ANNUAL REVIEW OF ASTRONOMY AND ASTROPHYSICS

EDITORIAL COMMITTEE (1971)

ANNUAL REVIEW OF ASTRONOMY AND ASTROPHYSICS

LEO GOLDBERG, *Editor*
Harvard College Observatory

DAVID LAYZER, *Associate Editor*
Harvard College Observatory

JOHN G. PHILLIPS, *Associate Editor*
University of California, Berkeley

VOLUME 9

1971

ANNUAL REVIEWS INC.
4139 EL CAMINO WAY
PALO ALTO, CALIFORNIA 94306

ANNUAL REVIEWS INC.
PALO ALTO, CALIFORNIA, USA

Standard Book Number 8243–0909–X
Library of Congress Catalog Number: 63–8846

FOREIGN AGENCY
Maruzen Company, Limited
6, Tori-Nichome, Nihonbashi
Tokyo

PRINTED AND BOUND IN THE UNITED STATES OF AMERICA BY
GEORGE BANTA COMPANY, INC.

PREFACE

Volume 9 of the *Annual Review of Astronomy and Astrophysics* was planned by the Editorial Committee at a meeting at the University of California, Santa Cruz on 17 May, 1969. Dr. Joseph Wampler of the Lick Observatory was an invited participant and made many valuable suggestions. The Committee believes that the current volume should attract wide interest because it deals with so many timely and important subjects, and we are grateful to the authors for their excellent contributions. If other interesting topics are conspicuous by their absence, it is usually not that we have overlooked them but, as we have reminded our readers previously, because new work in progress makes a delay of a year or two seem highly desirable.

After this volume had gone to the printer, we learned with sorrow and dismay of the sudden death of Mrs. Joann Huddleston, who had served as Assistant Editor of the series from Volume 1 on. The high quality of her editorial work, the skillful and tactful management that brought out every volume on schedule, and the kind and friendly assistance she gave to editors and authors were deeply appreciated. Much of the heavy editorial work on manuscripts and proofs for this volume had been performed by Mrs. Huddleston.

We thank Mrs. Jacqueline Handley of the staff of Annual Reviews, Inc. for carrying on the work to completion so capably, and we are also indebted to the George Banta Company for an outstanding printing job.

THE EDITORIAL COMMITTEE

CONTENTS

REPRINTS

The conspicuous number (2014 to 2026) aligned in the margin with the title of each review in this volume is a key for use in the ordering of reprints.

The sale of reprints from all Annual Reviews volumes was initiated in July 1970. Reprints of most articles published in the *Annual Reviews of Biochemistry* and *Psychology* from 1961 and the *Annual Reviews of Microbiology* and *Physiology* from 1968 are now maintained in inventory.

Available reprints are priced at the uniform rate of $1 each postpaid. Payment must accompany orders less than $10. The following discounts will be given for large orders: $5–9, 10%; $10–24, 20%; $25 and over, 30%. All remittances are to be made payable to Annual Reviews Inc. in US dollars. California orders are subject to sales tax. One-day service is given on items in stock.

For orders of 100 or more, any Annual Reviews article will be specially printed and shipped within 6 weeks. Reprints that are out of stock may also be purchased from the Institute for Scientific Information, 325 Chestnut Street, Philadelphia, Pa. 19106. Direct inquiries to the Annual Reviews Inc. reprint department.

The sale of reprints of articles published in the Reviews has been expanded in the belief that reprints as individual copies, as sets covering stated topics, and in quantity for classroom use will have a special appeal to students and teachers.

METEORITES AND THE EARLY SOLAR SYSTEM   2014

EDWARD ANDERS

Enrico Fermi Institute and Department of Chemistry, University of Chicago, Chicago, Illinois[1] and Physikalisches Institut, Universität Bern, Switzerland

CONTENTS

[1] Permanent address.

INTRODUCTION

Meteorites contain far more information about the early solar system than was once believed. Chondrites, as the most primitive meteorites, have of course long served as the basis of the "cosmic" (better, "solar-system") abundance curve, on the assumption that they represent well-mixed rubble from the inner solar system. But during the last decade another view has gained ascendancy: that chondrites are an original and relatively unaltered condensate from the solar nebula. If this is true, then chondrites contain a unique archeological record of physical and chemical conditions in the solar nebula. It is the job of the meteoriticist to decipher this record—to translate structure, composition, and mineralogy into temperature, pressure, time, and chemical environment.

By a fortunate coincidence, observational and theoretical work on proto-stars has advanced to the point where detailed estimates of some of these same parameters can be made. To be sure, the meteoritic evidence is some-times ambiguous and at best applies to only one solar system. Nonetheless it appears that comparisons between meteoritic and astronomical observations are becoming increasingly fruitful.

The origin of meteorites has been vigorously debated during the last decade. The prevailing view at present is that most (or all) meteorites come from the asteroid belt. But a cometary origin of some types (e.g. carbo-naceous chondrites) remains an intriguing possibility. Thus meteorites contain clues to conditions at 2 to 4 au, perhaps also to those in the giant planet zone.

In this paper, I shall review the meteoritic evidence as of late 1970, em-phasizing those properties that relate to the solar nebula. The paper begins with a glossary followed by brief sections on the classification of chondrites and the composition of primordial matter. The next section, comprising the major part of the paper, reviews the processes that may have influenced the chemistry and mineralogy of chondrites. The following section discusses the time scale of meteorite evolution. The last section sums up the principal conclusions and reviews the implications for planetary cosmogony.

Any decipherment of an archeological record involves some subjectivity. I shall try to follow the rules set down in my previous reviews (Anders 1964, 1968): (a) any interpretation must be consistent with all available evidence; and (b) of several alternative interpretations, the simplest one is to be pre-ferred. It is not obvious that the latter rule (Occam's principle) is applicable to meteorites. Authorities as eminent as Urey (1967) and Suess (1965) be-lieve that meteorites had a very complicated history; if this view is accepted then Occam's principle is not a valid guide. The reader is therefore referred to the above reviews for a different view of the same subject. A more detailed discussion along the lines of the present review may be found in Wood's (1968) book.

GLOSSARY

Chondrites.—Chondrites are stony meteorites containing *chondrules*, millimeter-sized silicate spherules that appear to be frozen droplets of a melt. They consist largely of *olivine* [(Mg, Fe)$_2$SiO$_4$], *pyroxene* [(Mg, Fe)SiO$_3$], and *plagioclase feldspar* [solid solution of CaAl$_2$Si$_2$O$_8$ and NaAlSi$_3$O$_8$]. In the more primitive chondrites, glass is often found in place of crystalline feldspar. (Special names exist for the endmembers of the Fe-Mg solid solutions: MgSiO$_3$ = *enstatite*; FeSiO$_3$ = *ferrosilite*; Mg$_2$SiO$_4$ = *forsterite*; Fe$_2$SiO$_4$ = *fayalite*. They are abbreviated En, Es, Fo, and Fa. We shall use them only in cases of absolute necessity.)

Chondrules are embedded in a ground-mass or *matrix*. In the less primitive chondrites, the matrix is somewhat more fine grained than the chondrules, but otherwise has the same mineralogy and composition. Millimeter-sized particles of *metal* (nickel-iron with 5–60% Ni) and *troilite* (FeS) are also present. In the more primitive chondrites, the matrix is very fine grained (to ∿10^{-6} cm) and richer in Fe^{2+} than the chondrules.

CLASSIFICATION OF CHONDRITES

Of the 4 principal meteorite classes (chondrites, achondrites, irons, and stony irons), chondrites are undoubtedly the most primitive, containing all nonvolatile elements in approximately solar proportions. Nonetheless, there are well-defined differences among the 5 subclasses of chondrites. At last count, some 50-odd elements showed such abundance differences (Larimer & Anders 1970), mostly subtle (10 to 50%), but occasionally large (factors of 2 to 1000). A few of these differences may have been established in the meteorite parent bodies, but the majority seem to date back to the solar nebula. Let us review the classification of chondrites, and then explore the chemical differences one by one.

The most up-to-date classification of chondrites is that of Van Schmus & Wood (1967). It retains the traditional subdivision into five major groups on the basis of chemical criteria: ratio of metallic to total iron; Fe/Si and Mg/Si ratios; degree of oxidation of iron as measured by Fe^{2+}/(Fe^{2+}+Mg^{2+}) ratio in silicates; etc (Table 1). In addition, it uses 10 petrographic criteria to subdivide each chemical group into a maximum of 6 "petrologic types" reflecting the extreme range of chondrule-matrix intergrowth. Chondrites with sharply delineated chondrules and submicron-sized, opaque matrix are classified as types 2 or 3; those with less distinct chondrules and coarser-grained matrix, as progressively higher types, 4 to 6. A separate category, type 1, is reserved for the chondrule-free Type I carbonaceous chondrites (≡ C1).

An important distinction is only subtly hinted at in Table 1. Whereas all 3 classes of ordinary chondrites are *isochemical*, i.e. have essentially constant major element contents from types 3 to 6, C and E chondrites are not. The

TABLE 1. Classification of chondrites[a]

Chemical group	Petrol. types	$\dfrac{Fe_{metal}}{Fe_{total}}$	Fe/Si	Mg/Si	$\dfrac{Fe^{2+}[b]}{(Fe^{2+}+Mg^{2+})}$	Known falls
E enstatite	3–6	0.80±0.10	0.83±0.32	0.79±0.06	0–2%	17
H ⎫	3–6	0.63±0.07	0.83±0.08	0.96±0.03	16–20%	413
L ⎬ ordinary	3–6	0.33±0.07	0.59±0.05	0.94±0.03	22–26%	504
LL ⎭ ("O")[c]	3–6	0.08±0.07	0.53±0.03	0.94±0.03	27–31%	49
C carbo- naceous	1–4	low	0.83±0.08	1.05±0.03	33%	36

[a] After Van Schmus (1969).

[b] Atom percentage of Fe^{2+} in principal ferromagnesian silicate [$(Mg, Fe)SiO_3$ in enstatite chondrites and $(Mg, Fe)_2SiO_4$ in all others]. Values given refer to petrologic types 4 to 6; those in types 1 to 3 cover a wider range and generally extend down to 0.

[c] The symbols *H*, *L*, and *LL* refer to total iron content, and stand for high, low, and low-low, respectively. We shall find it convenient to use the collective symbol *O* for these classes.

larger standard deviations in major element ratios reflect systematic compositional variations from the lower to the higher petrologic types.

Which of the properties in Table 1 were established in the solar nebula and which in the meteorite parent bodies? Though we have no foolproof criterion for deciding where a given feature arose, circumstantial evidence suggests that most of the *chemical* properties date back to the nebula (Larimer & Anders 1967, 1970, Keays et al 1971). On the other hand, some of the *textural* properties such as the increasing recrystallization from types 2 to 6 seem to have developed in the meteorite parent bodies by sustained heating (thermal metamorphism; Wood 1962a, Dodd 1969). Yet there exist certain correlations between chemical composition and petrologic type, and at least some of these apparently predate metamorphism. For this reason, the petrologic classification, though ostensibly based on a postnebular feature, will be useful in our discussion.

COMPOSITION OF PRIMORDIAL MATTER

It is almost an axiom that the solar nebula was well mixed in an isotopic and elemental sense (Suess 1965). Certainly no *isotopic* differences have yet been found that might be attributed to incomplete mixing of material with different nucleosynthetic histories (Anders 1964, Arnold & Suess 1969). *Elemental* differences have of course been observed (Table 1). All are explicable by chemical fractionations, however, and though the locale of such fractionation cannot be strictly specified, the pressures inferred in some cases ($10^{-4\pm2}$ atm; Larimer & Anders 1970) are more appropriate to the solar nebula than to an interstellar cloud. Moreover, it is unlikely that initial compositional differences could have persisted throughout the enormous (10^{13}-fold) contraction from interstellar to nebular densities, even if interstellar grains had not been vaporized in the process. Thus we are probably justified in assuming that the solar nebula once had completely uniform elemental composition.

Granted this assumption, we can attempt to deduce this composition. Goles (1969) has likened the search for this "primordial" composition to the search for the Holy Grail. The analogy is apt insofar as fervor is concerned, though medieval bloodshed has been replaced by mere acrimony.

Most authors regard $C1$ chondrites as the closest available approximation of primordial solar-system matter. This point of view was opposed by Urey (1964, 1967) and Arnold & Suess (1969), on the grounds that the abundance of iron in $C1$ chondrites was some 5–10 times higher than that in the solar photosphere. Anders (1964) reviewed the problem at some length and suggested that the photospheric value was in error because it led to various implausible consequences.[2] The issue was resolved by Garz & Kock (1969) who discovered a tenfold error in the oscillator strengths of the Fe I lines used for the solar abundance determinations. Excellent agreement now exists between the meteoritic value and all solar values, whether based on allowed or forbidden lines of Fe I or Fe II in the photosphere, or on coronal lines (Garz et al 1969, 1970, Baschek et al 1970).

Even with the iron problem resolved, there remains the legitimate question just how closely $C1$ chondrites resemble primordial solar-system matter. Anders (1971) has considered the problem in detail, on the basis of nuclear abundance systematics, cosmic-ray and solar abundances, and cosmochemical fractionation processes. He concludes that the maximum difference between $C1$ and primordial abundances is a factor of 1.5 for groups of 10 or more elements and factors of 2 to 5 for individual elements. Obvious exceptions are highly volatile elements: H, C, O, N, and noble gases. It seems that $C1$ chondrites (or abundance tables based thereon, e.g. Cameron 1968) closely approximate the composition of primordial solar-system matter.

CHEMICAL PROCESSES IN THE EARLY SOLAR SYSTEM

Separation of solids from gases is perhaps the most efficient cosmochemical fractionation process known (Suess 1965). And there is good evidence for such separations during the formation of the meteorites and terrestrial planets. In terms of Brown's (1950) classification, these bodies consist largely of "earthy" material. The Earth, for example, contains only $\sim 10^{-4}$ its cosmic complement of "ices" (e.g. H_2O) and $\sim 10^{-7}$ to 10^{-12} of "gases" (e.g. noble gases). The depletion seems to have been governed by

[2] Urey (1966, 1967) has accused me of "ignoring" or "neglecting" this problem although I devoted 11 pages of my 1964 review to it (pp. 644–650, 700–703).

volatility and chemical reactivity rather than by molecular weight, because Ne (mol wt 20) is some 8 orders of magnitude more depleted than H_2O (mol wt 18) (Urey 1954).

During the last few years it has become apparent that many other, more subtle fractionations are also due to volatility. Let us therefore examine the condensation sequence of a cosmic gas, and use it as a frame of reference for interpreting the observed fractionations in meteorites.[3]

CONDENSATION SEQUENCE OF A COSMIC GAS

Condensation sequences have been calculated by Urey (1952a, 1954), Lord (1965), Larimer (1967), Gilman (1969), and Clark et al (1971). For an element E condensing in pure form, the basic condition for incipient condensation is that the partial pressure p_E equal the vapor pressure p_v. At that point, the gas is saturated in E. Since the gas consists largely of H_2, p_E can be approximated by $(2E/H)p_t$, where E and H are the abundances of E and hydrogen in the gas, and p_t is the total pressure. The vapor pressure, in turn, is given by log $p_v = -A/T+B$, where A and B are constants varying from element to element. For elements condensing as compounds, the situation is similar, though the equations are more complex.

Two condensation diagrams based on the work of Larimer (1967 and unpublished) are shown in Figure 1. They were calculated for a total pressure of 10^{-4} atm (a value appropriate to the asteroid belt) and Cameron's (1968) cosmic abundances. Other pressures and abundances are discussed in the references cited above. The upper diagram refers to the situation just discussed: condensation of pure elements and compounds. It applies to conditions of rapid cooling, where substances condense in successive layers without interdiffusion or alloy formation. The lower diagram, on the other hand, assumes complete diffusional equilibration, with formation of alloys and solid solutions to the limit of solubility. The effect of such alloy formation is to raise the condensation temperatures of minor elements, and to widen their condensation intervals. Of course, all intermediate situations are possible.

A cooling gas of cosmic composition should thus condense in the following order. Below 2000°K, some highly refractory compounds of Ca, Al, Mg, and Ti appear, followed by magnesium silicates and nickel-iron at ~1350 to ~1200°K and alkali silicates at 1100 to 1000°K. Up to this point, some 90% of chondritic matter has condensed. Only H, C, N, O, S, and some volatile trace elements still remain in the gas phase. At 680°K sulfur begins to condense on solid Fe grains by the reaction $Fe+H_2S \rightleftarrows FeS+H_2$, followed by

---

[3] Arrhenius & Alfvén (1971) have proposed a radically different alternative: that condensation took place not from the solar nebula, but from a partially ionized and excited low-density gas that was gradually added to the circumsolar region. This model offers novel and promising explanations for the abundance patterns of mercury and noble gases. However, its few remaining predictions happen to agree with those of the equilibrium condensation model. Detailed evaluation of the Arrhenius-Alfvén model will have to await its further development.

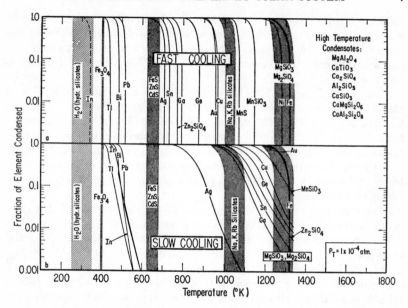

FIGURE 1. Condensation sequence of a gas of cosmic composition (Larimer 1967, Anders 1968, with minor revisions). Slow-cooling sequence assumes diffusional equilibrium between grain surfaces and interiors, with formation of solid solutions, while fast-cooling sequence corresponds to deposition of pure elements and compounds, without interdiffusion. Shaded areas represent condensation or chemical transformation of major constituents. The formation range of hydrated silicates is poorly known.

Pb, Bi, Tl, and In. Any remaining Fe reacts with H_2O at 400°K to give Fe_3O_4. Finally, water is bound as hydrated silicates at some temperature between 400 and 250°K.

Both diagrams assume that nucleation is instantaneous, so that the vapor never becomes supersaturated. This need not necessarily be so. Blander and his associates have pointed out that iron vapor, in particular, may supersaturate greatly, in which case the condensation would take a rather different course (Blander & Katz 1967, Blander & Abdel-Gawad 1969, Blander 1971). We shall return to this alternative at appropriate times.

Next we shall discuss individual fractionation processes, in the order of falling temperature.

FRACTIONATION OF REFRACTORY ELEMENTS

Significantly, the elements which head the condensation sequence all happen to be fractionated among the major chondrite classes. The variation in Mg/Si ratio has been known for a decade (Urey 1961, DuFresne & Anders 1962) and has in fact been used as a classification criterion (Table 1). More recently, similar variations have been found for Al, Ca, Ti (Ahrens et al 1969), Cr, Hf, La and lanthanides, Sc, Th, U, Y, and Zr (Larimer & Anders

1970). When the data are normalized to Si, the abundances consistently increase from enstatite to ordinary to carbonaceous chondrites in the average
ratio 0.56:0.73:1.00. All these elements are known to be highly refractory,
and would be expected to concentrate in an early condensate or a volatilization residue.

Larimer & Anders (1970) have suggested that the trend from C to E
chondrites reflects progressively greater loss of an early condensate ($MgAl_2O_4$,
$CaTiO_3$, Mg_2SiO_4, etc). The degree of loss cannot be uniquely determined
from the data, because Si, the element used for normalization, is itself fractionated, and so are all other elements that might serve in its place. Assuming
that $C1$ chondrites represent primordial composition, they deduce minimum
losses of 14 and 27% of the total condensable matter from ordinary and
enstatite chondrites.

There is some evidence that material of the composition of this early
condensate was present in the early solar system. Most $C3$ chondrites contain a few chondrules or irregular inclusions (up to several cm in diameter)
of about the composition of the early condensate: rich in Ca, Al, Ti, Mg;
poor in Si (Christophe 1968, 1969, Keil et al 1969, Marvin et al 1970). And
astronomical evidence also points to a fractionation of this sort. Herbig
(1970a) noted that the interstellar gas is deficient in these same elements, and
suggested that this is due to partial condensation of refractories. Such condensation cannot have taken place at the low densities of interstellar space.
Herbig therefore suggests that a major part of interstellar matter has once
passed through solar nebulae. It would seem that partial condensation of refractories is a fairly universal feature, not limited to our own solar system.

There is some evidence that the least volatile metals (Ir, Os, Re) were
likewise fractionated with the early condensate (Larimer & Anders 1970).
Their abundances in iron meteorites vary by 3 orders of magnitude and
are *anti*-correlated with those of equally reducible but more volatile elements,
e.g. Pd and Ni (Yavnel 1970). Similar though much smaller variations have
been observed in chondrites (Mueller et al 1971).

The *mechanism* of the fractionation is still uncertain. Material rich in
refractories can be produced either by partial condensation or by partial
volatilization. The former would be expected if all interstellar grains were
completely vaporized during collapse of the solar nebula (Cameron 1962);
and the latter, if they were not. Once produced by whatever process, such
material might settle toward the median plane of the nebula, thus becoming
separated from the gas.

The fate of the "lost" refractories invites speculation. Some authors have
proposed that the Earth and Moon are enriched in refractory elements
(U, Th, Ca, Sr, and Al) relative to chondrites (Gast 1968, 1971, Ringwood &
Essene 1970). This might mean that the refractory material, being the oldest
solid phase in the solar system, was preferentially accreted into asteroids and
planets. However, the reality of this enrichment is in some doubt. Thermal
history calculations by Hanks & Anderson (1969) suggest that the Earth is

not enriched in U and Th relative to C chondrites. The same conclusion has been reached by Larimer (1971) from consideration of crustal abundances of 12 elements.

METAL-SILICATE FRACTIONATION

Chondrites.—This fractionation manifests itself by variations in Fe/Si ratio among major chondrite classes (Table 1). That this is not merely a fractionation of *iron* but of *metal phase* is shown by the parallel variation in other siderophile (=iron-loving) elements, such as Ni, Co, Au (Larimer & Anders 1970). There has been much argument whether this fractionation represents loss of *silicate* from iron-poor material (e.g. L chondrites; Urey & Craig 1953) or loss of *metal* from iron-rich material (e.g. $C1$ or H chondrites; Anders 1964). Now that the discrepancy between the solar and the $C1$ chondrite abundance of iron has been eliminated, there is little reason not to take the latter view.

Larimer & Anders (1970) have attempted to infer from chemical data various details of the metal-silicate fractionation: specifically, direction, extent, temperature, and redox conditions. The "raw" Fe/Si and Ni/Si ratios suggest that many E chondrites have been enriched in metal relative to C chondrites (Table 1). However, these ratios are misleading, because appreciable amounts of Si were lost during refractory-element fractionation. When a correction for this effect is applied, all chondrites turn out to have lower Fe/Si ratios than do $C1$ chondrites. Thus the fractionation seems to have been unidirectional, involving only loss, but no gain, of metal. The maximum loss seems to have been $\sim 50\%$.

The temperature during fractionation was estimated from abundances of siderophile elements of increasing volatility, using the condensation diagram in Figure 1. It turns out that Ir, Pd, Ni, Co, Au, and Ge are systematically less abundant in L chondrites than in H chondrites, by a mean factor of 0.64, while Ga and S are about equally abundant in both classes. This implies that the first group of elements had already condensed on the metal when it was fractionated, while the latter group had not. The temperature during fractionation must thus lie below 1050°K, the temperature for condensation of 90% of the Ge at 10^{-4} atm. A 100-fold change in pressure shifts this temperature by $\pm 50°$K. For E chondrites, a corresponding value of $\leq 985 \pm 50°$K was derived.

Planets.—Metal-silicate fractionations also seem to have occurred during formation of the terrestrial planets, as inferred by Noddack & Noddack (1930) and Urey (1952b) from the variations in density. Urey (1967) has estimated the following Fe/Si ratios for the planets:

	Mercury	Venus	Earth	Mars	Moon
Fe/Si	~ 3.0	~ 1.0	~ 1.0	0.6	0.3

These ratios must be regarded as approximate, because they are based on equations of state and planetary radii that are not always well known. Bullen

(1969) finds, for example, that identical compositions for Mars and the Earth are still within the realm of possibility. Nonetheless, there is no doubt that Mercury and the Moon differ substantially from the other three planets.

Some uncertainty also arises from the assumption made in these calculations that planets and ordinary chondrites have silicate phases of identical composition, including Fe^{2+} content. This neglects possible differences in oxidation state and refractory-element content. Indeed, Ringwood (1959, 1966) has argued that the density variations are due solely to differences in degree of oxidation, while Levin (1957, 1963) postulated formation of metallized silicates at high pressures. However, neither hypothesis accounts for the high density of Mercury or for the variation in Fe/Si among various chondrite classes (Table 1). Since these data clearly indicate that a metal-silicate fractionation took place in some parts of the solar system (0.4 and 2–4 au), there seems to be little reason to assume that this process was inoperative between 0.7 and 1.5 au.

Mechanism.—A variety of mechanisms have been proposed for the metal-silicate fractionation, based on differences in density, brittleness, volatility, magnetic susceptibility, etc (Urey 1952b, 1956, Wood 1962b). The last two seem especially promising. At higher pressures (e.g. $\geq 10^{-2}$ atm), iron condenses ahead of Mg_2SiO_4, $MgSiO_3$. If accretion is rapid, the innermost part of a growing planet may consist largely of nickel-iron. Such "inhomogeneous accretion" has recently been proposed for the terrestrial planets (Turekian & Clark 1969). With pressures increasing toward the center of the solar nebula, one would expect metal contents to increase correspondingly, in accordance with observation.

For the chondrites, a different mechanism suggests itself, which may or may not be applicable to the planets. The Ni content of the metal during metal-silicate fractionation seems to have been about 6.4 and 5.2% for *O* and *E* chondrites (Larimer & Anders 1970). The ferromagnetic Curie points for these alloys, 900 and 940°K, lie only slightly below the upper temperature limits of ≤ 1050 and $\leq 985°K$ inferred from the Ge and Ga contents. Perhaps that is a coincidence but then it is hard to understand why the temperature had to fall 400° after the refractory-element fractionation began. More probably, the relationship is causal, the fractionation being triggered by the appearance of ferromagnetism in the metal grains (Wood 1962b).

Harris & Tozer (1967) estimated that ferromagnetism would enlarge the capture cross section of metal grains by a factor of 2×10^4, causing them to aggregate rapidly to larger clumps. Separation from silicates might then be effected not only by differences in magnetic susceptibility but also by differences in size. Harris & Tozer's conclusions have been challenged by Banerjee (1967) on the grounds that particles in the size range considered by them (10^{-4} to 10^{-3} cm) would be multidomained and thus have only a negligible magnetic moment. At 950°K, only particles between 1.8×10^{-6} and 3.3×10^{-5} cm would have moments large enough to permit significant magnetostatic attraction. However, this happens to be just the size range of Fe_3O_4 particles in primitive chondrites [$\sim 2 \times 10^{-6}$ cm in Tieschitz (*H3*) and Renazzo (*C2*); Wood 1962a; $1-10 \times 10^{-5}$ cm in Orgueil (*C1*); Kerridge 1970]. Though we have no data on the original particle size in more recrystallized chondrites, it is not unreasonable to suppose that it was in the same range. In any case, even a rather inefficient process would suffice to account for the fractionations in Table 1.

It would be interesting to know what became of the iron-poor material complementary to the inner planets. It may have been incorporated in the outer planets, expelled into interstellar space, or both.

FRACTIONATION OF VOLATILES; TEMPERATURE AND PRESSURE IN THE SOLAR NEBULA

Urey (1952a,b, 1954) was the first to note that volatile metals (Hg, Cd, etc) might serve as "cosmothermometers" to determine the accretion temperature of meteorite parent bodies and the Earth. A careful analysis of the data then available showed no clear-cut evidence of significant depletion. Urey therefore concluded that both the Earth and meteorite parent bodies had accreted at a temperature of $\sim 300°$K, not far from the present black-body temperature at 1 au. This concept of "cold accretion" became firmly established in planetary cosmogony.

But the situation has changed in subsequent years. Precision measurements by neutron activation analysis showed that nearly all elements considered volatile by Urey were in fact underabundant; some by factors as large as 10^3, many others by smaller but nonetheless definite factors, e.g. 2 to 10 (Anders 1964, Larimer & Anders 1967). The new data led to a different and much more detailed picture of accretion conditions in the solar nebula.

Abundance patterns.—At last count, some 31 volatile elements appeared to be fractionated in chondrites. The simplest fractionation pattern is shown by carbonaceous chondrites (Figure 2).

The elements are arranged in somewhat unconventional order, based on their depletion pattern in ordinary chondrites. Abundances decrease consistently from *C1* to *C3*, and by constant *factors* as shown by the parallelism of the curves.[4] This parallelism extends to the last 5 elements on the graph, though their absolute depletion is very much greater.

---

[4] Apparent exceptions occur at Au, Cl, and Br, but past experience suggests that at least the last two may be due to sampling or analytical errors. Some points, especially for *C1*'s, are based on only one or two measurements. The original version of this graph (Figures 16a, b in Anders 1964) contained many more irregularities and gaps, which gradually disappeared in subsequent versions (Figure 1, Larimer & Anders 1967; Figure 3, Anders 1968). In view of this secular trend, the irregularities at Cl and Br can perhaps be set aside for the time being.

The situation for Au is unclear. The overall fractionation is smaller than that of the other elements in Figure 2, and may be due to some combination of sampling and analytical errors, metal fractionation, etc. Moreover, Au is near the top of the condensation sequence in Figure 1, where less complete depletion may be expected.

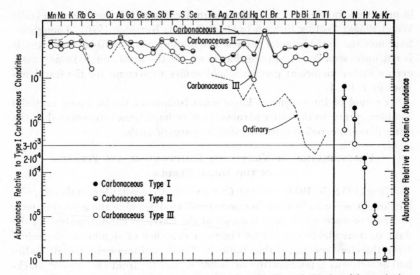

FIGURE 2. Volatile elements are underabundant in *C2* and *C3* chondrites relative to *C1*'s, by nearly constant factors. Even for the strongly depleted elements C to Kr, relative abundances stand in approximately the same ratio, 1:0.6:0.3. Dashed line: mean abundances in ordinary chondrites. (Data from Larimer & Anders 1967, Anders 1968, with minor revisions.)

Partial volatilization or condensation cannot account for this pattern. The condensation curves in Figure 1 show clearly that there is no one temperature at which all elements will be condensed to the same fractional degree. Partial loss of a major phase (metal, sulfide, etc) can also be ruled out, because the elements in Figure 2 differ greatly in geochemical character, and do not all reside in the same phase.

To account for these trends, it seems necessary to postulate that *C* chondrites are a mixture of two components: a high-temperature, volatile-free component, and a low-temperature, volatile-rich component (Wood 1963, Anders 1964). To explain the observed mean relative abundances (*C1*:*C2*:*C3* = 1.00:0.55:0.32), one must postulate ~100, ~55, and ~32% low-temperature material in the three types.

There exists some circumstantial evidence in favor of this model. Broadly speaking, the minerals of carbonaceous chondrites can be divided into two groups: high-temperature (olivine, pyroxene, metal) and low-temperature (hydrated silicates, $MgSO_4$, S, FeS, Fe_3O_4). Significantly, the *observed* content of low-temperature minerals in the three types agrees closely with the *postulated* content of low-temperature fraction, and so does the ratio of matrix to chondrules. *C1*'s consist almost entirely of a fine-grained matrix of low-temperature minerals; *C3*'s consist mainly of chondrules of high-temperature minerals. Furthermore, there is direct experimental evidence that chondrules and metal particles are depleted in volatile elements such as Na, Mn, and Ga, relative to matrix (Schmitt et al 1965, Chou & Cohen 1970). This suggests that the low-temperature fraction may be identical with the matrix, and the high-temperature fraction, with chondrules plus metal particles. (In much of our subsequent discussion we shall use the shorter term "chondrules" for "chondrules plus metal particles of comparable size.")

Formation of chondrules and matrix.—Each class of chondrites has its own characteristic brand of chondrules and matrix. This suggests that these two components are genetically related, not independently drawn from distant sources. Several possibilities exist for the formation of chondrules and matrix in the nebula. Wood (1958, 1962b, 1963) suggested that they represent material from two different condensation paths. Condensation of vapor to solid would give submicron-sized smoke (= matrix), while condensation of vapor to liquid would give droplets that might grow to millimeter-sized chondrules by coalescence. With $\sim 10^5$ times the radius of matrix grains, chondrules would have only 10^{-10} times the specific surface area and hence be far less efficient collectors of volatiles under conditions where surface area is the governing factor (Figure 1a). Similarly, they would take 10^{10} times longer to reach diffusional equilibrium, and would thus be correspondingly less efficient where such equilibration is the determining factor (Figure 1b).

Qualitatively Wood's model is most satisfactory, as it accounts for the main structural components of chondrites and their chemical differences. Quantitatively it is in some trouble, however. Condensation of magnesium silicates and nickel-iron from a cosmic gas requires pressures of 10 to 10^3 atm (Suess 1963, Wood 1963), far higher than those expected in the solar nebula. One alternative is that the primordial condensate was entirely dust, and that some fraction of it was remelted to chondrules by electric discharges or shock waves (Whipple 1966, Cameron 1966). At pressures outside the liquid field of the phase diagram (i.e. <100 atm) molten droplets are of course unstable with respect to vapor, but if the remelting events are of a purely local scale, droplets may freeze before they have completely evaporated. Indeed, petrographers have agreed for over a century that the texture of chondrules indicates rapid crystallization.

Another alternative is that condensation did not proceed under equilibrium conditions but involved supercooling and metastability. Blander & Katz (1967) have argued that liquids are easier to nucleate than solids; thus much of the silicate condensing from the gas even at low pressures might be a metastable liquid rather than a stable solid. Once crystallization is initiated by some chance event, the droplet would of course freeze rapidly.

Temperatures and pressures.—Whatever the chondrule-forming process, it is instructive to compare the abundance patterns in chondrites with the predictions of the condensation diagram (Figure 1). For each of the two fractions, one can determine a "condensation temperature," defined as the temperature at which the solid grains ceased to equilibrate with the gas. (Under conditions of slow cooling, such equilibration would normally terminate with accretion. In that case, these temperatures may be interpreted as accretion temperatures.) If condensation did proceed according to Figure 1, temperatures derived from different elements should be concordant, and consistent with other evidence.

It appears that the high-temperature fraction is best explained in terms of Figure 1a; and the low-temperature fraction, in terms of Figure 1b. This is not surprising in view of the differences in grain size noted above.

The high-temperature fraction of C chondrites contains magnesium silicates and nickel-iron but is depleted in all elements from Mn on down. This is readily understood in terms of Figure 1a (but not 1b), for a condensation temperature of 1150–1250°K. The low-temperature fraction,[5] on the other hand, contains its full complement of even the highly volatile metals Pb, Bi, Tl, and In. By either diagram, this implies an accretion temperature of <400°K. A value of the same order can be independently estimated from the presence of magnetite and hydrated silicates in C chondrites. The formation temperature of hydrated silicates is not well known and can only be roughly estimated as 300–350°K. But the formation temperature of Fe_3O_4 is 400°K in a cosmic gas, and is independent of pressure (Urey 1952b).

The ordinary chondrites (dashed line in Figure 2) show a somewhat different trend, implying different condensation and accretion temperatures. [Only mean abundances are plotted here. Actually, the range of variation is quite large, chondrites of lower petrologic types generally being richer in volatile elements (Figure 4, Anders 1968; Figure 1, Keays et al 1971).] Nine elements from Cu to Se parallel the $C2$ and $C3$ curves. Though the three segments are not particularly flat in this region, their parallelism suggests that they reflect the same process. Any deviation from horizontality is probably due to errors in the $C1$ data used for normalization. It therefore seems likely that the ordinary chondrites, too, are a mixture of two components. The mean depletion factor between Cu and Se (using the latest data, J. C. Laul, private communication, and omitting Au; cf Footnote 4) is 0.24, which suggests a matrix content of 24%.

The remaining elements deviate from the $C2,3$ pattern. Mn, Na, K, and Rb are essentially undepleted, while Cs and the elements from Te to Tl show progressively greater depletion. The first trend suggests a lower condensation temperature of the high-temperature fraction, e.g. 1000°K vs 1150 to 1250°K, or a higher pressure. The second trend suggests a higher accretion temperature of the low-temperature fraction, within the region of partial condensa-

[5] It is perhaps misleading to speak of the low-temperature fraction in the singular. If accretion took place over a range of temperatures, the composition of the low-temperature fraction must vary accordingly. Indeed, the abundance of the highly volatile elements H and N drops more rapidly from $C2$ to $C4$ than one would expect from the decrease in matrix content. Presumably this reflects increasing accretion temperatures.

TABLE 2. Accretion temperatures of meteorites and planets from content of volatile metals[a]

(Total pressure $= 10^{-4}$ atm)

	Class	Temperature (°K)[b]		
		Bi	Tl	In
Mezö-Madaras	L3	—	464	464
Krymka	L3	483	—	474
Khohar	L3	490	480	479
Tennasilm	L4	*469*	*466*	501
Barratta	L4	489	499	486
Fukutomi	L4	505	504	491
Goodland	L4	510	498	—
Homestead	L5	*508*	*498*	529
Farmington	L5	521	*500*	534
Modoc	L6	554	—	533
Bruderheim	L6	556	551	548
L Chondrites, weighted average		542	539	533
Eucrites, average		470	450	480
Earth		480, 570	450, 530	490, 560
Moon		540, 660	500, 600	510, 580

[a] From Keays et al (1971), Anders et al (1971).

[b] Italicized values disagree with In-based temperatures, and are deemed less reliable. Perhaps they reflect unforeseen complexities in the condensation behavior of Bi and Tl. Where two values are given, the second refers to $P_t = 10^{-2}$ atm, a more appropriate value for the Earth and Moon.

tion. Unfortunately, condensation curves are available for only 4 of the 13 strongly depleted elements (Ag, Pb, Bi, and Tl), owing to lack of thermodynamic data or complexity of chemical equilibria. The curve for a fifth element, In, has been estimated indirectly (Keays et al 1971).

Attempts to determine accretion temperatures from Bi, Tl, and In contents have been made by Larimer (1970) and Keays et al (1971). Data for 11 L chondrites are shown in Table 2. Though there are minor discrepancies among the 3 cosmothermometers, all 3 give fairly similar results. Accretion temperatures lie in a surprisingly narrow, 100-degree interval centered on 510°K, and increase with petrologic type. The mean value, weighted according to the frequency of the 4 petrologic types, is 540°K. Temperatures derived from Pb contents lie in the same range (Larimer 1970).

The above temperature estimates are based on pressure-dependent condensation curves and may thus be in error if the pressure differed greatly from the assumed value of 10^{-4} atm. However, at least for the L chondrites the uncertainty in pressure is only 2 orders of magnitude in either direction, corresponding to a temperature error of $+80$, -60 degrees (Keays et al

1971). At pressures of 10^{-2} or 10^{-6} atm, calculated Bi-Tl correlation curves fit the data noticeably less well than at 10^{-4} atm. Moreover, all chondrites contain FeS, whose formation temperature (680°K) is pressure independent. At a pressure of 10^{-2} atm, the condensation curves of Bi and Tl are shifted upward sufficiently to imply accretion temperatures *greater* than 680°K for the more strongly depleted Type 6's. Such meteorites then should contain no FeS, contrary to observation. Similarly, a lower limit of 10^{-6} to 10^{-7} atm can be inferred from the absence of Fe_3O_4 (formation temperature, 400°K) from all but the least recrystallized and least depleted Type 3's. Thus the pressure in the region of the L chondrites apparently was $10^{-4\pm2}$ atm. It will be of interest to extend this treatment to other chondrite classes. A somewhat roundabout estimate for eucrites[6] gave a similar result, 470°K (Anders et al 1971).

It is remarkable that temperatures as high as 500°K prevailed during accretion of the meteorite parent bodies, because the present blackbody temperature at 2.8 au is only 170°K. Thus a powerful, transient heat source must have been present. One possibility is the Hayashi stage of the Sun; another is the collapse stage of the nebula which would release vast amounts of gravitational potential energy over a short span of time (Cameron 1962, 1963, 1969). Indeed, the temperatures inferred for the dust shells or disks surrounding protostars are of the right order: $\leq$1850 to 350°K for VY Canis Majoris (Herbig 1970b) and $\sim$850°K for R Monocerotis and other infrared stars (Low & Smith 1966, Low 1969). Perhaps it will soon be possible to compare time, temperature, and pressure estimates from meteorites with observational and theoretical data for protostars.

Planets.—There is some indication that the bulk of the accretion in the inner solar system took place between 500 and 700°K. Only 5–10% of the ordinary chondrites are Type 3's which accreted below 500°K, and few if any meteorites have the FeS-free, but otherwise chondritic composition expected for material accreted above 680°K. The Earth, too, seems to contain substantial amounts of FeS (Anderson et al 1970), and resembles O4–O6 chondrites in the abundance of volatile elements (Anders 1968, Larimer 1971). There are two limiting interpretations for this pattern: accretion at a constant temperature of $\sim$560°K (at 10^{-2} atm; Table 2) or at falling temperatures from >700 to $\sim$350°K.[7] In the latter case, a large part of the

---

[6] Eucrites are the most common variety of Ca-rich achondrites. They consist largely of pyroxene ($\sim$70%) and feldspar ($\sim$30%), with mere traces of metal and sulfide. Relative to chondrites, they are enriched in Ca, Al, U, Th, etc by factors of 5–10. They are somewhat similar to terrestrial and lunar basalts, and were presumably derived from chondritelike material by magmatic processes.

[7] We are continuing to use the term accretion temperatures although strictly speaking, these are temperatures of equilibration with the nebula. For a large planet, the former are much higher, owing to the release of gravitational potential energy

volatiles may have been brought in as a thin veneer of carbonaceous-chondrite-like material, swept up by the Earth in the final stage of accretion (Anders 1968, Turekian & Clark 1969).

The Moon seems to be depleted in volatiles by another 2 orders of magnitude relative to the Earth (Ganapathy et al 1970, Anders 1970, Anders et al 1971). In terms of the preceding two models, this may imply either accretion at a higher temperature (620°K at 10^{-2} atm; Table 2) or a much lower accretion efficiency in the final stages of formation, when volatile-rich material was being swept up. Ganapathy et al (1970) and Anders (1970) have suggested that such inefficient accretion might result if the Moon formed near the Earth. Singer & Bandermann (1970) have shown, however, that no such effect can arise in the absence of gas. Concentration of dust by the Earth would offset the Moon's lower capture cross section resulting from its orbital motion. However, it is generally believed that gas was present during accretion, and this would prevent any significant concentration of dust in the Earth's neighborhood.

Material balance.—The fate of the lost volatiles remains largely undetermined. A few of the most volatile elements (In, Bi, Tl, Cs, and Br) are occasionally enriched in O3 chondrites beyond cosmic levels. Such enrichment might be expected in the final stages of accretion when the gas phase had become enriched in volatiles left behind by the previously accreted O4's–O6's. An even more striking enrichment is shown by mercury, thought to be the most volatile of all metals (Urey 1952b, 1954). It is enriched in C chondrites by 1–2 orders of magnitude above cosmic levels and is only slightly depleted in O chondrites (Reed & Jovanovic 1967, Ehmann & Lovering 1967). Larimer & Anders (1967) have tried to explain the "mercury paradox" by postulating the existence of a relatively involatile mercury compound, but Arrhenius & Alfvén (1971) have proposed an attractive alternative. If condensation took place from a partially ionized plasma, mercury, because of its high ionization potential, would be neutralized ahead of most other elements, thus becoming available for condensation at an earlier stage.

Some part of the volatiles may have been lost with the gases during dissipation of the solar nebula. Hence it seems unlikely that the average composition of all meteorites is close to cosmic. Meteorites apparently did not evolve in a closed system.

Other models.—Alternative explanations have been offered for the trace-element depletion in ordinary chondrites. Tandon & Wasson (1968) suggested that the most volatile elements were contained largely in a third component, which becomes increasingly abundant in the lower petrologic

---

during accretion. This has little effect on composition, though, because escape of volatiles during accretion of an Earth-sized planet is unlikely.

types. Blander & Abdel-Gawad (1969) proposed instead a multicomponent model, according to which each chondrite contains a variety of materials covering a range of grain sizes and condensation temperatures. Neither model predicts a simple relationship between abundance and temperature, let alone concordance of temperatures derived from different elements (Table 2). There are other objections to these more complex models (Keays et al 1971), and it seems that there is no need at present to go beyond the simple two-component model.

STATE OF OXIDATION

Chondrites differ strikingly in degree of oxidation, as reflected by the partition of iron between metal (plus sulfide) and silicates (Table 1). A convenient index of oxidation level is the iron content of the principal silicates, olivine and pyroxene. It is usually given as the atom percentage of Fe^{2+} in the mineral, using abbreviations Fa and Fs for the iron endmembers of the solid solution. Thus Fa_{18} means $Fe^{2+}/(Fe^{2+}+Mg^{2+})_{olivine} = 0.18$.

Each of the 5 chondrite classes covers a narrow, characteristic range of iron contents (Table 1, Column 6), separated by well-marked hiatuses. The equilibrated members of each class fall wholly within this range. The unequilibrated chondrites contain minerals of widely varying iron contents, but their *mean* iron content generally falls close to that of their equilibrated counterparts (Dodd et al 1967).

Ordinary chondrites.—Silicates condensed from a hot cosmic gas are expected to have very low iron contents, less than 1%. For this reason some authors have assumed that the high Fe^{2+} contents of ordinary chondrites were established in a noncosmic environment, i.e. the meteorite parent bodies or a fractionated region of the solar nebula (Suess 1964, 1965, Ringwood 1966, Wood 1967a). Blander & Katz (1967), on the other hand, have pointed out that high Fe^{2+} contents might arise even by condensation from the solar nebula if the nucleation of metallic iron were kinetically inhibited. Under these conditions, the partial pressure of iron vapor remains orders of magnitude above the equilibrium value, causing a proportionate increase in the Fe^{2+} content of the silicates.

A simpler alternative is equilibration of the condensate with the nebular gas upon cooling. Under equilibrium conditions, iron-bearing olivine is produced by the reaction:

$$MgSiO_3 + Fe + H_2O \rightleftarrows \tfrac{1}{2}Mg_2SiO_4 + \tfrac{1}{2}Fe_2SiO_4 + H_2$$

The iron content of the pyroxene, in turn, is coupled to that of the olivine by the exchange equilibrium:

$$\tfrac{1}{2}Mg_2SiO_4 + FeSiO_3 \rightleftarrows MgSiO_3 + \tfrac{1}{2}Fe_2SiO_4$$

Larimer (1968a) has considered these equilibria in detail and predicts the following olivine compositions in a solar gas of $H_2/H_2O = 500$:

T (°K)	1400	1200	1000	800	700	600	500
Fa (%)	0.6	1	2	6	10	25	87

The olivine compositions in equilibrated ordinary chondrites (Fa_{16} to Fa_{31}, Table 1) thus seem to point to equilibrium near 600°K. Reaction rates are probably fast enough to permit establishment of equilibrium at these low temperatures. The equilibration process involves transfer of iron from metal to silicate grains, which can proceed by solid-state diffusion if the grains are in contact or by vapor-phase transport if they are not. For $t = 10$ years and a grain diameter of 10^{-5} cm (probably representative of matrix particles) the diffusion mechanism may be effective to ~630°K. The limit for vapor-phase transport is less certain, but should be of the same order if volatile iron species such as $FeCl_2$ were present (Larimer & Anders 1967).

If this dust is to be transformed into chondrules of the same Fe^{2+} content, it is obviously necessary that no appreciable reduction take place during remelting. This appears to be feasible. Reduction requires that a sufficient number of H_2 molecules from the gas phase impinge on the surface of the molten chondrule. A minimum half-life for Fe^{2+} reduction can be estimated on the assumption that every impact is effective and that any iron reduced at the surface is instantly replaced by fresh Fe^{2+} from the interior. At $P = 10^{-4}$ atm and $T = 1200$°K, this minimum half-life is ~10 sec for 1 mm chondrules, of the same order as the freezing time (seconds to minutes). But the actual half-life is likely to be orders of magnitude longer and thus the average degree of reduction will remain slight. Individual chondrules should of course vary, depending on their cooling times, and such variations have in fact been observed in primitive chondrites ($Fa_{\sim 0}$ to Fa_{90} in $O3$'s, Van Schmus 1969; $Fa_{0.1}$ to Fa_{69} in $C2$'s, Wood 1967a).

It thus appears that the oxidation state of ordinary chondrites was established directly in the solar nebula, by equilibration of the primary condensate with the gas at ~600°K.

Enstatite chondrites.—These meteorites are much more strongly reduced. They contain almost iron-free $MgSiO_3$ ($Fs_{0.1-2}$) and minerals such as graphite, FeSi, TiN, and CaS, which are stable only under very reducing conditions. It appears that these facts cannot be explained by temperature alone; the gas phase must have been more reducing than cosmic. Larimer (1968b) investigated the thermodynamics in detail and noted that the requisite degree of reducing power could be achieved most simply by a slight increase in C/O ratio, from the solar value of ~0.6 to about 0.9 or 1.0. It is well known from stellar spectra and thermodynamic calculations (Suess 1962, Tsuji 1964, Dayhoff et al 1964, Dolan 1965, Morris & Wyller 1967) that drastic chemical changes take place when the C/O ratio in a hot cosmic gas approaches unity. The bulk of the oxygen is tied up as CO and metal oxides, while H_2O virtually disappears. Since the reducing power of the gas depends on the H_2/H_2O ratio, various reduced species become prominent.

Larimer suggested that the increased C/O ratio might be achieved by a slight local enrichment of dust relative to gas, if the dust contained some carbon. Direct condensation of carbon from a solar gas (as graphite or a solid solution in metallic iron) requires rather specialized, though not impossible conditions (Urey 1953, Eck et al 1966). Yet the high C/O ratio would have to persist from $\sim$1000°K (where silicates otherwise begin to take up significant amounts of Fe^{2+}) to 600–700°K (temperature of metal-silicate fractionation and chondrule formation). It is not clear that prolonged survival of large, compositionally peculiar regions in the solar nebula is plausible.

Alternative models for the enstatite chondrites have been proposed, involving either reduction of carbon-rich dust in the parent body (Ringwood 1966) or condensation from a cosmic gas supersaturated in Fe vapor at pressures greater than 10^{-2} atm (Blander 1971). All of these models need to be explored further.

Oxidation state during metal-silicate fractionation.—Conditions during this fractionation can be estimated from a plot of Ni/Mg vs Fe/Mg for a group of cogenetic meteorites. The slope gives Ni/Fe in metal and the intercept, Fe/Mg in silicate during fractionation. The latter ratio can of course be related to temperature.

Larimer & Anders (1970) have determined these quantities for O, E, and C chondrites, on the assumption that all members of each class are cogenetic. The Fe^{2+}/Mg^{2+} ratio for O chondrites is 0.14 ± 0.02, corresponding to $T \approx 700 \pm 15$°K. The Fe^{2+}/Mg^{2+} ratio of 0.14, slightly lower than the present value, suggests that the fractionation happened shortly before the final stage of oxidation. The ratios for E and C chondrites (0.00 ± 0.07 and 0.14 ± 0.08) have larger errors and hence cannot be used for meaningful temperature estimates.

ORGANIC MATTER

Carbonaceous chondrites contain up to 4% carbon. Most of it is in the form of an insoluble, aromatic polymer somewhat similar to coal or to humic acids in soils. The remainder consists of Mg, Fe, and Ca carbonates and a variety of organic compounds. Ordinary chondrites contain smaller amounts of carbon (0.01 to 2%) in an ill-defined chemical form, presumably again an aromatic polymer (Hayes & Biemann 1968).

The nature and origin of meteoritic organic matter has been reviewed by Hayes (1967), Vdovykin (1967), and Levin (1969). From the standpoint of the present review, the principal question is whether the organic matter represents a primary condensate from the solar nebula or a secondary product.

Urey (1953) investigated the thermodynamics of carbon condensation from the solar nebula, and showed that simple, nonvolatile substances (graphite, Fe_3C) could condense only under specialized and somewhat improbable conditions. At moderate temperatures ($<$500–600°K at $<10^{-3}$

atm), methane is the dominant form of carbon, but is far too volatile to condense in the inner solar system ($T_c = 48°K$ at 10^{-3} atm). Urey therefore suggested that the primary condensate consisted of "complex, tarry carbon compounds" similar to those in carbonaceous chondrites. Such compounds might form as metastable intermediates in the transformation of CO, the stable high-temperature form of C in the solar nebula, to CH_4, the stable low-temperature form.

Much detailed information on the organic compounds in meteorites has subsequently become available. We can therefore try to reconstruct the processes that led to the formation of these compounds, and see what relationship they bear to the primary condensation process.

The study of organic compounds in meteorites has been somewhat hampered by contamination, but the isotopic composition of H, S, and C shows that the major part of the organic matter is indigenous (Briggs 1963). Surprisingly, the compound distribution is far from random. A few classes or structural types dominate: for example, straight-chain hydrocarbons are far more abundant than their branched-chain counterparts, although the straight-chain configuration is only one of 10^3 to 10^5 structural possibilities for molecules with 15 to 20 carbon atoms. Apparently some highly selective synthesis process was at work.

A mechanism of the right degree of selectivity is the Fischer-Tropsch reaction: hydrogenation of CO in the presence of an iron or cobalt catalyst. This reaction, which is used industrially, produces mainly straight-chain or slightly branched hydrocarbons of the general formula C_nH_{2n+2}, e.g.

$$10CO + 21H_2 \rightarrow C_{10}H_{22} + 10H_2O$$

The hydrocarbon distribution resembles that in meteorites (Figure 3) and in certain oil shales and natural gases (Studier et al 1968, Friedel & Sharkey 1968). When NH_3 is present in the gas mixture, various nitrogen compounds are produced, including some of biological significance (adenine, guanine, cytosine; the constituent bases of DNA and RNA), and other structurally similar ones (ammeline, melamine) without such significance (Hayatsu et al 1968). The same compounds have been found in carbonaceous chondrites (Hayatsu 1964).

A further argument for the Fischer-Tropsch reaction comes from the isotopic differences between carbonate carbon ($=$ bonded to O, e.g. $MgCO_3$) and organic carbon ($=$ bonded to H, e.g. C_xH_y) in carbonaceous chondrites. Clayton (1963) showed that the carbonate carbon was some 7 to 8% richer in ^{13}C than the organic carbon (Briggs 1963). This trend was confirmed by Krouse & Modzeleski (1970) and Smith & Kaplan (1970). Although fractionations of this magnitude are theoretically possible under equilibrium conditions at very low temperatures ($\leq 0°C$), they are not observed in nature. Urey (1967) therefore proposed that the two types of carbon came from two unrelated reservoirs. It turns out, however, that the Fischer-Tropsch reaction gives an isotopic fractionation of just the right sign and magnitude at 400°K (but not

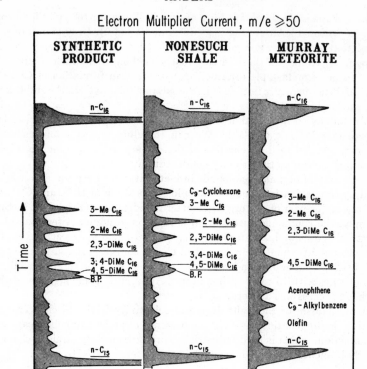

FIGURE 3. Gas chromatogram of hydrocarbons in the range C_{15} to C_{16} (unpublished work by M. H. Studier and R. Hayatsu). *Synthetic product:* from CO and D_2 by a Fischer-Tropsch reaction. *Nonesuch shale:* terrestrial oil shale, about 1.1 AE old (hydrocarbon fraction courtesy W. G. Meinschein); *Murray meteorite:* $C2$ chondrite. Only 6 of the $\sim 10^4$ isomeric hydrocarbons with 16 C atoms are present in appreciable abundance; 5 of these (underlined) are common to all three samples. Identifications were confirmed by mass spectrometry (Me = methyl; B.P. = branched paraffin).

500°K), owing to a kinetic isotopic effect (Lancet & Anders 1970). The temperature of $\sim 400°K$ agrees fairly well with the accretion temperature of $C2$ chondrites estimated from their volatile element content.

An alternative mechanism frequently proposed for the prebiotic synthesis of organic compounds on Earth is the Miller-Urey reaction: irradiation of CH_4, NH_3, and H_2O by UV, γ rays, etc (Miller 1953; see Lemmon 1970 for a recent review). However, this reaction involves random linkup of free radicals, and therefore lacks the required selectivity. It produces *all* hydrocarbon isomers in comparable amounts, without preference for straight-chain isomers, and does not yield many of the heterocyclic or aromatic compounds found in meteorites. Moreover, the Miller-Urey reaction does not give significant carbon isotope fractionations.

Although the Fischer-Tropsch reaction seems capable of accounting for much of the evidence there is as yet no certainty that conditions in the solar nebula were appropriate for this reaction. To be sure, a variety of suitable catalysts (nickel-iron, Fe_3O_4, hydrated silicates) were undoubtedly present, and so were at least two of the three reactants (NH_3, H_2). Problems arise for the third reactant, CO, however. In a cooling solar nebula it should be abundant down to $\sim 600°K$. But below this temperature, it becomes unstable, first with respect to CH_4, and only later (at temperatures 100 to 150° lower) with respect to higher hydrocarbons. Thus, if equilibrium were maintained, most of the CO would be converted to CH_4 before formation of heavier hydrocarbons became possible.

In order to resolve this difficulty, Lancet & Anders (1970) postulated that the hydrogenation of CO might have been very sluggish until 400°K, when magnetite began to form. At that temperature, both CH_4 and heavier hydrocarbons are thermodynamically possible products; the latter actually form more readily for mechanistic reasons. In support of this suggestion they noted that significant amounts of organic compounds are found only in meteorites containing magnetite.

The survival of any heavy hydrocarbons formed also presents problems. These compounds are unstable with respect to CH_4 in the presence of excess H_2. Studier et al (1968) have shown that this reaction does not proceed to a perceptible degree during several days at 500°K and $P = 10$ atm. It is not known, however, whether this is sufficient to ensure survival of hydrocarbons at the lower temperatures, lower pressures, and longer times characteristic of the solar nebula.

Likewise, the reaction has not yet been carried out at the high H_2/CO ratios ($\sim 10^3$) and low pressures ($\leq 10^{-3}$ atm) appropriate to the solar nebula, because the carbon content of such a system would be too small for present analytical techniques. Of the two variables, pressure seems to be the more critical one. Studier et al (1968) were able to synthesize hydrocarbons at $H_2/CO = 500$ and $P = 10$ atm, but saw no detectable reaction in gas mixtures of lower H_2/CO ratio during several days at 10^{-2} or 10^{-3} atm.

Another unsolved problem is the origin of the aromatic polymer. Calculations by Dayhoff et al (1964) show that aromatic hydrocarbons of high molecular weight might form metastably in systems of low H/C ratio, and this has been confirmed experimentally by Eck et al (1966). A similar process seems to be involved in the transformation of terrestrial plant remains to coal, and might also occur if the hydrocarbons first formed were reheated in the meteorite parent body after accretion. Some features of the hydrocarbon distribution in meteorites actually seem to call for such reheating, e.g., the predominance of aromatic over aliphatic hydrocarbons in the range C_6 to C_{10} (Studier et al 1968). However, no material resembling the meteoritic polymer has yet been synthesized in the laboratory.

In summary, the Fischer-Tropsch reaction (perhaps augmented by radiation chemistry and reheating) seems to be a likely mechanism for producing the organic compounds in meteorites. Whether this reaction is indeed the primary condensation process sought is less certain. One cannot yet rule out the possibility that the primary condensate was made by some other process (Miller-Urey?), and transformed to Fischer-Tropsch material by secondary reactions in the meteorite parent bodies. Such transformations are in fact observed on Earth. The hydrocarbon distribution in some crude oils, coals, and oil shales is strikingly similar to a Fischer-Tropsch distribution (Figure 3) although these hydrocarbons are obviously made from biological material by subterranean degradation (Friedel & Sharkey 1963, 1968). Apparently the Fischer-Tropsch pathway is kinetically favored over competing mechanisms in a wide range of conditions. Thus the Fischer-Tropsch compounds in meteorites may be of secondary origin.

This problem may soon be resolved by radio astronomers. The organic molecules of growing complexity discovered in interstellar clouds seem to require formation in an environment far denser than the cloud itself, i.e. stellar envelopes or "solar nebulae" (Herbig 1970a,c). These compounds almost certainly were made by the "primary" process of carbon condensation. As the list of positive (and negative!) identifications grows longer, it may provide a decisive clue to the nature of the primary process.

PRIMORDIAL NOBLE GASES

Some meteorites contain remarkably large amounts of noble gases whose isotopic composition clearly rules out a radiogenic or cosmogenic origin. Such gases have been called primordial or trapped (Gerling & Levskii 1956, Zähringer 1962, Suess et al 1964; see Pepin & Signer 1965 for a review). It has gradually become apparent that there exist two varieties of primordial gas. The first, called solar, contains the noble gases in nearly solar proportions, and is characterized by He and Ne relatively rich in the light isotope: $He^3/He^4 = (3.79 \pm 0.40) \times 10^{-4}$; $Ne^{20}/Ne^{22} = 12.5 \pm 0.2$ (Jeffery & Anders 1970, Black & Pepin 1969). Solar gas is found sporadically in meteorites of all classes, is always associated with shock and brecciation, and seems to represent trapped solar-wind particles (Wänke 1965). Some authors have proposed that the trapping took place before accretion of the meteorite parent bodies (Pellas et al 1969, Lal et al 1969), but the discovery of a similar gas component in lunar soil and breccias suggests that the implantation is an ongoing process in the solar system.

The second, "planetary" component seems to be of much more ancient origin. It is disproportionately enriched in the heavier noble gases (not unlike the Earth's atmosphere), and contains lesser amounts of the light isotopes of He and Ne: $He^3/He^4 = (1.43 \pm 0.40) \times 10^{-4}$; $Ne^{20}/Ne^{22} = 8.2 \pm 0.4$. It is found only in "primitive" meteorites, e.g. chondrites and ureilites. [The latter, though classified as achondrites, are the least differentiated members of that class. Their high C content and O^{18}/O^{16} ratio links them to the carbonaceous chondrites, and some authors believe that they are derived from them by a simple alteration process (Anders 1964, Vdovykin 1970).]

There is evidence for several additional types of neon, with Ne^{20}/Ne^{22} ≤ 3.4, 11.25 ± 0.25, and ≥ 14 (Black & Pepin 1969, Black 1970). The first two are always associated with planetary neon, but the second is not quite firmly established: it may be a mixture of the 8.2 and 12.5 components. The third component seems to be associated with solar gas, but its reality is not yet settled.

We shall be concerned only with the planetary gas, as it alone seems to date from the nebular stage. Three properties need to be explained: amounts, elemental ratios, and isotopic ratios.

The amounts of planetary gas show a most remarkable correlation with petrologic type, among both the ordinary and carbonaceous chondrites (Zähringer 1966, Marti 1967, Mazor et al 1970). Absolute abundances of Ar, Kr, Xe rise by 3 orders of magnitude from $O6$ to $C1$ chondrites or ureilites. (He and Ne show a different behavior: they are found only in $C1,2$'s and a few $O3$'s.) Elemental ratios of the heavier gases (Ar/Kr, Ar/Xe) remain constant within a factor ≤ 6 over the entire range.

A rather common view is that planetary gases reflect some kind of solubility equilibrium between nebula and solid phase, perhaps modified by adsorption effects. The equilibrium solubility of these gases may be expected to follow Henry's law, the amount of dissolved gas at a given temperature being proportional to its partial pressure. Unfortunately there exist no published data on equilibrium solubility of noble gases in meteoritic minerals at the pertinent temperatures (~ 300 to $600°K$; see above). Kirsten (1968) has measured solubilities in enstatite at $1773°K$, near its freezing point. The distribution coefficient for Ar, for example, was 2×10^{-5} cm^3 STP/g atm. To account for the amounts of planetary Ar in $O5,6$ chondrites, the distribution coefficient at lower temperatures would have to be some $10^{5\pm2}$ times greater (if $P_t = 10^{-4\pm2}$ atm).

The only pertinent experimental data are some unpublished measurements by M. S. Lancet (quoted by Jeffery & Anders 1970). The distribution coefficient of Ar in magnetite is about 1 cc STP/g atm at $500°K$. This is some 5×10^4 times higher than Kirsten's value for enstatite at $1773°K$, and only slightly less than the value required to account for the ^{36}Ar content of $O5,6$ chondrites. Solubilities for He and Ne are markedly lower, so that magnetite equilibrating with "solar" noble gases should absorb them in approximately planetary proportions.

The question is far from settled, however. No solubility data are available for Kr and Xe, or for meteoritic minerals other than magnetite. It is not at all certain that these solubilities will be of the right order to account for the evidence. Xenon, in particular, is a problem. The observed amount of Xe in $C3$ chondrites, for example, requires a distribution coefficient of $\sim 10^{5\pm2}$ cm^3 STP/g atm, corresponding to a number density in the solid some $3 \times 10^{5\pm2}$ higher than in the gas. This implies a rather strong bond between Xe and the host lattice, corresponding to a heat of solution of about -20 to -25 kcal/mole. Perhaps adsorption on a growing crystal was involved as a first step. The density of atoms in an adsorbed layer can approach that in a liquid, and even if only a fraction of the adsorbed atoms are trapped in the growing crystal, quite high gas concentrations can result.

An interesting alternative has been proposed by Jokipii (1964): enrichment of the heavier gases by ambipolar diffusion. In a partially ionized solar nebula, the solar magnetic field would retard ions relative to neutral atoms. Thus the heavier noble gases, with their lower ionization potentials, would be preferentially retained in the inner solar system. Calculated enrichment factors indeed agree rather well with terrestrial atmospheric abundances. This process can also lead to highly efficient trapping of the retained gases, if the ions are accelerated and implanted in solid grains.

The *isotopic* differences between solar and planetary gas have been the subject of much speculation (Pepin & Signer 1965, Black & Pepin 1969, Mazor et al 1970, Black 1970). Some of the obvious possibilities, such as mass fractionation by gravitational escape (Krummenacher et al 1962; Kuroda & Manuel 1970) or diffusion from a solid lattice (Zähringer 1962), seem rather unlikely, because the elemental and isotopic ratios do not show the expected correlations (Pepin 1967, Jeffery & Anders 1970). The existence of a neon component with $^{20}Ne/^{22}Ne < 3.5$ (Black & Pepin 1969) is a particularly strong argument against these models, because production of this component from solar neon with $^{20}Ne/^{22}Ne = 12.5$ requires an implausibly high fractionation factor, 10^{-12}.

No wholly satisfactory explanation is yet available. The most plausible mechanism is production of 2.6 year ^{22}Na in meteoritic silicates by a charged-particle irradiation in the solar nebula, followed by *in situ* decay to ^{22}Ne. The amount of ^{22}Na produced, and hence the final $^{20}Ne/^{22}Ne$ ratio, should vary with the composition of the mineral (Jeffery & Anders 1970). The variations in $^3He/^4He$, on the other hand, might represent a secular increase in the ^3He abundance at the solar surface (E. Schatzmann, private communication quoted by Eberhardt et al 1970). The ratio would have to change from $\sim 1.4 \times 10^{-4}$ about 4.6 AE ago to $\sim 4 \times 10^{-4}$ in recent times. [The "recent" ratio is not well defined; variations were predicted owing to differential acceleration of ^3He and ^4He (Geiss et al 1970a) and have been confirmed by measurements on Apollo 11,12 lunar soils and Al-foil solar-wind collectors (Geiss et al 1970b).]

CHRONOLOGY OF THE SOLAR NEBULA

It has been clear for more than a decade that the active period of planet formation lasted no more than a few hundred million years—a few percent of the age of the solar system (Urey 1957). The obvious challenge to meteoriticists was to determine the length of this interval with greater accuracy, and to resolve individual events within this interval. At first, all such efforts were based on "extinct" radionuclides with half-lives of 10^7 to 10^8 years, e.g.

16 Myr ^{129}I and 82 Myr ^{244}Pu (Reynolds 1968, Hohenberg et al 1967). But more recently, some spectacular advances were made in the ^{87}Rb–^{87}Sr dating method (Papanastassiou & Wasserburg 1969). In spite of the long half-life of ^{87}Rb (47 AE), improved mass-spectrometric techniques now permit resolution of time differences as small as ± 2 Myr.

The picture still is quite fragmentary, owing to the lack of a common reference point. All of the methods used measure time intervals between two events, but the events themselves differ from method to method.

Rubidium-87.—Seven eucrites (Ca-rich achondrites of very low Rb content) fall on a well-defined isochron (= line of constant age) on a ^{87}Sr/^{86}Sr vs ^{87}Rb/^{86}Sr plot, the maximum deviation being 6×10^{-5}. This implies that if these meteorites were derived from material with chondritic Rb/Sr ratio (0.25), they must have formed within ≤ 4 Myr of each other (Papanastassiou & Wasserburg 1969).

This result would be trivial if all these meteorites came from a single lava flow. However, this is rather unlikely. Two of the seven meteorites (Stannern and Moore County) differ from the others in composition or texture. Moreover, the radiation ages of these meteorites suggest that they originated in at least four and perhaps seven separate impacts (Heymann et al 1968). Thus it seems that the 4 Myr isochronism applies to an event of more than local scale. It is interesting in this connection that at least one asteroid, Vesta, has a reflection spectrum suggestive of a eucritic composition (McCord et al 1970).

The Rb/Sr method has also provided an estimate for the duration of metamorphic heating in chondrite parent bodies. An analysis of the *H*6 chondrite Guareña showed that it had a higher ^{87}Sr/^{86}Sr ratio than the eucrites when its minerals ceased to exchange Rb and Sr (Wasserburg et al 1969). The time required to evolve Guareña Sr from eucritic Sr is 74 ± 12 Myr, and it probably reflects the duration of metamorphism in this meteorite.

Iodine-129.—In the decade since its inception, the ^{129}I/^{129}Xe dating method has gone through several stages of refinement, as certain pitfalls were recognized and averted. In principle, this method dates the time when the meteorite began to retain ^{129}Xe from the decay of ^{129}I. But the various iodine-bearing sites in the meteorite differ greatly in their retentivity, the more labile ones retaining less ^{129}Xe to begin with and being most vulnerable to losses in subsequent reheating events. Manuel et al (1968) have used data on low-retentivity sites to estimate cooling rates of chondrite parent bodies. The results, 5 to 9°C/Myr, are reasonably consistent with metallographically determined cooling rates, 2 to 10°C/Myr (Wood 1967b).

However, for determining the chronology of the early solar system, data for high-retentivity sites are most pertinent. Podosek (1970) has measured (or reinterpreted) such data on 17 meteorites. A variety of chondrite classes and petrologic types were represented: $C(2, 4)$; $E(4, 5)$; $H(5)$; $L(4, 6)$; $LL(3, 6)$; as well as aubrites, a nakhlite, and a silicate inclusion from a hexahedrite.[8] Most of these meteorites began to retain ^{129}Xe within 6 to 8 Myr of each other. Errors were often only about 1 Myr or less, and thus at least some of these differences seem to be real.

It is difficult to generalize, however, because only a few members of each class have been measured. A safe conclusion is that the parent bodies of many meteorite classes cooled to xenon-retention temperatures (1000–1500°K?) within 6–8 Myr of each other. Since cooling rates must have varied with size of the body, the small spread in cooling times implies an even smaller spread in formation times: a few million years or less.

Actually, the time scale may have been considerably shorter still. Hanks & Anderson (1969) have shown that the existence of 3.6 AE old rocks on the Earth implies core formation at the very beginning of the Earth's history, in order to allow time for removal of the gravitational potential energy released. This in turn requires a high initial temperature and hence a short accretion time: 0.5 Myr or less. Though this time scale does not necessarily apply to the formation of asteroids and planetesimals, it does strengthen the case for rapid evolution of the inner solar system. [Hartmann (1970) has estimated a time scale of 10^7–10^8 years for the accretion of 10–100 km planetesimals.]

Podosek & Hohenberg (1970) noted that three primitive chondrites—Renazzo ($C2$), Chainpur ($LL3$), and Manych ($L3$)—did not give precise isochrons. They suggest that some of the constituents of these meteorites have retained the I–Xe record of an earlier stage, but there are other alternatives. At least one of these meteorites has been severely reheated by shock (Wood 1967b). This might have redistributed I and Xe so as to smear out any initial correlations.

The I–Xe method has not yet been able to establish the relative age of chondrules and matrix. The matrix of Chainpur appears to be 7.4 Myr older than the chondrules; but data on Renazzo suggest an age difference of 16 Myr in the reverse direction. In view of the ambiguities surrounding these two meteorites, the question must be regarded as open.

Aluminum-26.—This nuclide ($t_{\frac{1}{2}} = 0.74$ Myr) would be produced in fairly high abundance in nucleosynthetic processes, and has therefore been proposed as a heat source for the melting of asteroids (Urey 1955, Fish et al 1960). A slight (0.4 to 0.6%) enrichment of ^{26}Mg in Al-rich meteorite phases was tentatively attributed by Clarke et al (1970) to ^{26}Al, but a later study by Schramm et al (1970) showed no perceptible enrichment to a 2σ upper limit

---

[8] *Aubrites:* highly reduced Ca-rich achondrites, consisting largely of $MgSiO_3$. *Nakhlites:* a rare type of Ca-rich achondrite, consisting largely of *diopside* [CaMg $(SiO_3)_2$], with some Mg^{2+} and Fe^{2+} substituting for Ca^{2+} and Mg^{2+}]. *Hexahedrites:* highly reduced irons containing 5.5% Ni.

of $\leq 0.05\%$. All meteorites in this study were highly evolved, i.e. achondrites or Type 6 chondrites. This suggests that no detectable amounts of ^{26}Al were present when the feldspar in these meteorites ceased exchanging Mg with other silicate phases, but does not preclude the possibility that significant amounts of ^{26}Al were present several million years earlier.

Superheavy elements.—Xenon from chondrites is enriched in the heavier isotopes (^{131}Xe–^{136}Xe) in proportions suggestive of heavy-element fission (see Rowe 1968 for data and references to earlier work). Yet the isotopic ratios do not correspond to the fission spectrum of any known nuclide, and the amounts are far larger than expected from spontaneous fission of 4.51 AE ^{238}U or 82 Myr ^{244}Pu.

The abundance of this fission component parallels that of the most volatile elements, Pb, Bi, Tl, In, Hg, Xe, etc, being highest in the most primitive chondrites. Anders & Heymann (1969) therefore suggested that the progenitor of the fission component was a superheavy element with atomic number Z-112 to 119, because these elements (congeners of the Hg–Fr sequence) are the only volatile ones between $Z = 88$ and >130. Indeed, an "island of stability" centered on $Z = 114$, $A = 298$ has been predicted from nuclear systematics. (See Seaborg 1968 for a general review of superheavy elements, and Dakowski 1969 and Rao 1970 for a further discussion of the meteorite problem.)

Nucleosynthesis.—The presence in meteorites of extinct radionuclides shows that the last nucleosynthetic event which contributed matter to the solar system occurred no more than a few hundred Myr before the formation of the solar system. The simple picture that all solar-system matter was synthesized in a single event was abandoned long ago. Almost certainly several successive stages were involved; and there have been many attempts to determine the number, duration, and relative importance of such stages from the abundances of radioactive nuclides. Early models generally assumed "continuous" nucleosynthesis over a long period of time (10–20 AE), at a constant or slowly declining rate (Fowler & Hoyle 1960, Kohman 1961). More recent efforts favor "prompt" synthesis of the major part of the elements at an early stage (Dicke 1969, Unsöld 1969), followed by continuous synthesis and a "last-minute" event just before the formation of the solar system. Hohenberg (1969) has estimated that the three stages contributed 81–89, 0–8, and 11–13% of the total r-process material ever produced. Prompt synthesis began 8.0 to 8.8 AE ago, while last-minute synthesis took place 176 to 179 Myr before the onset of ^{129}Xe retention in chondrites. Schramm & Wasserburg (1970) have shown, however, that these estimates are still model dependent to a considerable extent. The *mean* age of the elements might indeed be close to Hohenberg's estimate of 8 AE, but the time-dependence of the nucleosynthesis function is not well determined, and thus total times as high as 200 AE cannot be excluded. The interval between cessation of nucleosynthesis and onset of Xe retention is better defined, and probably lies between 75 and 250 Myr.

SUMMARY

Most of the chemical differences among chondrites, involving some 55 elements, can be explained by a small number of fractionation processes in a cooling solar nebula.

A. Ordinary Chondrites (2 to 4 au)

1. An early condensate, containing about one third of the refractory elements (Ca, Al, Ti, U, Th, lanthanides, Pt metals, etc), was lost at $\geq 1300°$K.

2. After condensation of the remaining material to grains of $10^{-5} - 10^{-6}$ cm, 20 to 50% of the nickel-iron was lost. The fractionation seems to have taken place below 1050°K, perhaps around 700°K, and may have been based upon the ferromagnetic properties of the metal grains.

3. When the Fe^{2+} content of the silicates had risen to $\sim 20\%$ by equilibration with the nebular gas ($\sim 600°$K), perhaps 75% of the condensate was remelted to millimeter-sized droplets by local heating events (electric discharges), on a time scale of seconds to minutes. Volatiles were lost from the remelted material.

4. The unremelted, fine-grained material continued to take up volatiles from the nebula (Pb, Bi, Tl, In, etc) and accreted together with the remelted material. Accretion took place in a 100° interval centered on $510^{+80}_{-60}°$K, at a pressure of $10^{-4\pm2}$ atm.

5. Within a few million years after accretion, chondrites were reheated to $\leq 1200°$K by some unknown heat source, perhaps Al^{26}. They then cooled, traversing the 1000 to 700°K interval at rates of 2 to 10°C/Myr.

B. Enstatite Chondrites (2 to 4 au, perhaps inner fringe of asteroid belt)

1. As above, but loss approached 50%.

2, 3. As above, but in a more reducing gas phase ($C/O \geq 0.9$?). No fully satisfactory explanation for the more reducing gas composition is available. A gas-dust fractionation is a possibility.

4. As above. The total pressure may have been higher.

5. As above, but temperature and cooling rate are less well determined.

C. Carbonaceous Chondrites [outer fringe of asteroid belt, or original source region of comets (~ 5 to ~ 40 au?)]

1, 2. These processes seem to have been largely inoperative in the source region of the carbonaceous chondrites. Only a slight metal-silicate fractionation ($\leq 10\%$) is evident among members of this group.

3. There is no evidence for remelting in $C1$'s ($\leq 10\%$ is possible, though). $C2$'s to $C4$'s show an increasing degree of remelting (≤ 40 to $\sim 70\%$).

4. Accretion seems to have taken place from ~ 450 to $\sim 300°$K. Hydrated silicates, magnetite, and organic compounds are present in meteorites that accreted below 400°K ($C1$'s, $C2$'s, some $C3$'s). The pressure may have

been slightly lower than that in the region of ordinary chondrites, but probably not less than $\sim 10^{-6}$ atm.

5. Reheating was much less severe than for ordinary chondrites, especially in the case of $C1$'s and $C2$'s.

D. *Inner Planets (0.4 to 1.5 au)*

1. Uncertain, and possibly small. Earth and Moon may be slightly ($\leq 50\%$) enriched in refractories.

2. Uncertain, and possibly small. Earth and Venus may be slightly (~ 10 to $\sim 30\%$) enriched in Fe. Extreme values of Fe/Si in Mercury and Moon may be due to special factors: early condensation of Fe at 0.4 au and preferential accretion; fractionation of metal between planet and satellite.

3. Remelting may have affected the major part of planetary matter. The depletion of alkalis in the Earth and Moon suggests contents of ~ 85 and $\sim 95\%$ remelted, alkali-free material.

4. Abundance pattern of volatiles in Earth is similar to that in ordinary chondrites. This is consistent with accretion either at a constant temperature near 550°K or at falling temperatures, the last few percent of the accreted material being carbonaceous-chondrite-like. The 100-fold greater depletion of the Moon implies either formation at higher temperatures (620°K) or a much smaller content of the volatile-rich, carbonaceous-chondrite-like component (0.03%).

The principal conclusion of this paper is that substantial chemical fractionations took place in the inner solar system. These processes were independent of each other, and thus the known meteorite types represent only a limited sampling of the possible range of planetary materials. The proper building blocks for the construction of planetary models are not the few known classes of meteorites, but the four components that behaved independently during chemical fractionations: early condensate, metal, remelted dust, and unremelted dust. Thus at least eight degrees of freedom are available for the composition of a planet: one each for the amounts and formation temperatures of the four components.

ACKNOWLEDGMENTS

I am indebted to Professor Johannes Geiss for his hospitality during my stay at Berne. This work was supported in part by AEC contract AT(11-1)-382, NASA grant NGL 14-001-010, and the Kommission für Weltraumforschung der Schweizerischen Naturforschenden Gesellschaft.

LITERATURE CITED

Ahrens, L. H., von Michaelis, H., Erlank, A. J., Willis, J. P. 1969. *Meteorite Research*, ed. P. M. Millman, 166–73. Dordrecht: Reidel. 940 pp.

Anders, E. 1964. *Space Sci. Rev.* 3:583–714

Anders, E. 1968. *Accounts Chem. Res.* 1: 289–98

Anders, E. 1970. *Science* 169:1309–10

Anders, E. 1971. *Geochim. Cosmochim. Acta* 35. In press

Anders, E., Ganapathy, R., Keays, R. R., Laul, J. C., Morgan, J. W. 1971. *Proc. Apollo 12 Lunar Sci. Conf.*, *Geochim. Cosmochim. Acta Suppl. 2.* In press

Anders, E., Heymann, D. 1969. *Science* 164:821–23

Anderson, D. L., Sammis, C., Jordan, T. 1970. Preprint, *Composition of the Mantle and Core*

Arnold, J. R., Suess, H. E. 1969. *Ann. Rev. Phys. Chem.* 20:293–314

Arrhenius, G., Alfvén, H. 1971. *Earth Planet. Sci. Lett.* 10:253

Banerjee, S. K. 1967. *Nature* 216:781

Baschek, B., Garz, T., Holweger, H., Richter, J. 1970. *Astron. Ap.* 4:229–33

Black, D. C. 1970. Paper submitted for 1970 Nininger Meteorite Award

Black, D. C., Pepin, R. O. 1969. *Earth Planet. Sci. Lett.* 6:395–405

Blander, M. 1971. *Geochim. Cosmochim. Acta* 35:61–76

Blander, M., Abdel-Gawad, M. 1969. *Geochim. Cosmochim. Acta* 33:701–16

Blander, M., Katz, J. L. 1967. *Geochim. Cosmochim. Acta* 31:1025–34

Briggs, M. H. 1963. *Nature* 197:1290

Brown, H. 1950. *Ap. J.* 111:641–53

Bullen, K. E. 1969. *Ann. Rev. Astron. Ap.* 7:177–200

Cameron, A. G. W. 1962. *Icarus* 1:13–69 Ibid 1963. 1:339–42

Cameron, A. G. W. 1966. *Earth Planet. Sci. Lett.* 1:93–96

Cameron, A. G. W. 1968. *Origin and Distribution of the Elements*, ed. L. H. Ahrens, 125–43. Oxford: Pergamon. 1178 pp.

Cameron, A. G. W. 1969. *Meteorite Research*, ed. P. M. Millman, 7–15. Dordrecht: Reidel. 940 pp.

Chou, C. L., Cohen, A. J. 1970. *Meteoritics* 5:188

Christophe Michel-Lévy, M. 1968. *Bull. Soc. Fr. Minéral. Cristallogr.* 91:212–14

Christophe Michel-Lévy, M. 1969. *Meteorite Research*, ed. P. M. Millman, 492–99. Dordrecht: Reidel. 940 pp.

Clark, S. P., Turekian, K. K., Grossman, L.

1971. The Early History of the Earth. To be published in *Proc. Symp. in Honor of Francis Birch*. McGraw-Hill

Clarke, W. B., deLaeter, J. R., Schwarcz, H. P., Shane, K. C. 1970. *J. Geophys. Res.* 75:448–62

Clayton, R. N. 1963. *Science* 140:192–93

Dakowski, M. 1969. *Earth Planet. Sci. Lett.* 6:152–54

Dayhoff, M. O., Lippincott, E. R., Eck, R. V., 1964. *Science* 146:1461–64

Dicke, R. H. 1969. *Ap. J.* 155: 123–34

Dodd, R. T. 1969. *Geochim. Cosmochim. Acta* 33:161–203

Dodd, R. T., Van Schmus, W. R., Koffman, D. M. 1967. *Geochim. Cosmochim. Acta* 31:921–51

Dolan, J. F. 1965. *Ap. J.* 142:1621–32

DuFresne, E. R., Anders, E. 1962. *Geochim. Cosmochim. Acta* 26:1085–114

Eberhardt, P. et al 1970. *Geochim. Cosmochim. Acta Suppl.* 1:1037–70

Eck, R. V., Lippincott, E. R., Dayhoff, M. O., Pratt, Y. T. 1966. *Science* 153: 628–33

Ehmann, W. D., Lovering, J. R. 1967. *Geochim. Cosmochim. Acta* 31:357–76

Fish, R. A., Goles, G. G., Anders, E. 1960. *Ap. J.* 132:243–58

Fowler, W. A., Hoyle, F. 1960. *Ann. Phys.* 10:280–302

Friedel, R. A., Sharkey, A. G. Jr. 1963. *Science* 139:1203–5

Friedel, R. A., Sharkey, A. G. Jr. 1968. *US Bur. Mines, Rep. Invest. 7122.* US Dep. Interior

Ganapathy, R., Keays, R. R., Laul, J. C., Anders, E. 1970. *Geochim. Cosmochim. Acta Suppl.* 1:1117–42

Garz, T., Kock, M. 1969. *Astron. Ap.* 2:274–79

Garz, T. et al 1969. *Nature* 223:1254–55

Garz, T., Richter, J., Holweger, H., Unsöld, A. 1970. *Astron. Ap.* 7:336–39

Gast, P. W. 1971. To be published in *Proc. Symp. in Honor of Francis Birch*. McGraw-Hill

Gast, P. W. 1968. *History of the Earth's Crust*, ed. R. Phinney, 15–27. Princeton: Princeton Univ. Press. 244 pp.

Geiss, J., Hirt, P., Leutwyler, H. 1970a. *Solar Phys.* 12:458–83

Geiss, J., Eberhardt, P., Bühler, F., Meister, J., Signer, P. 1970b. *J. Geophys. Res.* 75:5972–79

Gerling, E. K., Levskii, L. K. 1956. *Dokl. Akad. Nauk USSR* 110:750–53

Gilman, R. C. 1969. *Ap. J.* 155:L185–87

Goles, G. G. 1969. *The Handbook of Geochemistry* 1: Chap. 5, 123. Berlin: Springer

Hanks, T. C., Anderson, D. L. 1969. *Phys. Earth Planet. Interiors* 2:19–29

Harris, P. G., Tozer, D. C. 1967. *Nature* 215:1449–51

Hartmann, W. K. 1970. *Mém. Soc. Sci. Liège [8°], 5th Sér.* 19:215–27

Hayatsu, R. 1964. *Science* 146:1291–93

Hayatsu, R., Studier, M. H., Oda, A., Fuse, K., Anders, E. 1968. *Geochim. Cosmochim. Acta* 32:173–90

Hayes, J. M. 1967. *Geochim. Cosmochim. Acta* 31:1395–440

Hayes, J. M., Biemann, K. 1968. *Geochim. Cosmochim. Acta* 32: 239–67

Herbig, G. H. 1970a. *Optical Spectroscopy of Interstellar Molecules*. Presented at Am. Astron. Soc. Meet., Boulder, Colo.

Herbig, G. H. 1970b. *Ap. J.* 162: 557–70

Herbig, G. H. 1970c. Introductory Remarks to Symp. "Evolution des Etoiles avant leur Sejour sur la Sequence Principale." *Mém. Soc. Sci. Liège [8°], 5th Sér.* 19:13–26

Heymann, D., Mazor, E., Anders, E. 1968. *Geochim. Cosmochim. Acta* 32:1241–68

Hohenberg, C. M. 1969. *Science* 166:212–15

Hohenberg, C. M., Munk, M. N., Reynolds, J. H. 1967. *J. Geophys. Res.* 72:3139–77

Jeffery, P. M., Anders, E. 1970. *Geochim. Cosmochim. Acta* 34: 1175–98

Jokipii, J. R. 1964. *Icarus* 3:248–52

Keays, R. R., Ganapathy, R. Anders, E. 1971. *Geochim. Cosmochim. Acta* 35. In press

Keil, K., Huss, G. I., Wiik, H. B. 1969. *Meteorite Research*, ed. P. M. Millman, 217. Dordrecht: Reidel. 940 pp.

Kerridge, J. F. 1970. *Earth Planet. Sci. Lett.* 9:299–306

Kirsten, T. 1968. *J. Geophys. Res.* 73: 2807–10

Kohman, T. P. 1961. *J. Chem. Educ.* 38: 73–82

Krouse, H. R., Modzeleski, V. E. 1970. *Geochim. Cosmochim. Acta* 34:459–74

Krummenacher, D., Merrihue, C. M., Pepin, R. O., Reynolds, J. H. 1962. *Geochim. Cosmochim. Acta* 26:231–40

Kuroda, P. K., Manuel, O. K. 1970. *Nature* 227:1113–16

Lal, D., Rajan, R. S. 1969. *Nature* 223: 269–71

Lancet, M. S., Anders, E. 1970. *Science* 170: 980–82

Larimer, J. W. 1967. *Geochim. Cosmochim. Acta* 31:1215–38

Ibid 1968a. 32:1187–207

Ibid 1968b. 32:965–82

Larimer, J. W. 1970. Presented at Conf. Orig. Evol. Planets, Calif. Inst. Tech., Jet Propulsion Lab, Pasadena, Calif.

Larimer, J. W. 1971. *Geochim. Cosmochim. Acta* 35. In press

Larimer, J. W., Anders, E. 1967. *Geochim. Cosmochim. Acta* 31: 1239–70

Ibid 1970. 34: 367–88

Lemmon, R. M. 1970. *Chem. Rev.* 70:95–109

Levin, B. J. 1957. *Mém. Soc. Sci., Liège 4th Ser.* 18:186–97

Ibid 1963. *5th Ser.* 7:39–46

Levin, B. J. 1969. *Russ. Chem. Rev.* 38: 65–78

Lord, H. C. III. 1965. *Icarus* 4:279–88

Low, F. J. 1969. *Science* 164:501–5

Low, F. J., Smith, B. J. 1966. *Nature* 212: 675–76

Manuel, O. K., Alexander, E. C. Jr., Roach, D. V., Ganapathy, R. 1968. *Icarus* 9:291–304

Marti, K. 1967. *Earth Planet. Sci. Lett.* 2:193–96

Marvin, U. B., Wood, J. A., Dickey, J. S. Jr. 1970. *Earth Planet. Sci. Lett.* 7:346–50

Mazor, E., Heymann, D., Anders, E. 1970. *Geochim. Cosmochim. Acta* 34:781–824

McCord, T. B., Adams, J. B., Johnson, T. V. 1970. *Science* 168:1445–47

Miller, S. L. 1953. *Science* 117:528–29

Morris, S., Wyller, A. A. 1967. *Ap. J.* 150: 877–97

Mueller, O., Baedecker, P., Wasson, J. 1971. *Geochim. Cosmochim. Acta* 35. In press

Noddack, I., Noddack, W. 1930. *Naturwissenschaften* 18:758–64

Papanastassiou, D. A., Wasserburg, G. J. 1969. *Earth Planet. Sci. Lett.* 5:361–76

Pellas, P., Poupeau, G., Lorin, J. C., Reeves, H., Audouze, J. 1969. *Nature* 223:272–74

Pepin, R. O. 1967. *Earth Planet. Sci. Lett.* 2:13–18

Pepin, R. O., Signer, P. 1965. *Science* 149: 253–65

Podosek, F. A. 1970. *Geochim. Cosmochim. Acta* 34:341–66

Podosek, F. A., Hohenberg, C. M. 1970. *Earth Planet. Sci. Lett.* 8:443–47

Rao, M. N. 1970. *Nucl. Phys.* A140:69–73

Reed, G. W. Jr., Jovanovic, S. 1967. *J. Geophys. Res.* 72:2219–28

Reynolds, J. H. 1968. *Origin and Distribution of the Elements*, ed. L. H. Ahrens, 367–77. Oxford: Pergamon. 1178 pp.

Ringwood, A. E. 1959. *Geochim. Cosmochim. Acta* 15:257–83

Ibid 1966. 30:41–104

Ringwood, A. E., Essene, E. 1970. *Science* 167:607–10

Rowe, M. W. 1968. *Geochim. Cosmochim. Acta* 32:1317–26

Schmitt, R. A., Smith, R. H., Goles, G. G. 1965. *J. Geophys. Res.* 70:2419–44

Schramm, D. N., Tera, F., Wasserburg, G. J. 1970. *Earth Planet. Sci. Lett.* 10:44–59

Schramm, D. N., Wasserburg, G. J. 1970. *Ap. J.* 162:57–69

Seaborg, G. T. 1968. *Ann. Rev. Nucl. Sci.* 18:53–152

Singer, S. F., Bandermann, L. W. 1970. *Science* 170:438–39

Smith, J. W., Kaplan, I. R. 1970. *Science* 167:1367–70

Studier, M. H., Hayatsu, R., Anders, E. 1968. *Geochim. Cosmochim. Acta* 32:151–74

Suess, H. E. 1962. *J. Geophys. Res.* 67:2029–34

Suess, H. E. 1963. *Origin of the Solar System*, ed. R. Jastrow, A. G. W. Cameron, 154–46. New York: Academic. 176 pp.

Suess, H. E. 1964. *Isotopic and Cosmic Chemistry*, ed. H. Craig, S. L. Miller, G. J. Wasserburg, 385–400. Amsterdam: North-Holland. 553 pp.

Suess, H. E. 1965. *Ann. Rev. Astron. Ap.* 3:217–34

Suess, H. E., Wänke, H., Wlotzka, F. 1964. *Geochim. Cosmochim. Acta* 28:595–607

Tandon, S. N., Wasson, J. T. 1968. *Geochim. Cosmochim. Acta* 32:1087–109

Tsuji, T. 1964. *Ann. Tokyo Astron. Obs., Univ. Tokyo, 2nd Ser.* 9: No. 1

Turekian, K. K., Clark, S. P. Jr. 1969. *Earth Planet. Sci. Lett.* 6:346–48

Unsöld, A. O. J. 1969. *Science* 163:1015–25

Urey, H. C. 1952a. *Geochim. Cosmochim. Acta* 2:269–82

Urey, H. C. 1952b. *The Planets.* New Haven: Yale Univ. Press. 245 pp.

Urey, H. C. 1953. *Plenary Lectures, 13th Int. Congr. Pure Appl. Chem., London, 1953*, 188–214

Urey, H. C. 1954. *Ap. J. Suppl.* 6:147–73

Urey, H. C. 1955. *Proc. Nat. Acad. Sci. USA* 41:127

Urey, H. C. 1956. *Ap. J.* 124:623–37

Urey, H. C. 1957. *Progr. Phys. Chem. Earth* 2:46

Urey, H. C. 1961. *J. Geophys. Res.* 66:1988–91

Urey, H. C. 1964. *Rev. Geophys.* 2:1–34

Urey, H. C. 1966. *MNRAS* 131:199–223

Urey, H. C. 1967. *Quart. J. Roy Astron. Soc.* 8:23–47

Urey, H. C., Craig, H. 1953. *Geochim. Cosmochim. Acta* 4:36–82

Van Schmus, W. R. 1969. *Earth Sci. Rev.* 5:145–84

Van Schmus, W. R., Wood, J. A. 1967. *Geochim. Cosmochim. Acta* 31:747–65

Vdovykin, G. P. 1967. *Carbon Matter of Meteorites (Organic Compounds, Diamonds, Graphite).* Moscow: Nauka. 271 pp.

Vdovykin, G. P. 1970. *Space Sci. Rev.* 10:483–510

Wänke, H. 1965. *Z. Naturforsch.* 20a:946–49

Wasserburg, G. J., Papanastassiou, D. A., Sanz, H. G. 1969. *Earth Planet. Sci. Lett.* 7:33–43

Whipple, F. L. 1966. *Science* 153:54–56

Wood, J. A. 1958. *Smithsonian Ap. Obs. Tech. Rep. No. 10. ASTIA Doc. No. AD 158364*

Wood, J. A. 1962a. *Geochim. Cosmochim. Acta* 26:739–49

Wood, J. A. 1962b. *Nature* 194:127–30

Wood, J. A. 1963. *Icarus* 2:152–80

Wood, J. A. 1967a. *Geochim. Cosmochim. Acta* 31:2095–108

Wood, J. A. 1967b. *Icarus* 6:1–49

Wood, J. A. 1968. *Meteorites and the Origin of Planets.* New York: McGraw-Hill. 117 pp.

Yavnel, A. A. 1970. *Geokhimiya* 228–42

Zähringer, J. 1962. *Z. Naturforsch.* 17a:460–71

Zähringer, J. 1966. *Earth Planet. Sci. Lett.* 1:379–82

DWARF GALAXIES

PAUL W. HODGE

Astronomy Department
University of Washington, Seattle, Washington

For the purpose of this review, the term *dwarf galaxy* is used in its historically traditional way, that is, to refer to galaxies of small intrinsic size, small absolute luminosity, and low surface brightness. This definition excludes other types of small galaxies, such as the compact galaxies or the class of spiral galaxies that is small in intrinsic size but normal in other respects.

The dwarf galaxies discussed in this review are those objects that make up most of the faint end of the galaxy luminosity function. Although this luminosity function is still not well established for field galaxies, it is clear that in the nearby clusters and groups, dwarf galaxies are an important constituent in terms of numbers. Whether they are as common in the field as they are, for example, in the local group, is a matter of dispute, as well as a matter of importance concerning their origin (Zwicky 1957).

Virtually all of the dwarf galaxies so defined can be classified as either dwarf ellipticals or dwarf irregulars. Because the features and properties of these two classes are very distinctive, they are discussed below separately.

The luminosity limit of dwarf galaxies lies approximately in the range of absolute magnitude -15. Thus, the small close companions to Andromeda, NGC 205 and M32, are both near the borderline between dwarf and normal galaxies, while the Small Magellanic Cloud, with an absolute magnitude of -16, is too bright to be so considered. The faintest known dwarf galaxies have absolute magnitudes fainter than -10.

DWARF ELLIPTICAL GALAXIES

Most of the dwarf ellipticals that have been studied in detail are members of the local group of galaxies. These are primarily objects of extremely low surface brightness, making them so inconspicuous that they were not discovered until the middle of the 20th century. For these the most common absolute luminosities are approximately -10, and the combination of their low total brightness and low surface brightness has made them virtually undetectable beyond distances of about 250 kpc. Exceptions are the four companions of Andromeda, which have higher than average luminosities.

When the smallest and faintest of these objects were first discovered there was some controversy over whether they should be considered separate galaxies or merely unusually distant globular star clusters. Over the years, tradition has dictated that the linear size is the determining factor in this respect. The so-called *intergalactic tramp* globular star clusters are all approximately the same in linear diameter as the globular star clusters imbedded in our Galaxy, whereas the objects called *dwarf elliptical galaxies* are approximately an order of magnitude larger than normal globular star clusters. Very little else is different between the two objects, as their total luminosities and estimated masses overlap one with the other.

MEMBERS OF THE LOCAL GROUP

The local group of galaxies contains ten recognized elliptical galaxies that can be considered members of the dwarf class. Four of these, the four most luminous and least extreme in their dwarf characteristics, are companions to M31. The remaining six are all within 230 kpc of our local Galaxy.

Sculptor.—The first of the extreme dwarf galaxies near our local system to be discovered is the object called Sculptor, named after the constellation in which it lies. It was discovered by Shapley (1938) and its true nature was revealed by a study of it using the 60 inch telescope at Boyden Observatory (Shapley 1939) and the 100 inch telescope at Mt. Wilson (Baade & Hubble 1939). It is an extremely faint object, spread over more than a degree and difficult to photograph. Measurements of its brightest stars and of its RR Lyrae variables imply that its distance is 84 ($\pm$14) kpc (Hodge & Michie 1969).

Fornax.—The second of these objects to be discovered was found as a result of an exhaustive survey of Harvard plates carried out by Shapley and his co-workers (Shapley 1939). Shapley recognized that the Fornax system (see Figure 1) was clearly much more distant than the system in Sculptor, but pointed out that its total apparent magnitude was very nearly the same. The Fornax system, therefore, is intrinsically brighter than Sculptor. It differs from the other extreme dwarfs in having a somewhat higher central surface brightness, a brighter absolute magnitude, and a larger intrinsic size. It is in a sense intermediate in its character between the extreme dwarfs like the Sculptor system and the more luminous objects near M31. Its distance was found by Baade & Hubble (1939) to be 188 kpc, a figure based on an approximate and difficult photographic transfer of magnitudes to the Fornax system, where the brightest stars in the globular star clusters of the system were used as criteria.

Leo I.—Four additional extreme dwarf ellipticals were discovered in the 1950s as a result of the examination of the plates of the Palomar Sky Survey (Harrington & Wilson 1950, Wilson 1955). Leo I has not been studied extensively because of its proximity to the star Regulus, which makes photography of the system difficult for large telescopes with correcting lenses. Baade (1963) obtained plates of it with the 200 inch telescope, using di-

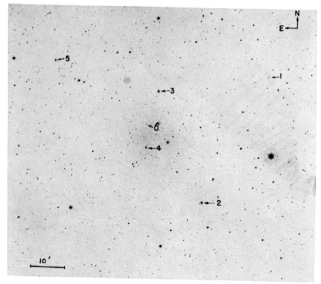

FIGURE 1. The Fornax dwarf galaxy, photographed by the ADH Schmidt telescope. The six globular clusters are identified.

aphragms in the optics to cut down reflection from Regulus. He estimated that the distance of the galaxy is approximately the same as that of the Leo II system for which Sandage (1961) quotes a distance modulus of $m - M = 21.8$, corresponding to 230 kpc.

The total apparent magnitude of Leo I has been measured by Holmberg (1958), who finds $m_V = 10.40$ and $CI = +0.87$. The dimensions from Holmberg's data are 12.0 by 9.5 min; these compare with Harrington & Wilson's estimate for the diameter of 17 min. Accepting the distance of 230 kpc leads to a value for the absolute visual magnitude of the system of $M_V = -11.4$, and the linear dimensions determined by Holmberg become 770 by 610 pc.

Leo II.—A nearly circular faint galaxy not far from Leo I was similarly discovered during the process of the Palomar Observatory–National Geographic Sky Survey. On the basis of Baade's (1963) plates and the original estimate by Wilson that the mean brightness of its 20 brightest stars averages $m_{pg} = 20.0$, the distance modulus is estimated to be $m - M = 21.8$, corresponding to 230 kpc. Apparently, the two Leo systems are at the same distance from us. The total apparent magnitude of Leo II has been measured by Holmberg (1958), who finds $m_{pv} = 12.04$ and $CI = 0.81$. From Holmberg's data and the assumed distance, we find the system to have an absolute visual magnitude of $M_V = -9.8$.

Ursa Minor.—The Ursa Minor dwarf elliptical galaxy is one of the most difficult to photograph. It was discovered in the Palomar survey (Wilson 1955), and Baade (1963) obtained an excellent series of plates with the 200 inch telescope. van Agt (1967, 1968) has used the RR Lyrae variables to derive an apparent distance modulus of 19.5.

Draco.—The last of the six extreme dwarf elliptical galaxies is a faint system discovered by Wilson (1955). This system is the most thoroughly studied, due to an extensive series of plates obtained by Baade and their virtually complete reduction by Miss Swope (Baade & Swope 1961) using photoelectric calibrations carried out by Arp. The magnitude at mean luminosity for the RR Lyrae variables is $V = 20.48$, indicating a true distance modulus of 19.14 and implying a linear distance of 68 kpc.

The dwarf companions of M31.—The Andromeda galaxy is apparently accompanied by four elliptical galaxies, all of which satisfy, to various degrees, our criteria for dwarf galaxies. The closest to the Andromeda spiral is the smallest and brightest of these, M32, which has an absolute magnitude of −15.5 and is therefore on the borderline between dwarf and normal elliptical galaxies. It differs very strikingly from the dwarf ellipticals described above in its comparatively steep luminosity gradient and high central surface brightness. It also differs in having a bright nucleus at the center, as well as a redder integrated color and probably a greater abundance of heavy elements (Spinrad 1966).

NGC 205 (see Figure 2), also a close companion to M31, has more of the appearance of a dwarf galaxy, which is due to its more gradual luminosity gradient and its smaller overall luminosity density. Nevertheless, its total integrated apparent magnitude is 8.9, indicating an absolute magnitude of −15.7. Thus, it also is a borderline case. It has a conspicuous nucleus, though not as bright as that of M32.

NGC 185, with an absolute magnitude of −13.9, is more akin to the extreme dwarf elliptical galaxies, both in its total luminosity and in its luminosity gradient. Its surface brightness is smaller than for NGC 205. Its nearby companion NGC 147 is the most similar in appearance to the extreme dwarf ellipticals, although its absolute magnitude is more than a magnitude greater than any of the latter. It is 2.4 kpc in diameter and has a total absolute magnitude of −13.6. There is no reported nucleus, which is also true for NGC 185, and in that respect both are more similar to the extreme dwarf ellipticals than to NGC 205 or M32. Distances to these four elliptical galaxies are currently based on the statement by Baade (1944) that they all are resolved at approximately the same apparent magnitude as the halo Population II stars of M31. They all have radial velocities that lie within only a few kilometers per second of each other (Humason, Mayall & Sandage 1956).

STELLAR CONTENT

Information on the general stellar content of dwarf elliptical galaxies is available only for those within the local group. This is because of the ex-

FIGURE 2. NGC 205, photographed at the U S Naval Observatory,
Flagstaff Station, by John Priser.

treme faintness of the systems, which do not at present permit study easily
by means of spectrophotometric analysis. For the local group galaxies our
information comes both from the integrated properties of the dwarfs and
from detailed measurements of the individual bright stars.

Integrated colors.—Some information of a gross nature can be inferred
from the integrated colors of a Population II stellar system. Baum (1959)
was one of the first to show a progression of color between dwarf galaxies and
giant. He showed from some preliminary measurements of the colors of
dwarf galaxies that they resemble the globular star clusters in integrated
colors more nearly than they resemble typical giant ellipticals. More recently published measures of the colors of dwarf ellipticals confirm this (de
Vaucouleurs 1961). Four of the six extreme dwarf elliptical galaxies near our
Galaxy have measured colors; they include the two Leo systems, whose
colors were measured photographically by Holmberg (1958), the Sculptor
system (Hodge 1966), and Fornax (de Vaucouleurs & Ables 1968). In all
cases the colors of the dwarf elliptical galaxies are approximately $B-V=0.6$

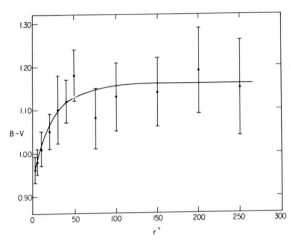

FIGURE 3. The variation of color with central distance
for NGC 185 (Hodge 1963).

to 0.7 mag. These results conform to the idea that the dwarf elliptical galaxies
are, at least in stellar content, nearly indistinguishable from the globular
clusters. The much redder colors of giant elliptical galaxies indicate the
preponderance in them of a mixture of stellar content, including higher metal
abundance giants (Spinrad 1962).

The integrated colors of the brighter dwarfs near M31 are redder than
for the extreme dwarfs, presumably for the same reason. NGC 147 and M32
are apparently made up primarily if not totally of old stars. For M32,
scanner observations indicate the presence of fairly metal-rich giants (Spinrad & Taylor 1971). The colors of NGC 185 (see Figure 3) and NGC 205,
however, are contaminated in their centers by the presence there of a Population I component. The brightest OB stars were first reported by Baade
(1944) and their effects on the colors of the systems have been detected by
de Vaucouleurs (1961) and Hodge (1963, 1970). In both cases the color in
the center is ∼0.2 mag bluer than the general color reached in the outer
regions.

Color-magnitude diagrams.—The nearby extreme dwarfs are close enough
for detailed color-magnitude diagrams to be derived for four of them, Draco
(Baade & Swope 1961), Ursa Minor (van Agt 1967), Sculptor (Hodge 1965),
and Leo II (Swope 1968). The latter lacks accurate photoelectric calibration
and so has not been published in detail.

The color-magnitude diagram of the Draco system (see Figure 4) is very
similar to that for globular star clusters of our Galaxy (Baade & Swope
1961). The horizontal branch is strongly populated on the red side, with
only a few stars delineating it on the blue side and it is thus similar in this

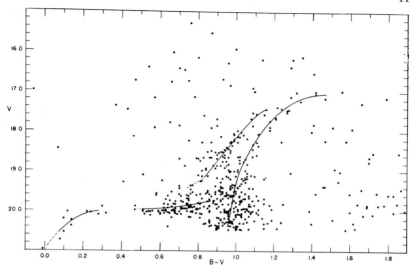

FIGURE 4. The color-magnitude diagram for the Draco
system (Baade & Swope 1961).

sense to moderately metal-poor globulars. The slope of the giant branch, though somewhat difficult to determine because of the presence of foreground stars, is quite steep. Baade & Swope estimate that the difference in visual magnitude between the horizontal branch and the giant branch at a color of $B - V = 1.4$ mag is approximately 3.0 mag, indicating that in this measure it is most similar to very low metal abundance globular star clusters (Sandage & Wallerstein 1960).

The Ursa Minor galaxy has a rather similar color-magnitude diagram (van Agt 1967). Its giant branch is also rather poorly defined, primarily because of a paucity of stars and confusion with field stars of similar color, but as best can be judged it is steep, with a slope like that for Draco. However, the horizontal branch of Ursa Minor is strongly populated on the blue side of the RR Lyrae gap instead of the red side, which actually makes it more similar to the extreme metal-poor clusters in this respect than Draco.

Leo II, on the other hand, has a color-magnitude diagram more similar to that of Draco, with a steep giant branch and with most of its horizontal-branch stars on the red end (Swope 1968). Unlike the color-magnitude diagram for Draco, however, that for Leo II does not show a split into two components.

For the Sculptor system, the lack of faint photoelectric standards allowed the construction of only the bright part of the color-magnitude diagram, down to but not including the level of the horizontal branch (Hodge 1965). In the steep but not extreme slope of the giant branch the Sculptor color-magnitude diagram closely resembles that of the galactic globular cluster M3. The top of the giant branch reaches colors of $B - V = +1.8$, the usual limit for globular star clusters. There are two stars that may be members of the system lying on the main sequence with absolute magnitudes of approximately -1 and -2, respectively. These are possible examples of the so-called *blue stragglers* of the type occasionally found in globular star clusters.

Luminosity functions.—Studies of the luminosity functions for the stars in dwarf elliptical galaxies show very striking similarities to those for globular star clusters. Shapley was the first to measure the luminosity function of one of them, and his results for Sculptor clearly show good agreement, when plotted differentially, with that for M3 (Shapley 1939, Hodge 1966). Results for the Draco and Ursa Minor systems, although more poorly defined because of the relatively smaller number of stars, also are in good agreement with that for M3.

Variable stars.—Various types of Population II variable stars have been found in most of the elliptical dwarf galaxies of the local group. For the four companions of Andromeda, Baade (1963) has mentioned the presence of unstudied long-period variables. For the closer dwarf ellipticals, five contain recognized RR Lyrae variables and a few somewhat unusual W Virginis variables. In Sculptor, the first of the systems to be studied in any detail, Baade & Hubble detected approximately 40 RR Lyrae variables (1939). In subsequent study of a much more extensive series of plates of the Sculptor system, Thackeray (1950) detected 216 variables, deriving accurate periods for 30 RR Lyrae variables. He estimated, on the basis of the size of his sample and the total size of the system, that the total number of RR Lyrae variables in Sculptor might eventually reach 700. More recent study by van Agt (1971) confirms this prediction of Thackeray. The mean period for the *ab* RR Lyrae variables found by Thackeray was 0.54 days, very similar to the mean for the variables of M3. This would suggest that Sculptor is of intermediate metallicity, rather than extremely metal poor, according to the apparent relationship between abundance estimates and RR Lyrae variables (Sandage & Wallerstein 1960).

The variable stars of the Ursa Minor system include 92 objects, all detected on a series of plates taken by Baade. According to van Agt (1967, 1968) the mean period of the *ab* RR Lyrae variables is 0.64 days. He noted a similarity to the variables of Omega Centauri, which has an almost identical period-amplitude relationship to that found for the 28 RR Lyrae variables studied in the Ursa Minor system.

In Draco, it is estimated that there are at least 260 variable stars detectable on Baade's plates. Baade & Swope (1961) measured 138 of these, 133 of which were RR Lyrae variables. The mean period was 0.61 days, and the properties of the variables were very similar to those of the RR Lyraes in M3.

Leo II contains nearly 200 variables (Swope 1968). Studies of 76 RR Lyrae variables showed a mean period of 0.59 days and a tendency for a smaller amplitude than for those in Draco.

The systems for which variable stars have been searched all contain objects that appear to be similar to short-period W Virginis stars. These objects, however, show distinct and puzzling dissimilarities with that class and have been the subject of considerable discussion (e.g., van Agt 1967). Most of them have periods between 1 and 2 days and are ~1 mag brighter than the RR Lyrae variables.

For the Leo II system Miss Swope (1968) reported twelve irregular variables, six of which were at the red end of the giant branch and are apparently similar to the irregular variables found in that position for globular clusters of our Galaxy. The other six were somewhat bluer than the RR Lyrae variables. No long-period variables have been found in any of these extreme dwarf systems, a somewhat surprising fact since they are found in galactic globular clusters and are reported for the brighter dwarfs in the Andromeda group. Also surprising is the lack of W Virginis or RV Tauri variables with periods greater than about 2 days. On the basis of the sample of globular clusters in our Galaxy, this lack suggests a significant and important difference between the two kinds of systems.

Globular star clusters.—Only one of the extreme dwarf galaxies near the local galaxy contains globular star clusters. The Fornax system contains six. Three of these were described by Shapley (1939) and three additional clusters have been discovered and discussed subsequently (Hodge 1961, 1969). Astrophotometer measures of the brightest stars of these clusters have shown that they all have the same maximum luminosity, within the measuring uncertainty of 0.1 mag. This would suggest, on the basis of arguments regarding the height of the giant branch of globulars of different metal abundance, that all of the globular clusters are uniform in age and chemical composition (Sandage & Wallerstein 1960). Three-color measures of most of these clusters have been carried out by several scientists (Demers 1969, Hodge 1969, de Vaucouleurs & Ables 1969). The colors are now known to be normal, falling within the range of colors found for halo globular clusters of our Galaxy. They are not, however, identical in color. Cluster 4, for example, is ~0.15 mag redder than the others, and cluster 1, the intrinsically faintest, is somewhat bluer than average. Spectroscopy of the brighter clusters of the Fornax system was carried out by van den Bergh (1969) who has found a significant range in metallicity, though the average heavy-element abundance of the systems is apparently very low.

Globular star clusters are also known for the four dwarf galaxies of the Andromeda group. Photometry of some of these has been carried out by Kron & Mayall (1960) and by Hodge (1971). They are in general normal in their colors, but high accuracy is very difficult to achieve, especially for those clusters imbedded in the bright portions of each galaxy. Baade (1944) reported four globular clusters each for NGC 147 and NGC 185, and eight for NGC 205. It is difficult, because of the proximity to the central regions of M31, to determine what clusters might belong to M32. More recent surveys of NGC 147 and NGC 185 have increased the number of probable globular clusters by approximately a factor of 2 (Hodge 1971).

   Total luminosities and masses.—Table 1 summarizes the known properties of the dwarf elliptical galaxies of the local group. Apparent magnitudes

TABLE 1. Data on dwarf elliptical galaxies

	Sculptor	Fornax	Leo I	Leo II	Draco	Ursa Minor
Ellipticity $1-b/a$	0.35 ± 0.05	0.35 ± 0.10	0.31 ± 0.07	0.01 ± 0.10	0.29 ± 0.04	0.55 ± 0.10
Core radius[a] r_c'	11.9	16	4.5	2.5	6.5	11.1
Limiting radius r_t'	53 ± 5	50 ± 6	13.9 ± 0.5	9.6 ± 1.5	26 ± 2	59 ± 25
Core radius[a], kpc	0.3	0.9	0.3	0.2	0.13	0.2
Limiting radius r_t, kpc	1.2 ± 0.1	3.1 ± 0.3	0.91 ± 0.04	0.65 ± 0.10	0.51 ± 0.04	1.2 ± 0.5
Computed tidal radius r_t, kpc	1.7 ± 0.4	7.0 ± 1.4	3.3 ± 1.5	3.0 ± 1.5	0.41 ± 0.15	0.42 ± 0.17
Absolute magnitude M_v	-10.9	-13.6	-11.4	-9.8	—	—
Mass $M/M_\odot$	3×10^6	2×10^7	4×10^6	10^6	1.2×10^5	10^5
Apparent distance modulus $m-M$	19.7 ± 0.4	21.4 ± 0.5	21.8 ± 0.6	21.8 ± 0.6	19.6 ± 0.3	19.5 ± 0.4
True distance modulus m_0-M	19.6 ± 0.4	21.3 ± 0.5	21.7 ± 0.6	21.7 ± 0.6	19.1 ± 0.3	19.1 ± 0.4
Distance, kpc	84 ± 14	188 ± 40	220 ± 50	220 ± 50	67 ± 10	67 ± 14
ρ (center), $M_\odot/pc^3$	7×10^{-4}	10^{-4}	5×10^{-3}	2×10^{-4}	2×10^{-3}	8×10^{-4}
Relaxation time,[b] y	7×10^{13}	2×10^{13}	4×10^{13}	8×10^{13}	7×10^{11}	2×10^{12}

   [a] As defined by King (1962).
   [b] Computed with $\overline{(v^2)}^{1/2}$ and v assumed equal to the circular orbital velocity at r_c.

of the extreme dwarf galaxies are extremely difficult to obtain, and are quite uncertain. Masses for these systems are also very difficult to determine, and the data in the table come primarily from indirect arguments based on tidal effects and on assumed mass-to-light ratios. It is clear that both in absolute luminosity and in mass, the extreme dwarf galaxies are not different from the brightest and most massive globular clusters of our Galaxy.

INTERSTELLAR MATERIAL

Interstellar gas.—There is no confirmed report of any interstellar gas, detected either by 21 cm radiation or by the presence of H II regions, for any of the dwarf elliptical galaxies in the local group. There is, however, a means of establishing an approximate estimate of the hydrogen mass associated with the Population I component of the galaxies NGC 185 and NGC 205. This involves arguments regarding the estimated mass both of dust (see below) and of the total Population I stellar content. For NGC 185, the total mass of gas has been estimated to be 4×10^4 $M_\odot$ (Hodge 1963). The gas mass for NGC 205 is calculated to be approximately twice that value (Hodge 1970).

Interstellar dust.—The six extreme dwarf elliptical galaxies near the local galaxy are free of any detectable interstellar extinction due to dust. Tests involving counts of background galaxies in all six cases indicate a very high degree of transparency for these systems (Hodge 1961a, b, 1962, 1963, 1964a,b). Figure 5 illustrates one of these cases.

Among the four elliptical companions of Andromeda, two anomalously contain conspicuous absorption regions. This has in fact led to the classification of both of these as peculiar galaxies (Sandage 1961). NGC 185 shows two large absorption regions spaced approximately symmetrically about its center. Both of these absorption regions are of a size typical for dark nebulae

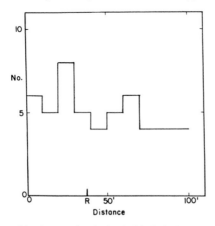

FIGURE 5. Counts of background galaxies behind the Fornax dwarf (*R* marks its radius), illustrating its transparency (Hodge 1961).

in the Milky Way of the Coal Sack type (Bok 1948), which have masses $\sim$10–100 $M_\odot$. Estimated masses of each of these systems have been calculated from their dimensions and profiles (Hodge 1963). If they are near the center of NGC 185, then it is calculated that the masses are 190 and 25 solar masses, respectively. Thus the total mass of dust in NGC 185 is approximately 200 $M_\odot$, which is a reasonable total considering the computed mass of Population I stars, approximately 2×10^5 $M_\odot$.

NGC 205 shows two large dust lanes and more than a dozen small faint ones (Hodge 1971). For the largest the measured color excess is $\Delta (B - V)$ $= +0.13$, implying a total absorption of $\Delta V = 0.4$ mag, under the conventional assumptions about the ratios of selective to total extinction. The total mass for the dust lanes in NGC 205 is comparable to but somewhat larger than that for NGC 185.

STRUCTURE

Star and luminosity distributions.—For the six extreme dwarfs, complete studies of stellar distributions based on star counts have been published (Shapley 1939, Hodge 1961a, b, 1962, 1963, 1964a, b, de Vaucouleurs & Ables 1968). The star counts have shown that the stars in these galaxies are so distributed that projected densities form smooth ellipses, as for giant elliptical galaxies. Any irregularities detected are probably due to observational effects.

Photometric studies of the luminosity distributions of only two of these extreme dwarfs have been published, for Sculptor (Hodge 1965) and for Fornax (de Vaucouleurs & Ables 1968, Hodge 1971). For Sculptor the photometry was completely consistent with the results of star counts. The surface brightness ranges in V from approximately 24 mag per sec^2 at the center down to 26 mag per sec^2 at the most distant measured spot. For the Fornax galaxy, de Vaucouleurs & Ables found a systematic difference between their surface measurements, made by the diurnal scanning technique, and the star counts previously published. However, Hodge, using offset photometric measures of Fornax surface brightness, found consistency with the star counts.

Luminosity distributions for the four elliptical companions of M31 have been determined by a number of investigators and techniques. From photoelectric and photography photometry, the surface brightness distribution over the face of NGC 147 was determined by Hodge (1971) at Mt. Wilson. The central brightness is $\sim$21.25 mag per sec^2 and the limit on the major axis beyond which no stars could be detected corresponded to a radius of 5.3 kpc.

Surface luminosity distribution measurements for NGC 185 were also carried out at Mt. Wilson (Hodge 1963). Knowledge of the structure of that system is based on 300 photoelectric measurements taken at different spots over the face of the galaxy. These measures were supplemented by isophotometry of a series of Lick and Palomar plates of different colors, leading to the conclusion that NGC 185 conforms to the general pattern for elliptical

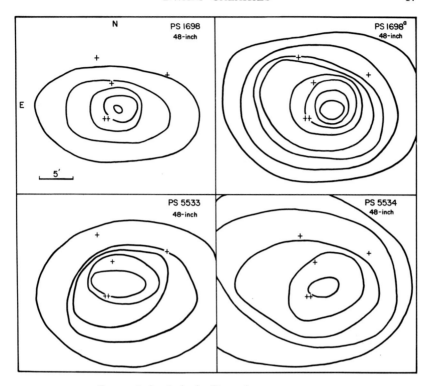

FIGURE 6. Isopleths for Draco from star counts on
Palomar Schmidt plates (Hodge 1964).

galaxies, with concentric and nearly identical ellipses forming its isophotes.
In the center, of course, there is photometric irregularity due to the presence
there of the system of dust clouds and OB stars.

A photometric study of NGC 205 was carried out by Hodge (1971) in
much the same way as that carried out for NGC 185. Previously this galaxy
was the subject of photometric studies by Redman & Shirley (1938), Richter
& Hogner (1963), and de Vaucouleurs (1958). Although the four studies do
not agree exactly in their determination of the luminosity and luminosity
profile for NGC 205, they find very similar shapes and luminosity distributions.

Photometric studies of M32 include many investigations, starting with
that of Hubble (1930). In shape this galaxy appears to be normal for an
elliptical galaxy, although details are difficult to establish because of the
intermingling of the isophotal data with that for M31.

Ellipticities.—In spite of early impressions to the contrary (Shapley
1939), most of the dwarf elliptical galaxies in the local group have fairly
high ellipticities. Because of their low surface brightness, they often appear

on plates to be circular or nearly circular in outline and this has led to fairly common usage of the term *dwarf spheroidal galaxies*. However, only one of the systems of the local group is found in actual fact to be nearly circular in outline. Table 1 gives data established for the various systems and shows that the ellipticity, defined as $1 - b/a$, where b is the minor axis and a is the major axis, is most commonly ~ 0.3. The Leo II system is approximately circular, while the Ursa Minor system has the very large ellipticity of 0.55. For the extreme dwarfs, it generally has not been possible to establish any variation of ellipticity with distance from the center, chiefly because of the relatively large uncertainties in the data. The exception to this is the Fornax system, which is found to have a variation of ellipticity with distance from the center that shows a typical curve, similar to those found for many giant elliptical galaxies (Liller 1960). The variation ranges from ~ 0.2 at the center to a maximum value of ~ 0.37 (Hodge 1961).

The brighter elliptical galaxies near Andromeda also exhibit a variation of ellipticity with distance from the center. For NGC 185, for example, the ellipticity varies from a value of approximately 0.18 in the central regions up to a maximum of 0.26 at a distance of 5′ from the center along the major axis (Hodge 1963). NGC 205, on the other hand, shows a variation from a value of 0.35 to as great a value as 0.53, with an apparent dip at a distance of 6′.5 along the major axis (Hodge 1971) and another possible dip at a distance of $\sim 1′$ (Richter & Hogner 1963). NGC 205 is anomalous in that its major axis is not oriented the same for different isophotes, varying in position angle over the range from $\theta = 320°$ up to $\theta = 350°$ (Hodge 1971). This is particularly visible in the case of the outer isophotes which show a twisting effect that has been likened to a spiral-arm pattern. Possibly this outer distortion is a tidal effect due to the proximity of M31 (van den Bergh 1968).

Star density and luminosity profiles.—All of the local group dwarf elliptical galaxies have been studied sufficiently that there are well-established data for each for either star density profiles or luminosity profiles (see Figure 7). The above cited references on the photometry of these systems give such profiles and discuss them in terms of standard elliptical galaxy interpolation formulae as well as dynamical models. Generally the observations do not fit the interpolation formulae, such as that of Hubble (1930) or that of de Vaucouleurs (1948). Departure from the formulae generally takes the form of a steeper decline of intensity or density in the outer parts of the system than predicted. This has been interpreted as most probably the result of tidal effects imposed on the outer parts of the galaxies by the neighboring massive galaxy (King 1962, Kinman 1962, Hodge 1966).

Actual intrinsic star densities within the dwarf elliptical galaxies are extremely low. This was immediately clear in the case of Fornax which even on its discovery plate showed many fewer stars per unit area near its center than in its globular star clusters. On a blue Schmidt plate that has an approximate faint limit of 19.9 the projected density of stars in the most dense portion of the Fornax galaxy is nine stars per square minute, while that for

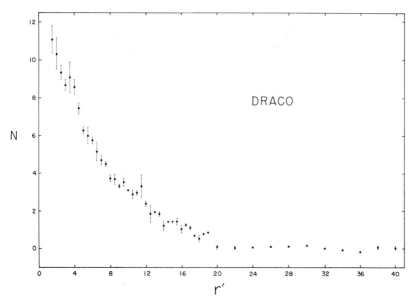

FIGURE 7. The star density profile for the Draco system (Hodge 1964).

the globular clusters averages eighty stars per square minute (Hodge 1961). Calculations of the central star density for each of the extreme dwarf galaxies (e.g. Figure 7) have been made on the assumption that the luminosity function is the same (except for a scale factor) as that for M3 (Sandage 1954) and based on King's (1962) formula for the spatial density in terms of projected densities. These calculations show that the dwarf elliptical galaxies have extremely low absolute density, $\sim 10^{-3}$ times the star density in the solar neighborhood, even at their centers. The range is from 5×10^{-3} to 10^{-4} $M_\odot$ per pc^3.

Dynamics.—The dynamical properties of the dwarf elliptical galaxies have been studied by fitting of models to the observed profiles (King 1962, Hodge & Michie 1969). Because of the very low star densities of the systems, relaxation times due to star-star encounters are extremely long, $\sim 10^{12}$–10^{13} years. Since this is very much longer than the probable age of the Universe, it was at first somewhat puzzling to find these systems to be in what appears to be a well-mixed state. However, two mechanisms can lead to a well-mixed star distribution in these systems in spite of their low densities: one is concerned with energy exchange between stars and star groups during the formation period (Hodge & Michie 1969), and the other is due to the stars being in a time-varying mean-field gravitational potential (Lynden-Bell 1967). Very good agreement is found between the observed profiles and models based on the Boltzmann equation (Michie 1963), when the models are made to take into account the tidal effects in the outer parts of the

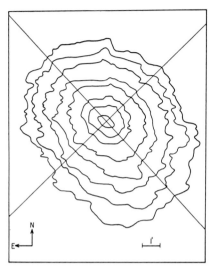

FIGURE 8. Photographic isophotes for NGC 185 (Hodge 1963).

galaxies. The models also allow a variety of types of orbits and the comparison shows that none of the six extreme dwarf galaxies of the local group has stars in exclusively isotropic orbits; elongated orbits are an important dynamical feature.

Although direct proof is lacking, a number of arguments point to the probability that the dwarf galaxies near our Galaxy are dynamically bound to it. For example, the radial velocity for the Fornax system, the only system with one known, is consistent with the idea of a closed orbit around our Galaxy (Hodge 1966). Furthermore, for Sculptor, Draco, and Ursa Minor, the agreement between observed and computed tidal radii is very good, while for the other three the computed limit is too large. These three are the most distant and therefore it can be concluded that they might have passed by our Galaxy at some time in the past. When the time for this transit is estimated, there is then not enough time for them to have been formed at very large distances from our Galaxy, and it is concluded that these systems probably formed in common with ours. It is also interesting that if the size of our Galaxy is expanded back to the time when its mean density was similar to the mean densities for these dwarf galaxies, we find the scale of our system and the distances to these galaxies to be in reasonable agreement.

Hodge & Michie (1969) inferred from such data that the dwarf galaxies came from the development of a few of the many fragments which formed during the early development of our protogalaxy. Only those early fragments in the outer regions would be expected to have survived the galactic tidal forces and to evolve into stellar systems. They concluded that the tidal force due to the main galaxy was a main determinant of the size, shape, density distribution, and mass in the outer regions of these systems, and that ~30% of the original mass of these systems was removed by tidal action. The total mass and the mean density, however, show no correlation with the tidal force and are presumably due to the properties of the original fragments from which they formed.

Ursa Minor has a computed tidal radius that is approximately one-third the observed limiting radius and if this datum is correct it must be true that a significant fraction of the total mass is now being pulled away from that system. It may be that the Ursa Minor system is making its final approach to our Galaxy, because it is difficult to escape the fact that if it is now approaching us, it is being disrupted drastically.

The dynamical properties of M32, one of the elliptical galaxies in the local group, was the subject of a very important high-dispersion spectroscopic study by Walker (1962). The nucleus of this small galaxy showed rotation similar to that found in the nucleus of M31. There is a steep increase in velocity with distance from the center to a maximum at a distance of only 2.5 sec of arc, where the velocity is 65 km per sec. Outward from this point, velocity decreases to zero at a distance of ~9 sec of arc from the center. From these data and from assumptions about the density gradient and the effects of noncircular motions, Walker found that the period of rotation of the nucleus must be ~6×10^5 years, its mass ~10^7 $M_\odot$, and its density ~5×10^5 $M_\odot$ per pc³. This nucleus is very similar dynamically to the nucleus of M31.

DWARF IRREGULAR GALAXIES

Irregular galaxies of small size and low surface brightness are fairly common in space in and near the local group (Zwicky 1957). In his catalog of dwarf galaxies, van den Bergh (1959) found the majority of the 222 dwarfs to be classifiable as dwarf irregular galaxies. Furthermore, the dwarf spirals of that catalog are primarily objects sufficiently small that the spiral pattern is either indistinct or incomplete, and the systems are very similar in appearance to dwarf irregulars. van den Bergh's survey was made on the Palomar Sky Survey charts and probably represents a fairly exhaustive sampling of the dwarf galaxies in the area in and near the local group. Karachentseva (1968) has also searched the Sky Survey for dwarf galaxies and has reported that the distribution of dwarfs in the sky is nonuniform and is evidently correlated with the distribution of the apparently brightest galaxies.

NEARBY DWARF IRREGULAR GALAXIES

Only two dwarf irregular galaxies satisfying the criteria outlined at the beginning of this review are clearly established members of the local group of galaxies. One is NGC 6822 (see Figure 9), studied exhaustively by Hubble (1925) and more recently by Kayser (1967) and Hodge (1971); the other is IC 1613, studied exhaustively by Baade (1963) and more recently by Sandage (1962), Ables (1968), and Hodge (1971). In addition, six small galaxies

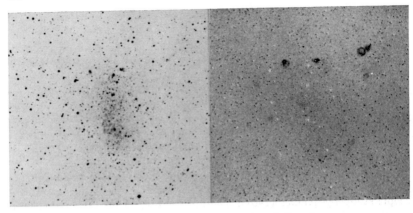

FIGURE 9. A comparison of a *V* plate (*left*) and an
Hα plate (*right*) of NGC 6822

might properly be considered members of the local group but have distances
that are either too poorly known or else somewhat too large for definite inclusion. These include the Leo A system (Hubble 1940), the Sextans A system (Hubble 1940, Zwicky 1942, 1957, Epstein 1964, Ables 1968), the
Pegasus system (Holmberg 1958), the Wolf-Lundmark-Melotte system
(Holmberg 1950), IC 10 (Roberts 1962), and GR 8 (Hodge 1967). Many of
these systems have been known in the literature by various names, and to
avoid confusion it is probably most proper at this time to designate them
according to the number assigned to them, based on their right ascensions,
in the *Reference Catalogue of Galaxies* (de Vaucouleurs & de Vaucouleurs
1964). The numbers for the above-named galaxies are, respectively, A0956,
A1009, A2326, A2359, IC0010, and A1256. The last of these is not in the
first edition of the *Reference Catalogue*, but from its position it would have a
designation as given here.

STELLAR CONTENT

The most exhaustive studies of dwarf irregular galaxies have concentrated on the two closest such objects, in the local group, NGC 6822 and
IC 1613, and most of our information on the stellar content of these galaxies
comes from these sources. In general these objects are quite similar in content to larger irregular galaxies, like the Magellanic Clouds, and to late-
type spiral galaxies. Resolved stars are primarily Population I objects.
Neutral hydrogen and H II regions indicate large amounts of gas, and
Cepheid variables are abundant.

Color-magnitude diagrams.—Colors and magnitudes for the brightest
stars have been measured and plotted for both NGC 6822 and IC 1613.

TABLE 2. Data on local group dwarf irregular galaxies

	NGC 6822	IC 1613
Diameter,		
arc min	$20' \times 10'$	$25'$
kpc	2.7×1.3	5
m_V	9.5	10.0
M_V	-14.7	-14.5
Distance, kpc	465	740
$M_H/M_\odot$	2×10^8	7×10^7
$M_T/M_\odot$	1×10^9	4×10^8

However, because of photometric problems in the faint sequence, that for IC 1613 has not yet been published (Baade 1963).

The color-magnitude diagram for NGC 6822 (see Figure 10) was obtained by Kayser (1967) with data based on photometry of photographic plates, primarily from the Palomar and Mt. Wilson telescopes, as calibrated by a photoelectric sequence obtained by Arp. The photometry of the galaxy was complete to apparent magnitude $V = 19.5$, equivalent to an absolute magnitude of $M_V = -5$. The diagram shows a conspicuous main sequence of bright blue supergiants, the brightest star having an absolute magnitude of $M_V = -9$. There is also a number of very red supergiants that reach to luminosities ~1 mag fainter than the blue. Many of these red stars show variations over a long time scale (1000 days). There is possibly a significant number of member stars in the Hertzsprung gap, though most are probably field stars. From the two-color photometry, Miss Kayser calculated that the average intervening reddening amounts to 0.27 mag.

Luminosity functions.—Miss Kayser (1967) derived the properties of the luminosity function for NGC 6822 from her measures of the magnitudes of stars in the system, after allowing for the galactic foreground. She found that in general the luminosity function for the most luminous stars is characterized by a smaller slope than in the case of our Galaxy. The slope of ϕ (M) in NGC 6822 is ~2 or 2.5, while in the local galaxy it is more nearly 3. This implies an overabundance of young stars in NGC 6822 as compared to the local galaxy, a result similar to that found for the Large Magellanic Cloud (Shapley 1931, Hodge 1961).

Luminosity functions for the other dwarf elliptical galaxies have not been published, except for GR 8, for which the luminosity function integrated over the top 5 mag totaled approximately 100 stars (Hodge 1967), indicating its extremely small total population.

Variable stars.—For NGC 6822, the study of Cepheid variables begun by Hubble (1925) has been carried out quite completely by Kayser (1967).

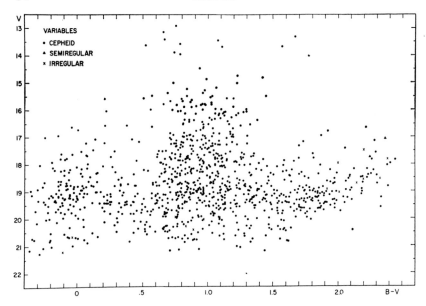

FIGURE 10. The color-magnitude diagram for the brightest stars in NGC 6822, including foreground stars, most of which are intermediate in color (Kayser 1967).

She discussed a total of 32 variable stars, 13 of which were classical Cepheids. The remainder include two red semiregular variables, one possible eclipsing variable, and a number of probable irregular variables. From the Cepheids, she found the period-luminosity relationship to have the form $\langle V \rangle = 23.35 - 2.97 \log P$.

The variable stars of IC 1613 have been described in general terms by Baade (1963). Of the 63 variables which he found on 100 inch Mt. Wilson plates, he was able to obtain periods and light curves for 36.

From the photometry available so far on IC 1613, the 25 Cepheid variables give a period-luminosity relationship with a somewhat shallower slope than is believed normal for Population I (Sandage 1962, Sandage & Tammann 1968). Baade derived a distance modulus for IC 1613 equal to 24.1 mag, based on Cepheids ranging in period from 2.4 to 146.35 days.

A long-period variable in IC 1613 appears to be a fairly normal object similar in all respects to those, for example, found in the Magellanic Clouds. Its period is 446 days and its absolute magnitude at maximum is $M_{pg} = -5.5$. Its amplitude is 1.8 mag, which Baade felt was unusually small, but which is not inconsistent with amplitudes for long-period variables in the Small Magellanic Cloud (Payne-Gaposchkin & Gaposchkin 1966) or in the Large Magellanic Cloud (Hodge & Wright 1968). It fits the wide maximum-luminosity-period relationship found to hold for the Magellanic Clouds.

The remainder of the variables include seven red irregular variables, a blue Hubble-Sandage variable of absolute magnitude approximately -4.4, a normal eclipsing binary with a period of 3.77 days, and a nova, which reached an absolute magnitude of -6.6 at its brightest detected level.

A lone exception to the normality that characterizes the variables of IC 1613 is an object that Baade called a "queer variable" (Baade 1963). It is extremely red and has a period of 28.687 days. Unlike most red variables, its light curve is quite regular. Baade described it as being like the light curve of a β Lyrae variable, only upside down, with most of the time spent at minimum. In absolute magnitude it is -5.5 mag at maximum, which places it about a magnitude above the period-luminosity curve for Cepheid variables. Its exact nature remains unknown.

Star clusters.—One of the peculiarities of the dwarf irregular galaxies in nearby space is their relatively small number of star clusters. Apparently the number of such clusters is not directly related, as one might think, to the total mass or luminosity of an irregular galaxy. For example, the Large Magellanic Cloud contains 1600 catalogued star clusters of all types (Hodge & Sexton 1966). On the other hand, IC 1613, with an absolute luminosity about 25 times smaller, contains none. Star clusters also are not very abundant in NGC 6822, in which Hubble found several objects which he thought might be identified as star clusters but which might instead be background galaxies.

INTERSTELLAR MATERIAL

H II regions.—Most of the dwarf irregular galaxies that have been detected contain conspicuous H II regions. Hubble identified 5 diffuse nebulae in NGC 6822 (Hubble 1925), and 16 H II regions in this galaxy were mapped from 48 inch Palomar-Schmidt Hα plates (Hodge 1969). A spectroscopic study by Peimbert & Spinrad (1970) was carried out for one of the brightest H II regions, Hubble No. V, and their absolute photoelectric calibration of line strengths allowed them to determine the chemical abundances in this object. Comparisons with line strengths in Orion and M17 showed that Hubble V is deficient in nitrogen and oxygen by factors of 6 and 1.7, respectively. Peimbert & Spinrad suggest that this deficiency is consistent with the apparent relationship between metal abundances and the total masses of galaxies, wherein the galaxies of small mass have a generally smaller heavy-element abundance. They found a ratio of the abundances of helium and hydrogen similar to that found for H II regions in our Galaxy and in other galaxies, and argue that this suggests that the helium is principally primeval. Assuming the H II region to be approximated by an isothermal sphere, they calculated the electron density to be $N_e = 4.1 \times 10^{-1}$ cm^{-3} and they found the total mass to be $M = 2.6 \times 10^4 \, M_\odot$.

IC 1613 contains several large H II regions, all concentrated in a small area of the outer part of the galaxy. This region also contains the brightest stars of the system. The largest emission object is ring shaped and $\sim$150 pc in diameter (Baade 1963).

The other small irregular galaxies in and near the local group contain few, if any, conspicuous H II regions, as shown by an unpublished Hα survey of several of them made recently at Kitt Peak. This paucity is apparently the result of the very small total mass and absolute magnitude of these objects. By extrapolation from the curve for Sc galaxies (Hodge 1967) it is clear that for galaxies of absolute magnitude less than -15 the number of H II regions is expected to be insignificant. For example, for the galaxy GR 8, which has an absolute magnitude of approximately -8, no H II regions exist whatever.

H I content.—Measurements of the neutral hydrogen content for both IC 1613 and NGC 6822 have been published (Volders & Hogbom 1961, Roberts 1970, Davies 1970). The total hydrogen mass for NGC 6822 is found to be $\sim$2$\times$10^8 $M_\odot$ and this is $\sim$11% of the total mass of the galaxy. Both of the recent investigations of the neutral hydrogen of NGC 6822 have shown that the gas extends over a vastly larger area than the stellar distribution outlines. For IC 1613 the total hydrogen mass is $\sim$7$\times$10^7 $M_\odot$, corresponding to $\sim$18% of the total galaxy mass. These represent large neutral hydrogen fractions, but not excessively large for irregular galaxies. Both are considerably smaller than the 32% found for the Small Magellanic Cloud (Hindman 1967).

Although their distances are more poorly known, the H I content has been studied for galaxies IC 10 (Roberts 1962) and Sextans A (A1009) (Epstein 1964).

Interstellar dust.—Dust in the form of dark nebulae is not conspicuous in any of the dwarf irregular galaxies and would, because of their irregular brightness distribution, be difficult to detect. The only available information on the dust content of these objects comes from comparing the three-color photometry of foreground galactic stars for NGC 6822 (Kayser 1967) with the reddening determined from the emission lines of the H II region Hubble V (Peimbert & Spinrad 1970). Miss Kayser found a color excess in the direction of NGC 6822 of $E(B-V)=0.27\pm0.03$. The reddening for the H II region, obtained by fitting the observed Balmer decrement to the theoretical one, was calculated to be $E(B-V)=0.37\pm0.04$. The difference of 0.10 mag is suggested by Peimbert & Spinrad to be due to absorption in the galaxy NGC 6822 and in the H II region itself. This represents a comparatively small amount of dust and may be a further indication that the dust content of dwarf galaxies is relatively low. Further research is needed to establish whether a correlation exists, but the evidence suggests that there is a range extending from the relatively dust-rich galaxies such as the local galaxy, through the Large Magellanic Cloud (which contains apparently only moderate amounts of dust), the Small Magellanic Cloud (which is apparently relatively dust-free), to the dwarf galaxies, which contain very small amounts of dust. More quantitative data are needed to establish whether this is a real or only an apparent effect.

STRUCTURE

Luminosity distributions.—Detailed photometric studies of the luminosity distributions for NGC 6822 (Hodge 1971), IC 1613 (Ables 1968, Hodge 1971), A1009 (Ables 1968), GR 8 (Hodge 1967), and the WLM system (Hodge 1971) are available. Some of the systems, for example NGC 6822, show primarily a barred structure, while others are more elemental in shape.

The barred structure of NGC 6822 is remarkable in that most of its H II regions and brightest stars are concentrated at the north end of the bar. Further, there are virtually no stars discernible and only one H II region lying exterior to the bar. The H I distribution suggests that there is gas extending in opposite directions from both ends of the bar in nearly equal amounts, almost like spiral arms. Only one of these gaseous appendages to the bar (which might be considered as uncondensed spiral arms) shows the presence of star formation activity at this time.

A study of the structure from the distribution of stars and luminosity in IC 1613 indicates the presence of eight small star groupings, similar to but somewhat looser than OB associations. From the maximum brightness of the main-sequence turnoff for these groups, it has been possible to trace back the recent evolutionary history of IC 1613, revealing that the apparent structure of this galaxy has changed drastically in a relatively short period of time (Hodge 1966).

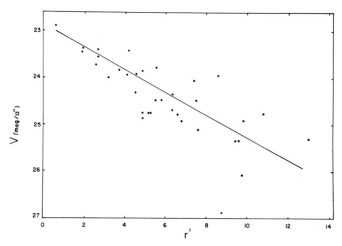

FIGURE 11. The mean photometric profile for IC 1613, determined from photoelectric measures of individual spots (*dots*). The line is a least-squares solution and shows the characteristic exponential decrease (Hodge 1971).

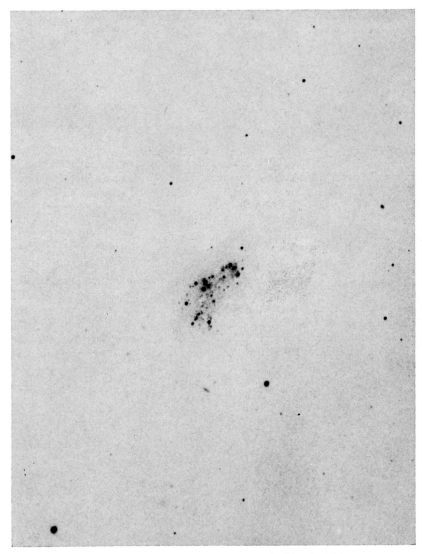

FIGURE 12. The dwarf GR 8, photographed by the Lick
120 inch telescope in blue light.

Density and luminosity profiles.—From the photometric studies of the
luminosity distribution in irregular dwarf galaxies and from star counts of
the brightest stars it has been possible to determine their density and
luminosity profiles. In all cases, within the limits of the observational errors
the dwarf irregular galaxies are characterized by an exponential decrease in

luminosity or in star density with distance from the center (see Figure 11). Thus they are similar in structure to giant irregular galaxies (de Vaucouleurs 1960). Even the galaxy GR 8 (see Figure 12), which has a mean dimension of only ~100 pc, seems to show this exponential character in its outer parts. Although the physical explanation of the exponential law of luminosity distribution for these objects has only been explored in a preliminary way (Freeman 1970) it is clearly an important clue to their dynamical properties and history. A remarkable correlation seems to exist between the slopes of the exponential decrease and the absolute magnitude of the galaxies, a correlation which seems to hold true equally well for the dwarf irregulars in the local group and for those studied photometrically in the Fornax cluster of galaxies (Hodge, Pyper & Webb 1965).

ASSOCIATION OF DWARFS WITH OTHER GALAXIES

As Members of Clusters

The local group.—It is clear from the derived luminosity function for galaxies in the local group that in terms of numbers the dwarf galaxies dominate the system. From the theoretical considerations outlined above, it is believed that the dwarfs are associated with individual giant galaxies of the local group, such as the Milky Way galaxy and M31. If such is the case then we may have a relatively complete sample of dwarf galaxies for our Galaxy and, by analogy, we might expect that there are approximately six or more dwarf elliptical galaxies of the Sculptor type surrounding M31 that we cannot detect. In this case, and under the assumption that the other galaxies of the local group are too small to have a significant family of such objects, we would estimate that the local group consists of about fifteen dwarf elliptical galaxies, four or five dwarf irregular galaxies, two larger irregular galaxies, a moderate-sized spiral, two giant spirals, and possibly a giant elliptical or two (the Maffei objects). Though the dwarf galaxies dominate in numbers, they are insignificant in terms of mass, most of the mass of the local group lying in the giant members.

Other nearby groups.—Dwarf members have been found for many of the nearby loose groupings of galaxies that are close enough to allow resolution of the brightest stars (see Figure 13). For example, Holmberg (1950) has catalogued dwarf members, all irregular galaxies, for the groups of galaxies centered around M81 and M101. Similar dwarf galaxies have been studied in the Leo group (see Figure 14) (Hodge 1964) and there are some interesting objects in the NGC 1023 group (Sandage 1961).

The Virgo Cloud. Since Reaves' (1956) pioneering survey of the dwarf members of the Virgo Cloud of galaxies, there have been several studies of dwarf members of this and other large clusters (Reaves 1962, Chester & Roberts 1964). For these fairly distant groups it is difficult to tell the differ-

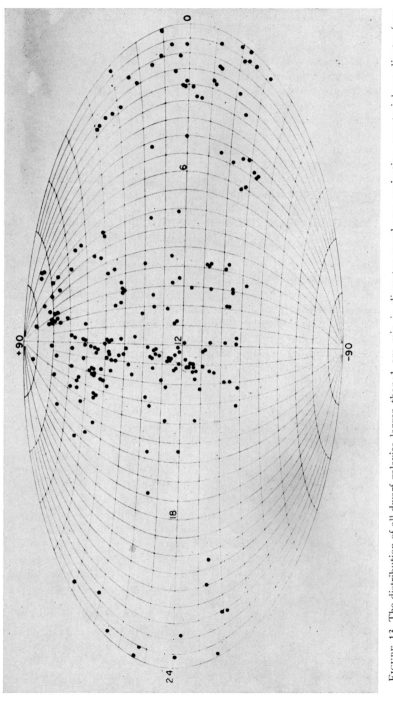

FIGURE 13. The distribution of all dwarf galaxies larger than 1 arc min in diameter, equal area projection, equatorial coordinates (van den Bergh 1966).

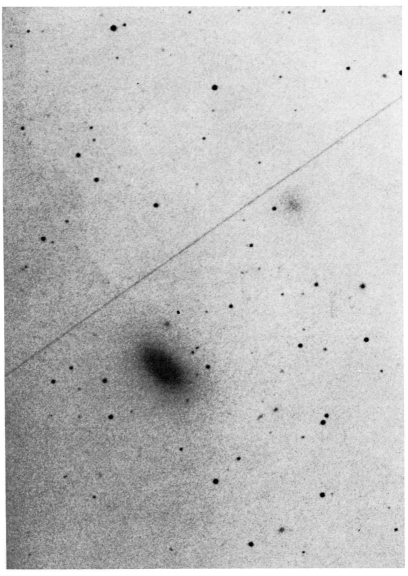

FIGURE 14. Giant and dwarf galaxies in the Leo group. The bright elliptical is NGC 3377 and the faint object above it and to the right is a dwarf irregular system. The line is a meteor trail. Palomar 48 inch photograph.

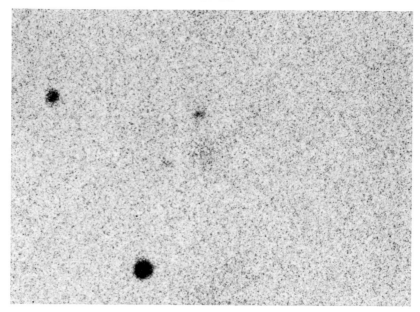

FIGURE 15. A dwarf of the Fornax cluster of galaxies, from a highly enlarged Palomar Schmidt plate in blue.

ence between dwarf elliptical and dwarf irregular galaxies because of the lack of bright enough stars in the latter, whose resolution would be a determining factor in identification. For the brighter objects of intermediate luminosity, on the order of absolute magnitude −15, the distinction is more straightforward than for the true dwarf systems. Reaves' observations of dwarf galaxies in the Virgo Cloud primarily concerned objects similar to NGC 185 and NGC 205 and probably did not include objects similar to the Sculptor system. From Reaves' data, Ambartsumian (1961) has suggested that the dwarfs and giants appear to have the same distribution within the cluster, which implies a lack of equipartition of energy for the galaxies of the cluster. This cloud, however, has a complicated structure and is a difficult object for which to apply this particular test.

The Fornax cluster.—Surveys of the dwarf members of the cluster of galaxies in the constellation of Fornax (see Figures 15, 16) (which should not be confused with the Fornax galaxy of the local group) have been carried out in some detail, including photometry of 50 of the individual dwarf members (Hodge 1959, 1960, Hodge, Pyper & Webb 1965). From their luminosity profiles it was possible to separate those which were dwarf elliptical galaxies from those which, from their exponential profiles, were apparently irregular dwarfs. Absolute (B) magnitudes of the objects range from −13.8 to almost −16. In size, the galaxies range from a radius of 0.8 kpc to 3.6 kpc; these fall

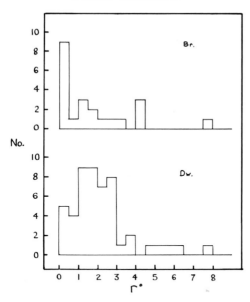

FIGURE 16. The difference between the radial distributions of the bright (Br.) and dwarf (Dw.) galaxies in the Fornax cluster (Hodge, Pyper & Webb 1965).

within the observed range for similar galaxies of the local group. The dwarfs make up a large percentage of the catalogued number of galaxies in this cluster, which, besides the fifty dwarfs, includes six spirals, four large irregular galaxies, and fourteen bright ellipticals. The spatial distributions of the giant galaxies and the dwarf galaxies were sufficiently different to suggest the presence of a tendency for the equipartition of galaxies of different mass.

More distant clusters.—As clusters are examined at greater and greater distances, the search for dwarf galaxies is more difficult and generally is limited to the objects of brighter absolute magnitude. For example, the Coma cluster has been searched by Reaves (1966) for dwarfs, using plates taken with the Palomar 200-inch telescope. He found objects in the luminosity range from absolute magnitude $M_V = -14$ to -16. These objects have sizes in the range 2 to 4 kpc. Kwast (1966) has examined the dwarf galaxies of the Hydra cluster and Rudnicki & Baranowska (1966) have examined them in three other more distant clusters.

There do not seem to be isolated groups of dwarf galaxies (van den Bergh 1959, 1966). Karachentseva (1967, 1968) has examined the relationship between the dwarf galaxies and the type of bright galaxy with which they appear to be associated, concluding that Sculptor-type systems are preferentially found as companions to normal giant ellipticals while there is an apparent close association of irregular dwarfs with giant spirals.

SUMMARY

Dwarf galaxies make up a large percentage of the galaxies in the local neighborhood and probably in the Universe, though their total mass is insignificant in comparison with that of the rarer giant galaxies. Local group dwarfs have in the past been most useful because of their proximity and the ease with which it has been possible to study their stellar and interstellar content. Their intrinsic interest centers around the question of their origin and its possible dependence upon the mode of formation of our Galaxy.

LITERATURE CITED

Ables, H. 1968. *PhD thesis*. Univ. Texas
Ambartsumian, V. 1961. *Astron. J.* 66:536
Baade, W. 1944. *Ap. J.* 100:147
Baade, W. 1963. *Evolution of Stars and Galaxies*. Cambridge: Harvard
Baade, W., Hubble, E. P. 1939. *PASP* 51:40
Baade, W., Swope, H. 1961. *Astron. J.* 66:300
Baum, W. A. 1959. *PASP* 71:106
Bok, B. J. 1948. *Harvard Monogr.* 7:53
Chester, C., Roberts, M. 1964. *Astron. J.* 69:635
Davies, R. D. 1970. Paper presented at *IAU Symp. No. 44, Uppsala, Sweden*
Demers, S. 1969. *Ap. Lett.* 3:175
de Vaucouleurs, G. 1948. *Ann. Ap.* 11:247
de Vaucouleurs, G. 1958. *Ap. J.* 128:465
de Vaucouleurs, G. 1960. *Ap. J.* 131:574
de Vaucouleurs, G. 1961. *Ap. J. Suppl.* 5:233
de Vaucouleurs, G., Ables, H. 1968. *Ap. J.* 151:105
de Vaucouleurs, G., Ables, H. 1969. *Ap. J.* 151:105
de Vaucouleurs, G., de Vaucouleurs, A. 1964. *Reference Catalogue of Galaxies*. Austin: Univ. Texas
Epstein, E. E. 1964. *Astron. J.* 69:490
Freeman, K. C. 1970. *Ap. J.* 160:811
Harrington, R. G., Wilson, A. G. 1950. *PASP* 62:118
Hindman, J. V. 1967. *Aust. J. Phys.* 20:147
Hodge, P. W. 1959. *PASP* 71:28
Hodge, P. W. 1960. *PASP* 72:188
Hodge, P. W. 1961a. *Astron. J.* 66:83, 249, 384. 1961b. *Ap. J. Suppl.* 6:235
Hodge, P. W. 1962. *Astron. J.* 67:125
Hodge, P. W. 1963. *Astron. J.* 68:237, 470
Hodge, P. W. 1964a. *Astron. J.*, 69, 438, 853. 1964b. *Sci. Am.* 210:78 (May)
Hodge, P. W. 1965. *Ap. J.* 141:308, 142:1390
Hodge, P. W. 1966. *Ap. J.* 144:869. 1966. *Galaxies and Cosmology*. New York: McGraw-Hill
Hodge, P. W. 1967. *Ap. J.* 148:719
Hodge, P. W. 1969. *Astron. J.* 72:129. 1969. *PASP* 81:875
Hodge, P. W. 1970. Paper presented at *IAU Symp. No. 44, Uppsala, Sweden*
Hodge, P. W. 1971. In preparation
Hodge, P. W., Michie, R. W. 1969. *Astron. J.* 74:587
Hodge, P. W., Pyper, D., Webb, C. J. 1965. *Astron. J.* 70:559
Hodge, P. W., Sexton, J. 1966. *Astron. J.* 71:363
Hodge, P. W., Wright, F. 1968. *Ap. J. Suppl.* 17:467

Holmberg, E. 1950. *Medd. Lunds Astron. Obs.* 2:128
Holmberg, E. 1958. *Medd. Lunds Astron. Obs.* 2:No. 136
Hubble, E. P. 1925. *Ap. J.* 62:409
Hubble, E. P. 1930. *Ap. J.* 71:131
Hubble, E. P. 1940. *Sci. Mon.* 51:398
Humason, M. L., Mayall, N. U., Sandage, A. R. 1956. *Astron. J.* 61:97
Karachentseva, V. 1967. *Astrophysics (USSR)* 3:225
Karachentseva, V. 1968. *Contrib. Obs. Bjurakan* 39:61
Kayser, S. 1967. *Astron. J.* 72:134
Kinman, T. 1962. *PASP* 74:424
King, I. 1962. *Astron. J.* 67:471
Kron, G., Mayall, N. U. 1960. *Astron. J.* 65:581
Kwast, T. 1966. *Acta Astron.* 16:45
Liller, M. 1960. *Ap. J.* 132:306
Lynden-Bell, D. 1967. *MNRAS* 136:101
Michie, R. 1963. *MNRAS* 125:127
Payne-Gaposchkin, C. H., Gaposchkin, S. 1966. *Smithsonian Contrib. Ap.* 9:1
Peimbert, M., Spinrad, H. 1970. *Astron. Ap.* 7:311
Reaves, G. 1956. *Astron. J.* 61:69
Reaves, G. 1962. *PASP* 74:392
Reaves, G. 1966. *PASP* 78:407
Redman, R. O., Shirley, E. G. 1938. *MNRAS* 98:613
Richter, N., Hogner, W. 1963. *Astron. Nachr.* 287:261
Roberts, M. S. 1962. *Astron. J.* 67:431
Roberts, M. 1970. Paper presented at *IAU Symp. No. 44, Uppsala, Sweden*
Rudnicki, K., Baranowska, M. 1966. *Acta Astron.* 16:55, 65
Sandage, A. R. 1954. *Astron. J.* 59:162
Sandage, A. R. 1961. *The Hubble Atlas of Galaxies*. Washington: Carnegie
Sandage, A. R. 1962. *Problems of Extragalactic Research*, ed. G. McVittie, 359. New York: Macmillan
Sandage, A. R., Tammann, G. A. 1968. *Ap. J.* 151:531
Sandage, A. R., Wallerstein, G. 1960. *Ap. J.* 131:598
Shapely, H. 1931. *Harvard Bull.* 881:1
Shapley, H. 1938. *Harvard Bull.* No. 908:1
Shapley, H. 1939. *Proc. Nat. Acad. Sci. US* 25:565
Spinrad, H. 1962. *Ap. J.* 135:715
Spinrad, H. 1966. *PASP* 78:367
Spinrad, H., Taylor, B. 1971. *Ap. J.* In press
Swope, H. 1968. *Astron. J.* 73:5204
Thackeray, A. D. 1950. *Observatory* 70:144

van Agt, S. 1967. *Bull. Astron. Neth.* 19:275

van Agt, S. 1968. *Bull. Astron. Neth. Suppl.* 2:237

van Agt, S. 1971. In preparation

van den Bergh, S. 1959. *Publ. David Dunlap Obs.* 2:No. 5

van den Bergh, S. 1966. *Astron. J.* 71:922

van den Bergh, S. 1968. *J. Roy. Astron. Soc. Can.* 62:145, 219

van den Bergh, S. 1969. *Ann. Rep. Dir. Mt. Wilson and Palomar Obs.* 1969:129

Volders, L., Hogbom, J. 1961. *Bull. Astron. Inst. Netherlands* 15:307

Walker, M. 1962. *Ap. J.* 136:695

Wilson, A. G. 1955. *PASP* 67:27

Zwicky, F. 1942. *Phys. Rev.* 61:489

Zwicky, F. 1957. *Morphological Astronomy.* Be.lin: Springer

INFRARED SOURCES OF RADIATION[1]

2016

GERRY NEUGEBAUER AND ERIC BECKLIN

Hale Observatories, California Institute of Technology
Carnegie Institution of Washington

A. R. HYLAND

Mount Stromlo & Siding Spring Observatories
Research School of Physical Sciences, The Australian National Observatory

CONTENTS

INTRODUCTION

The infrared spectral region is generally recognized as including wavelengths between 1 μ and 1 mm. Within the past 5 years, this region has been increasingly explored in astronomical observations using ground-based telescopes in the available atmospheric windows from 1 to 20 μ and airborne

[1] This work was supported in part by National Aeronautics and Space Administration grant NGL 05-002-007.

telescopes—balloons, rockets, and aircraft—for wavelengths between 50 and 500 μ. A large number of objects, both galactic and extragalactic, have been found whose radiation is emitted almost exclusively in the infrared or whose infrared radiation deviates markedly from the predictions based on visual or radio data.

In a review of the infrared photometry of stars, H. L. Johnson (1966a) concentrated on the problems of stellar bolometric corrections and effective temperatures; his review stands as a comprehensive summary of the infrared colors of stars. The present review of infrared sources will, generally, not re-examine the data contained in Johnson's review. It will also not include the rapidly expanding field of high-resolution infrared spectroscopy nor observations of the solar system or the infrared background radiation; the infrared spectra of stars have been reviewed by Spinrad & Wing (1969).

I. INFRARED SURVEYS

Knowledge of infrared sources has been severely biased by selection effects. In particular, although much anomalous infrared emission is conspicuous at wavelengths longer than 5 μ, technical difficulties have severely limited unbiased infrared surveys at these wavelengths; objects have been selected for study on a variety of visual properties, a priori assumptions, or semiempirical relations.

A complete survey of the northern sky at declinations between -33 and $+80°$ was carried out at the California Institute of Technology to provide an unbiased sample of celestial 2.2 μ emitters; approximately 5600 sources brighter than 4×10^{-25} W m^{-2} Hz^{-1} were found (Neugebauer & Leighton 1969).[2] Grasdalen & Gaustad (1971) have shown that for declinations above 0°, 93% of the sources can be identified with stars in the Dearborn Catalog of Faint Red Stars (Lee et al 1947), a catalog listing stars of type K5 or later. The large majority of the identifications are thus late-type giant stars. On the order of 50 stars seen in the survey had 0.8–2.2 μ color temperatures in the vicinity of 1000°K; most of these objects, the reddest found, could not be associated with cataloged optical sources.

Surveys with IN plates can be competitive with the 2 μ survey in finding stars with color temperatures in the range 1000–2000°K. Hetzler (1937), Chavira (1967), Ackermann et al (1968), and Ackermann (1970) have conducted such surveys over limited areas of the sky; their results with respect to finding red stars were similar to those of the 2 μ survey.

Several long-wavelength sky surveys are now either in preparation or in partial operation (e.g., Low 1970a); airborne surveys at wavelengths ≥ 10 μ of small, essentially arbitrarily selected areas of the sky have been published by Hoffmann et al (1967) (300–360 μ), Feldman et al (1968) (10–30 μ), and Friedlander & Joseph (1970) (45–250 μ). The latter two groups report the observation of several apparently strong infrared sources but the positional accuracy is too low to allow confirmation and study by ground-based observers

[2] The designation IRC followed by a 5 digit number refers to an object in the catalog resulting from this survey.

II. CHARACTERISTICS OF INFRARED EXCESSES IN STARS

Many galactic infrared sources of interest are stars which show an excess of radiation, within a given infrared wavelength interval, over that which would be expected were the star to radiate like a blackbody with an effective temperature appropriate to its spectral type. For the purposes of this review, we will call this excess "the infrared excess."

Currently, almost all stellar infrared excesses are assumed to arise from thermal radiation from a dust shell or cloud surrounding a central star. The observed energy distribution generally has both a short-wavelength component exhibiting the spectrum of a normal star and a relatively smooth long-wavelength component. The division of energy between the stellar and circumstellar components covers a large range.

At one extreme, typified by μ Cep and other M supergiants (Section IIIA), the excesses are found in limited wavelength intervals only, e.g., between 8 and 14 μ, and, typically, contain less than 10% of the total luminosity. The energy distribution at wavelengths less than about 8 μ follows that of a cool blackbody ($T \leq 3500°K$) roughly appropriate to the stellar spectral type with little circumstellar reddening.

At the other extreme, most of the observed luminosity can be attributed to infrared radiation from a circumstellar shell. The contributions of the cool central star are essentially undetectable; presumably the shell either is totally opaque or has become an integral part of the atmosphere of the central star. Sometimes, infrared spectra show that the central star contributes energy to the 1–2.5 μ regions, but the continuum radiation at wavelengths less than 3.5 μ is heavily reddened, presumably by the circumstellar shell.

The evidence that these excesses are caused by circumstellar dust shells is largely circumstantial. For many late-type stars, and in particular those which are oxygen rich, the energy distribution of the observed excess between 8 and 14 μ is often strikingly similar to that expected from an optically thin cloud of silicate grains (Gillett, Low & Stein 1968, Woolf & Ney 1969). Theoretical investigations of molecular equilibria in circumstellar shells of cool stars (Gilman 1969) further indicated that the silicates Al_2SiO_5 and Mg_2SiO_4 are in fact the condensates most likely to form if the stars are oxygen rich; the magnesium silicates are more likely because of the low cosmic abundance of aluminum. The identification of silicates in circumstellar clouds was strengthened by Low & Krishna Swamy (1970), who observed an excess in α Ori near 20 μ which they identified with a second, very characteristic, peak in the emissivitiy of silicates. Silicate-like emission in the 8 to 14 μ interval has also been identified near the Trapezium region of Orion (Stein & Gillett 1969a), and in Comet Bennett 1969i (Maas et al 1970); silicate absorption has been identified in the interstellar medium by Hackwell et al (1970).

Further evidence that infrared excesses originate in circumstellar shells is the fact that, at wavelengths longer than 3 μ, the color temperatures of stars with a large fraction of their energy radiated as excess infrared energy are typically less than 1000°K. Since these temperatures are appreciably less than the temperatures of the photospheres of the central stars, the long-wavelength radiation must be produced by an additional, cooler source. For these stars, the observed radiation at wavelengths shorter than 3 μ is consistent, both in its intensity and wavelength dependence, with stellar radiation which has been attenuated by absorption in dust grains.

Polarization measurements have also corroborated the dust-shell hypothesis for late-type stars. From observations of 60 late-type stars, Dyck et al (1971) find a strong correlation between the average visual polarization and the infrared excess at 11 μ, which they relate to the scattering and absorption optical properties of circumstellar clouds of solid particles. The correlation was previously suggested in data obtained by Kruszewski et al (1969).

There has been much speculation about the origin of the inferred dust shells. In many cases there is evidence that the central star is highly evolved and is losing mass; thus some dust shells may have formed from material ejected from the central stars (Hyland et al 1969, Geisel 1970, Burbidge & Stein 1970, Gehrz & Woolf 1971).

An alternate explanation for the origin of some dust shells arises in the case of several stars with large infrared excesses which are believed to be evolving onto the main sequence; notable examples are the T Tauri stars (Section IVB). Larson (1969a) investigated the evolution of protostars with masses up to $5M_{\odot}$ and predicted their energy distributions, taking account of the radiative transfer in the protostellar dust clouds. He showed that, on the basis of his model, it is possible to interpret several T Tauri and related stars in terms of a protostar surrounded by the remnants of its original dust cloud. The properties of "cocoon" stars, young massive stars with very cool and large dust envelopes ($T \sim 100°K$, $R \sim 10^{17}$ cm), had previously been discussed by Davidson & Harwit (1967).

In summary, both evolved and young stars show quite similar infrared excesses which apparently can be attributed to dust shells. There is thus evidence that, in some cases, mass loss is the source of the shell, while, in other cases, the shell may be the remnant of a protostellar dust cloud.

Note that although the majority of the galactic objects with infrared excesses are consistent with dust-shell models, some stellar sources have been given a quite different interpretation. Specifically, the infrared excesses of some hot Be and Of stars have been interpreted in terms of free-free emission from ionized hydrogen (Section IVC).

III. INFRARED EXCESSES IN LATE-TYPE STARS

Many strong infrared sources are, according to their spectral characteristics, classified as late-type stars. Although all of the excesses in late-type stars can probably be explained by emission from circumstellar shells, there are differences between these excesses which depend on the spectral type of the central source. In the following we will discuss the observed excesses associated with various classes of late-type stars.

A. M SUPERGIANTS

The supergiant stars of classes M2 and later are strong emitters of infrared radiation in the 1–2 μ region as a consequence of their low photospheric temperatures. In addition, almost all observed M supergiants have excesses at wavelengths longer than 5 μ; unfortunately energy distributions beyond 3.5 μ have been obtained for only about 20 objects (Johnson 1967a, Gillett, Merrill & Stein 1971).

In 1968, Gillett, Low & Stein (1968) published spectrophotometry of late-type stars, including the supergiants μ Cep and α Ori, in the wavelength region from 2.8 to 14 μ and with a resolution $\Delta\lambda/\lambda = 0.02$. Woolf & Ney (1969) showed that the energy distributions could be satisfactorily explained in terms of photospheric radiation with additional excess emission at wavelengths greater than 8.4 μ (Figure 1), which they identified with the emissivity curves of silicate materials (Section II).

Gillett, Merrill & Stein (1971) have shown that the majority of M supergiants have excesses only at wavelengths $>5\mu$ which typically account for less than 10% of the star's total luminosity. Some M supergiants show a deficiency at 4.8 μ which is probably due to the extreme absorption by the fundamental vibration-rotation band of CO.

A few M supergiants exhibit larger excesses. For these stars, the excesses between 5 and 8 μ are similar to blackbody radiation of a shell at $\sim 600°K$, presumably due to emission from an optically thick silicate dust shell. As an example, in S Per the excess accounts for approximately 40% of the total luminosity and is evident at all wavelengths longer than 3.5 μ (Johnson 1967a); the ratio of the flux at 11.5 μ to that at 8.4 μ is still suggestive of the silicate bump.

More extreme examples are provided by NML Cyg (Neugebauer et al 1965) and VY CMa, two of the brightest stars observed at 10 μ (Figure 2) (Johnson, Low & Steinmetz 1965, Hyland et al 1969, Low et al 1970); both are extremely strong OH microwave emitters (Wilson & Barrett 1968, Eliasson & Bartlett 1969) (Section VE). The former is fainter than 18th mag visually; its extreme visual faintness is partially due to its location in a strongly obscured region of Cygnus. VY CMa is of particular interest since it is bright enough in the visible to have been studied in detail; it has long been known to be a complex peculiar irregular M variable. Both sources show strong infrared polarization (Forbes 1967, Hashimoto et al 1970).

Spectral classifications of NML Cyg range from M6 to M8 (Ford & Rubin 1965, Wing et al 1967, Pesch 1967, Johnson 1968, Herbig & Zappala 1970). Johnson (1968) has argued for a luminosity of $\sim 5 \times 10^4$ $L_\odot$ based on a distance of 500 pc obtained from estimates of the interstellar reddening; the luminosity based on the strength of the 2.3 μ CO bands is similar (Low et al

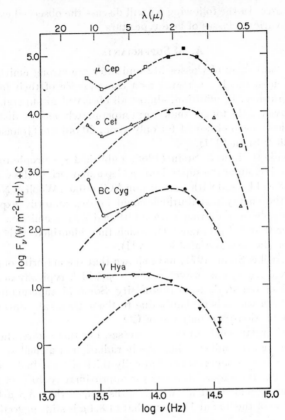

FIGURE 1. Log flux density vs log frequency for the late-type stars μ Cep ($C=27.5$), o Cet ($C=26.0$), BC Cyg ($C=26.0$) and V Hya ($C=24.0$). Dashed curves represent blackbodies which approximately define the photospheric energy distributions. Open symbols at $\lambda < 1$ μ are from Johnson et al (1966) and Lee (1970). Open symbols at 3.5, 4.8, 8.5, and 11.5 μ are from Gehrz et al (1970) and Dyck et al (1971). Filled symbols represent unpublished photometry at California Institute of Technology.

1970). Herbig & Zappala (1970), however, concluded from high-resolution optical spectra that the luminosity of NML Cyg is closer to that of a giant than a supergiant.

    After correction for interstellar reddening, the energy distribution of NML Cyg is almost identical to that of VY CMa. Furthermore, the structure and strength of the 2.3 μ CO bands in the two stars are remarkably similar (Low et al 1970, Hyland et al 1971). There are, however, minor differences between NML Cyg and VY CMa. VY CMa is somewhat earlier in spectral type (M4–5) and is more luminous than NML Cyg (Bidelman 1954, Wallerstein 1958, Herbig 1969, Humphreys 1970). At 2 μ VY CMa exhibits

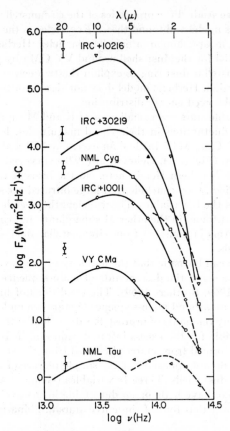

FIGURE 2. Log flux density vs log frequency for carbon-type infrared stars IRC +10216 ($C=27.0$) and IRC+30219 ($C=27.0$), M Mira-type infrared stars IRC +10011 ($C=26.0$) and NML Tau ($C=23.0$), and M supergiant-type infrared stars NML Cyg ($C=26.0$) and VY CMa ($C=24.0$). Solid curves represent blackbodies of the temperatures given in Table 1; for IRC+30219 $T=700°$K. Dashed curves represent the approximate photospheric energy distributions which have been reddened by circumstellar absorption. Filled symbols represent unpublished photometry at California Institute of Technology. The open symbols are data taken from Becklin et al (1969), Hyland et al (1969), Low et al (1970), and Hyland et al (1971).

the effects of steam absorption to a larger degree than NML Cyg (Low et al 1970, Hyland et al 1971). In the 8–14 μ region, weak silicate emission is found in VY CMa but not in NML Cyg (Stein et al 1969a, Gillett, Stein & Solomon 1970).

The parameters of a circumstellar spherical dust-shell model applied to NML Cyg and VY CMa are given in Table 1. Hyland et al (1969) and Low et al (1970) have studied the properties of the circumstellar dust absorptions required by this model and found that the particles in the shell are generally larger than the most common interstellar particles. Herbig (1969, 1970) has refined the model for the dust shell around VY CMa by postulating an expanding circumstellar dust ring to explain the two emission peaks observed in the OH emission. Herbig's model does not significantly alter the theoretical fit to the observed energy distribution.

The Ic variable stars are a small class of K and M supergiant stars which show irregular fluctuations in their visual magnitudes. Following their discovery that BC Cyg (M3.5 Ia) had an excess at 11.4 μ about twice as large as that of μ Cep (Figure 1), Gehrz et al (1970) surveyed all 12 Ic variables visible from the Northern Hemisphere. On the basis of the observed 3.5 to 11.4 μ color indices, they conclude that large infrared excesses are a common characteristic of the Ic variables. The wavelength dependence of the excesses is similar to that found in the other M supergiants and can be satisfactorily explained in terms of emission from circumstellar dust shells composed of silicate materials.

A good case can be made that the majority of the Ic variables are highly evolved massive stars, the descendants of main-sequence O stars (Stothers & Chin 1968, 1969, Stothers 1969). The conditions of high ionization and radiation pressure around main-sequence O stars are undoubtedly sufficient to blow away any remnants of protostellar dust clouds; thus, if circumstellar dust is responsible for the excess infrared radiation in Ic variables, it was most probably formed from material ejected from the stars themselves.

A few stars do not appear to fit the simple picture of all cool supergiant stars acquiring dust shells. Three Ic variables (Gehrz et al 1970), as well as α Her and α Sco, have little or no detectable long-wavelength excesses. It may be significant that four of these five stars are binaries with early-type companions.

B. LATE-TYPE GIANTS

M and S Mira variables.—Almost all observed M and S type Mira variables have infrared excesses which peak around 11 μ (Gillett, Low & Stein 1968, Woolf & Ney 1969, Gillett, Merrill & Stein 1971) as illustrated in Figure 1 for o Cet. The excesses are generally smaller than those observed in M supergiants; typically, the percentage of the total luminosity attributable to emission by dust grains is less than 5%. The excesses are commonly larger for the M Miras than for the S Miras; this difference may be due in part to the difference in the oxygen-to-carbon ratio which affects the condensation equilibrium of the silicate particles.

Although only a small fraction of the total luminosity of the better known, visually bright Mira variables is attributable to the infrared excess, there are several Mira variables whose luminosities are dominated by their infrared radiation. One such source is NML Tau (Neugebauer et al 1965) (Figure 2). Spinrad & Wing (1969) have reviewed the infrared spectral data which led to its identification as a Mira variable; spectral scans around 2 μ show that NML Tau is oxygen rich (Hyland et al 1971).

A second example is IRC+10011[3] (Figure 2). On the basis of spectro-photometry from 2.8–5.1 μ, Gaustad et al (1969) suggested that IRC+10011 is a very cool Mira variable. Two-micron spectra show the presence of strong H_2O absorption and CO bands, typical of the unreddened Mira variables (Hyland et al 1971). At wavelengths less than 3.5 μ, however, the energy distribution shown in Figure 2 is characteristic of a temperature much lower than that expected from Miras. Hyland et al (1971) interpret these data as a sign of extreme reddening by a thick circumstellar dust shell with a temperature around 700°K.

Both NML Tau and IRC+10011 are highly variable. Visually, NML Tau varies by more than 2 mag with a period of $\sim$400 days (Wing et al 1967, Cannon 1966, 1967). At 2 and 10 μ the amplitude is generally less; although the continuum becomes redder when the 2.2 μ flux is at a minimum, the resultant color temperature change is small. For IRC+10011, on the other hand, the flux variations at longer wavelengths are much smaller than at shorter wavelengths. This difference in behavior is interpreted as resulting from the difference in the thickness of the dust shells (Table 1). It is interesting that both NML Tau and IRC+10011 are OH radio emission sources (Wilson et al 1970).

Carbon stars.—Visually, carbon stars are among the reddest stellar objects; unfortunately, infrared energy distributions have been obtained for only a few (Mendoza & Johnson 1965, Gillett, Merrill & Stein 1971). As seen from the example of V Hya in Figure 1, the infrared excesses in carbon stars are a smooth function of wavelength; no emission peak similar to that of the M stars is observed in the 8–14 μ region.

Several correlations of the infrared excesses with optical properties of carbon stars do stand out in the data of Gillett, Merrill & Stein (1971). Stars with large 11.0 μ excesses show long-period variability whereas stars with small excesses have irregular variations; the latter show an apparent absorption around 5 μ, while little 5 μ absorption is observed in the former. Hyland & Frogel (1971) also show that long-period variables exhibit only weak CO and CN absorption bands around 2.2 μ, whereas these bands are conspicuous in the shorter-period and irregular variables.

One model which explains these features is thermal re-emission by a dust shell composed of graphite particles; pure graphite does not have an emission peak around 8–14 μ. According to Hoyle & Wickramasinghe (1962), Donn et al (1968), and Gilman (1969), graphite particles will condense in the atmospheres of cool carbon-rich stars at much higher temperatures, and thus much closer to the stellar surface, than can other condensates, such as silicates. That graphite particles can exist up to $\sim$2000°K may explain the ap-

[3] IRC+10011=CIT 3 (Ulrich et al 1966).

parent single-component composition since the particles could be, in effect, merely an extension of the star's atmosphere, and not, as in the case of the cool M stars, a separate cloud at a much lower temperature than the atmosphere. The correlations between the strength of the infrared excesses and the lack of photospheric absorption features at 2 μ could also be due to the intervening graphite shells.

The infrared object IRC +10216 (Becklin et al 1969) is an example of a carbon star whose shell is optically thick at all infrared wavelengths. At 5 μ, this object, located out of the galactic plane, is the brightest known source outside the solar system; visually, it is an extended object fainter than 18th mag. Its energy distribution from 1.2 to 20 μ roughly resembles that of a 650°K blackbody (Figure 2); no spectral features have been observed in the wavelength range from 1.5–14 μ. At 2.2 μ the source varies by as much as 2 mag with a period of ~600 days; at 10 μ the variation is ~0.8 mag. Becklin et al (1969) rejected the possibility that IRC +10216 is extragalactic and suggested a model of an evolved long-period variable star surrounded by an opaque dust shell with an angular diameter of ~0.5″ (Table 1). This size estimate has been confirmed by lunar occultation measures (Toombs 1971). Spectral scans at 1 μ by Miller (1970) supplied unequivocal evidence for the existence of CN bands; photographic spectra (Herbig & Zappala 1970) and independent scans by Oke (1970) confirm the presence of CN bands and the carbon-star nature of this source. The typical galactic distance of 200 pc adopted by Becklin et al (1969) leads to a luminosity of 2×10^4 $L_\odot$, similar to that proposed by Herbig & Zappala (1970) from a comparison with other long-period variable carbon stars.

It should be added that the spectrum of R Lep in the 8–14 μ region (Woolf & Ney 1969) is different from that of other cool stars and can be interpreted in terms of a 140°K blackbody. Broadband infrared photometry of the continuum such as done by Gillett, Merrill & Stein (1971) cannot isolate this spectral feature and suggests that further 8–14 μ spectrophotometry of carbon stars is necessary to determine the frequency of such cool blackbody excesses.

TABLE 1. Properties of late-type stars with extreme excesses

Object	Type of central star	T_{star} (°K)	$\dfrac{L_{shell}}{L_{total}}$	T_{shell} (°K)	R_{shell} (10^{15} cm)	M_{shell} (10^{28} g)
NML Cyg	Late M supergiant	2500	0.75	600	3.0	10
VY CMa	Late M supergiant	2500	0.75	600	3.0	10
IRC+10011	Late M Mira	1800	0.85	700	1.0	2
NML Tau	Late M Mira	2000	0.20	550	1.5	0.5
IRC+10216	Late carbon Mira	2000	>0.99	650	1.5	>10

Semiregular or nonvariable M giants.—In their extensive study of late-type stars, Gillett, Merrill & Stein (1971) observed a number of M giants ranging in spectral type from M2 III to M8e. Almost all stars M5 and later exhibit infrared excesses at 11 μ which the authors identify with emission from silicate particles. While a large range in the magnitude of the excesses is evident, even the extreme cases have much smaller excesses than those of most M Mira variables and M supergiants. The data of Gillett, Merrill & Stein (1971) also show clear evidence for absorption near 4.9 μ. In the case of normal M giants this is probably due to absorption by the fundamental vibration-rotation band of CO, although confirmation by medium-resolution spectrophotometry is required.

C. RV TAURI STARS

The RV Tauri stars are variable G to K type supergiants (Preston et al 1963), with atmospheric pulsations which appear to be closely connected to shockwave phenomena (Abt 1955, Preston 1962, 1964).

Observations of Gehrz & Woolf (1970) from 3.5–11.5 μ show that several RV Tauri stars have large excesses at wavelengths greater than 5 μ; the ratio of the 10 to 3 μ fluxes in AC Her is 2.5 times the same ratio in NML Cyg. The infrared excess increases to a maximum in the 8.4–11.5 μ region, with the excess at 8.4 μ comparable to or larger than that at 11.5 μ; this behavior is not characteristic of the emissivity of optically thin silicate material. Gehrz & Woolf (1970) suggest that the 10 μ emission could be due to a blackbody-like envelope of $\sim$300°K or to a material with high opacity in the 8–11 μ region. Some evidence that these excesses are, in fact, caused by circumstellar dust comes from the observation of large variable polarization (Serkowski 1970).

IV. INFRARED EXCESSES IN INTERMEDIATE AND EARLY-TYPE STARS

The early data of Johnson (1965) and Johnson et al (1966) demonstrated the existence of infrared excesses from some early-type stars. Although these generally can be interpreted in terms of circumstellar dust shells, several puzzling aspects remain. Moreover, it should be emphasized that in general, excesses are not conspicuous in early-type stars. Johnson (1967b) reviewed the infrared data on 85 stars with spectral type earlier than A2, and from a comparison of the 2.2 and 3.5 μ fluxes he concluded that "most early type stars, including supergiants, do not have significantly infrared radiating circumstellar shells (or late type stellar companions) and . . . their intrinsic infrared colors are in accord with those to be expected from stars of the appropriate effective temperature."

A. LUMINOUS YELLOW STARS

R CrB variables.—Two well-known hydrogen-deficient, carbon-rich variables R CrB and RY Sgr possess large infrared excesses at wavelengths beyond 1 μ (Stein et al 1969b, Lee & Feast 1969). The characteristics of the excesses, which comprise 40 and 80% respectively of the total luminosity of the stars at maximum light, can be attributed to blackbody radiation from a cloud of particles at a temperature close to 900°K; this is illustrated for RY Sgr in Figure 3. The data on both these stars support the model proposed by Stein et al (1969b) of absorption of the optical radiation and the subsequent thermal re-radiation by a cloud of graphite particles ejected during a previous deep minimum. The ejection velocities required by this hypothesis are less than 25 km/sec and are entirely reasonable. An apparent correlation of the variation in the visual magnitude of RY Sgr with its excess infrared radiation was reported by Lee & Feast (1969); although the correlation appears convincing, it needs further substantiation.

Another unusual hydrogen-deficient star v Sgr also exhibits a large infrared excess beyond 1 μ (Lee & Nariai 1967, Gehrz & Woolf 1971) which contains about 20% of the total luminosity of the star. Thus it is similar to, but not as marked as RY Sgr and R CrB.

F and G supergiants.—The hypothesis that circumstellar dust shells are accumulated during mass loss (Section II) has led to a search for large infrared excesses among stars with well-documented evidence for mass loss. Gillett, Hyland & Stein (1970) observed the four F and G supergiants α Cas, 89 Her, ϕ Her, and HD 8752 to 11 μ. All these stars have circumstellar shells and high rates of mass loss but, surprisingly, only 89 Her has significant excess infrared radiation. The colors of the other three stars and of δ CMa (Gehrz & Woolf 1970) are consistent with blackbody radiation at temperatures typical of F and G supergiants ($T \sim 6500$°K). It is clear that the entire question of the formation of dust particles in the extended atmospheres of such stars and the stability of circumstellar dust shells should be further investigated.

The infrared broadband colors and 2.0–2.5 μ spectra of 89 Her are remarkably similar to those of R CrB. This raises interesting questions regarding the chemical composition of the grains responsible for the infrared emission, since R CrB is carbon rich, while 89 Her has a normal oxygen-rich composition. No evidence is found for the existence of silicate grains in these stars, and it also seems unlikely that graphite particles are responsible for their common infrared properties.

The few Cepheids which have been observed show little or no excess infrared radiation. RS Pup (F8–K5), which is surrounded by a shell according to Westerlund (1961), and SV Vul (F7Ia–K0) probably show a little silicate emission (Gehrz & Woolf 1970).

Z CMa.—An interesting early-type star which is also an extreme infrared star is Z CMa whose energy distribution is shown in Figure 3. This irregular variable is associated with nebulosity and has been studied by Racine et al (1971) and Gillett & Stein (1971) at wavelengths from 0.36–20 μ. The recent

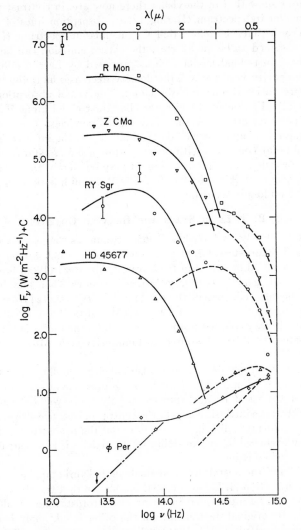

FIGURE 3. Log flux density vs log frequency for the early-type stars R Mon(C=31.0), Z CMa (C=29.5), RY Sgr (C=28.5), HD 45677 (C=27.0), and ϕ Per (C=27.0). Solid curves for R Mon, Z CMa, and HD 45677 are from the models of Larson (1969b). The solid curve for RY Sgr represents a 900°K blackbody. The solid curve for ϕ Per corresponds to optically thin free-free emission from ionized hydrogen; the broken curve corresponds to optically thick emission. Dashed curves represent the approximate photospheric emission. Data for R Mon and HD 45677 are from Low et al (1970) and Mendoza (1968), for Z CMa from Racine et al (1971) and Gillett & Stein (1971), for RY Sgr from Lee & Feast (1969), for ϕ Per from Johnson (1965), Johnson et al (1966), and Woolf et al (1970).

observations show that in the visual there now are, in contrast to previous descriptions of its spectrum (Herbig 1960), absorption lines of an early F star together with its characteristic P Cygni emission spectrum. If the central source is assumed to be an F2 star, the excess emission can be extremely well fit by a dust-cloud model such as proposed by Larson (1969); in this model the opacity varies as λ^{-1}, the density decreases as radius$^{-2/3}$, and the temperature varies throughout the cloud. The visual absorption and reddening in $(B—V)$ due to the circumstellar dust cloud alone indicate that the dust particles are larger than interstellar particles. The position of Z CMa in the HR diagram is well determined; when the infrared is included, it falls at a point consistent with the evolution of a 5 $M_\odot$ star from the main sequence towards the red-giant region. It may be merely an extreme case of the phenomenon exhibited by 89 Her. The P Cygni line profiles prove that extensive mass loss is occurring.

B. T Tauri Stars and Related Objects

Following closely on Poveda's (1965) arguments that very young stars, such as the T Tauri stars, might be accompanied by thick circumstellar dust clouds which would be bright in the infrared, Mendoza (1966) found that T Tauri stars do, in general, have large infrared excesses. Their total luminosity, including the infrared, exceeds the total luminosity expected from the visual data by a factor ranging from 1.3 to 6.6 (Mendoza 1968). For R Mon, an object in many ways related to the T Tauri stars, the factor is 58; as seen in Figure 3, almost 90% of the observed luminosity of R Mon is emitted beyond 1 μ.

Low & Smith (1966) developed a spherical circumstellar-dust-shell model specifically for R Mon. Partly because of the presence of emission lines, the exact spectral types of T Tauri stars cannot always be well determined from optical spectra; the intrinsic colors of the central sources are also often poorly determined. Thus it is difficult to test in detail the proposition that the optical energy absorbed by circumstellar dust particles is sufficient to account for the infrared emission.

The origin of the postulated dust shells in T Tauri stars is uncertain. As mentioned in Section II, Larson (1969a) has shown that the observed energy distribution can be interpreted in terms of a protostar surrounded by the remnants of its original dust cloud. On the other hand, analysis of optical emission lines by Kuhi (1964, 1966) indicates mass loss from T Tauri stars so that dust shells could perhaps be formed out of material from the stars themselves.

C. The Be Stars

In his study of bright stars earlier than A2, Johnson (1967b) found that all observed Be and Of emission-line stars have infrared excesses at 3.5 μ. The 3.5 μ fluxes range from 10 to 40% above those expected from the appropriate spectral class and 2.2 μ magnitude. Woolf et al (1970) investigated the possibility that both the infrared excesses and hydrogen line emission originate in a circumstellar shell of hot ionized gas. By combining observations of the brighter Be stars at 3.5, 5.0, and 11.5 μ with data from Johnson et al (1966) they showed that the nonstellar component of infrared radiation was generally consistent with free-free emission from ionized hydrogen. For ϕ Per, whose energy distribution is shown in Figure 3, they were able to derive values of the depth of the emitting gas in the line of sight as 3×10^{12} cm and of the electron density as 6×10^{11} cm^{-3}, values in reasonable agreement with optically derived values for Be stars. Since the hydrogen emission lines in Be stars are known to vary in strength, simultaneous photometric and spectroscopic observations should reveal any correlation between the infrared continuum and the emission-line strengths.

Possible variable excess radiation at 11.5 μ, which cannot be explained in terms of free-free emission from ionized hydrogen, has been reported for the Be stars 48 Per and κ Dra (Woolf et al 1970); observations separated by about 1 year showed that the 11.5 μ flux apparently changed by a factor of 10. The usual interpretation of the 11.5 μ excesses in terms of circumstellar dust cannot account for the transience of the excesses and the coexistence of dust and hot ionized gas within the Strömgren spheres of the Be stars.

The unusual Be star HD 45677 (Merrill 1928) is also relevant to this discussion. Numerous forbidden Fe II lines which appear under conditions of low excitation, and which have been suggested as good indicators of the presence of circumstellar dust (Geisel 1970), are present in the optical spectrum of HD 45677. Low et al (1970) found a two-component continuum, one with the colors and spectrum of an early B star, and the other with colors similar to those of a 600°K blackbody; the latter emits more than two thirds of the total luminosity of the system (Figure 3). The small reddening of the blue component (E $(B - V) \sim 0.2$) appears to preclude an optically thick spherical dust shell surrounding the B star, unless the dust particles are large enough to produce completely neutral absorption at optical wavelengths. An alternative proposal by Low et al (1970) is that HD 45677 is a binary system with one component surrounded by an optically thick dust shell with characteristics similar to the point source in Orion (Section VA). Velocity measurements are needed to test this proposal.

D. NOVAE

One prediction of the hypothesis that stellar material ejected during mass loss can condense into a circumstellar shell is that galactic novae would have large infrared excesses as a consequence of the matter ejected during their outbursts. These excesses are not expected to become apparent for up to 2 months following the optical outburst when the ejected material (moving typically at 1000 km/sec) encounters conditions suitable for condensation to occur. Hyland & Neugebauer (1970) and Geisel et al (1970) presented observations of Nova Ser 1970 which showed that, between 35 and 60 days following the visual maximum, radiation at wavelengths greater than 1 μ

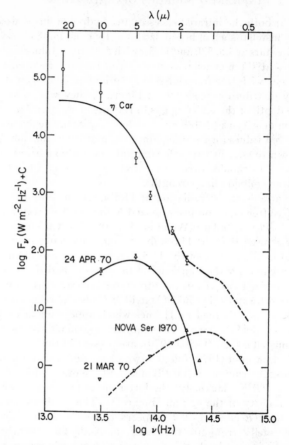

FIGURE 4. Log flux density vs log frequency for Nova Ser 1970 ($C = 25.0$) and η Car ($C = 26.0$). The solid curve for η Car is from the model of Larson (1969b). Solid and dashed curves for Nova Ser 1970 represent 900 and 4500°K blackbodies. Infrared data for η Car are from Westphal & Neugebauer (1969). Data at $\lambda < 1$ μ are from Rodgers & Searle (1967)—uncorrected for intrinsic reddening. Observations of Nova Ser 1970 are from Hyland & Neugebauer (1970) and Geisel et al (1970).

increased from an insignificant fraction to more than 90% of the total luminosity (Figure 4). By 50 days after the optical outburst, the total observed luminosity had decreased by a factor of 6 from its peak value, and then, remarkably, increased again to within a factor of 2 of its maximum. Geisel et al (1970) showed that the apparent infrared temperature decreased with time, approaching 600°K after 100 days, in accordance with ideas of a shell of dust cooling with expansion.

The most outstanding problem is that of the extensive secondary increase in the total luminosity. Geisel et al (1970) suggest that either the dust is heated by some unknown energy source independent of the optical emission of the nova, or the visual output of the central source actually increased while the observed visual brightness appeared to decrease by a factor of 100.

If the first alternative is true, the problem is to identify the source of energy. One possible explanation is that the dust particles are heated by collisions with the ejected stellar material. This mechanism appears to contain the ingredients necessary to produce the observed effects although detailed computations of the process are necessary to establish its magnitude. In the latter theory the apparent decrease is caused by an increase in the circumstellar dust opacity as the number or size (or both) of the particles increases. This idea has the advantage that the heating of the dust is similar to the mechanism proposed for most other galactic infrared stars. Such an increase in the optical output should produce observable spectroscopic effects.

Isolated photometric observations of Nova Del 1967, Nova Aql 1970, FU Ori, and RS Oph also suggest that an infrared phase may be a common feature of novae and novalike objects (Geisel et al 1970).

At 20 μ the brightest known celestial source outside the solar system is η Car, an unusual and novalike object whose continuum rises steeply into the infrared (Figure 4) (Rodgers & Searle 1967, Neugebauer & Westphal 1968, Westphal & Neugebauer 1969). Recent controversy has centered on the determination of the intrinsic reddening since it is sufficiently large to alter any interpretation of the origin of the ultraviolet and visible energy. Various authors have estimated the reddening correction with strongly contradictory results which are hard to reconcile unless the structure of η Car is extremely complex (Rodgers & Searle 1967, Pagel 1969a, Lambert 1969, Rodgers 1971).

The cool infrared component cannot be adequately respresented by blackbody radiation at a single temperature, but a slightly more detailed dust model can be used to describe the data. The best fit to the data using Larson's (1969b) model, which is described in connection with Z CMa (Section IVA), is included in Figure 4. Westphal & Neugebauer (1969) suggested the possibility that the infrared radiation is thermal re-radiation; Pagel (1969b) has shown there is sufficient ultraviolet flux to supply the infrared energy if his value of reddening is adopted. Early suggestions that the infrared rise be interpreted in terms of a synchrotron source (Searle et al 1965, Rodgers & Searle 1967, McCray 1967) were confronted with the need to explain a sharp decrease of the flux to 11 cm, a difficulty made apparently unsurmountable by the large 20 μ flux.

Westphal & Neugebauer (1969) computed the intrinsic luminosity of η Car on the basis of the observed fluxes, a distance of 1.5 kpc, Pagel's (1969a) reddening estimate, and a reasonable extrapolation of the infrared energy to longer wavelengths. The resulting $L = 5 \times 10^6 \, L_\odot$ is close to the value $L = 8 \times 10^6 \, L_\odot$ estimated for η Car in 1843, when it reached its maximum visual brightness. Westphal & Neugebauer (1969) were tempted to suggest that the luminosity of η Car has remained essentially constant since 1843, and that the energy has merely been redistributed in wavelength. In the light of the Nova Ser observations, it is not improbable that the luminosity may have decayed and subsequently increased.

V. INFRARED ASSOCIATED WITH SELECTED GALACTIC OBJECTS

A. H II REGIONS

Three well-known H II regions—the Orion Nebula, M8, and M17—have been studied in detail in the infrared; strong infrared sources are associated with all three. For example, Low & Aumann (1970) have found that at wavelengths from 50 to 300 μ the Orion Nebula and M17 each radiate between 10^5 and 10^6 $L_\odot$.

Observations.—The H II region most thoroughly studied in the infrared is the Orion Nebula; 1 to 4 μ excess radiation is generally emitted from the bright optical portion of the nebula. Approximately 1' north of the Trapezium there is an unresolved point source (Becklin & Neugebauer 1967) with an energy distribution like that of a 700°K blackbody and with no identified visible component; its observed luminosity is about 10^3 $L_\odot$ (Figure 5). At least one other similar, but weaker, unresolved object has been found in the same vicinity (Hilgeman et al 1968). Kleinmann & Low (1967) observed a large flux of 20 μ radiation from a source, about 30" in diameter, which was near the brighter point source and which they classified as an infrared nebula. Between 11.6 and 20 μ the energy distribution of this nebula resembles that of a 100°K blackbody (Figure 5); the measured luminosity between 2 and 25 μ is about 5×10^3 $L_\odot$. In the background around the brighter point source there is 2.2 μ radiation which exceeds that predicted from hydrogen recombination, and which is inconsistent with the blackbody radiation from the infrared nebula. It has not been determined whether this radiation is physically associated with either the point source or the infrared nebula or, indeed, whether the point source is physically associated with the infrared nebula.

Ney & Allen (1969) mapped the central 2' of the Orion Nebula at 11.6 and 20 μ and found another nebula much closer to the Trapezium stars with a total extent of about 1' (Figure 5); the maximum flux is about a factor of 100 above that expected from free-free emission. The color temperature derived from the 11.6 and 20 μ data is 220°K; since the brightness temperatures are less than 100°K, the nebula must be optically thin. This conclusion was confirmed by the 7 to 14 μ spectra of Stein & Gillett (1969a) which show that the spectral distribution of the radiation resembles that of the emission from silicate materials around cool stars. Finally, Low & Aumann (1970) have measured a flux of 0.9×10^{-8} W m^{-2} between 50 and 300 μ emitted from the central 8' of the Orion Nebula; they extrapolate this to a luminosity of 2×10^5 $L_\odot$ between 10 and 100 μ. Whether this radiation is emitted from a

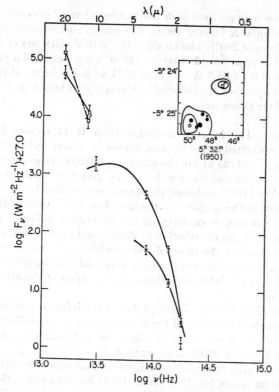

FIGURE 5. Log flux density vs log frequency for four infrared sources in the Orion Nebula along with their approximate position and extent. The curves drawn through the data of the two pointlike objects represent blackbodies at temperatures of 600°K(+) and 800°K(×). Photometry on the brighter southern pointlike source is from Becklin & Neugebauer (1967); data on the northern source are given by Hilgeman (1970). Observations of the Kleinmann-Low nebula (□) and the extended Trapezium source (O) at 11.5 and 20 μ with 26″ apertures are from Ney & Allen (1969). Contours of the extended sources are in intervals of 5×10^{-16} W m^{-2} Hz^{-1} ster^{-1} at 11.5 μ.

single source or from the nebula as a whole cannot be determined at this time.

A source whose energy distribution is generally similar to that of the radiation in the Trapezium region has been observed in a "starlike" optical condensation in M8, 13″ east of the O7 star Herschel 36 (Gillett & Stein 1970). The angular size of the infrared source is known only to be less than 16″.

M17 has been mapped at 10 μ by Kleinmann (1970) who finds two extended sources with flux densities of 1.7×10^{-22} and 1.2×10^{-22} W m^{-2} Hz^{-1}, coinciding closely in position with the maxima of the 2 cm radio continuum radiation (Schraml & Mezger 1969). The energy distributions roughly correspond to those of 200°K blackbodies; the optical thickness of each source is about 10^{-4} at 10 μ. Low & Aumann (1970) have measured a flux between 50 and 300 μ of 0.7×10^{-8} W m^{-2} from M17 with a 7' beam. If the distance of M17 is 1.5 kpc, then the infrared luminosity is at least four times greater than that of the Orion Nebula.

Discussion.—The infrared radiation from H II regions is generally thought to be emitted either by dust heated by visual and ultraviolet radiation from stars and gas within the nebula, or by dust and gas associated with the process of star formation and heated by gravitational contraction.

Ney & Allen (1969) endorsed the former explanation for the Trapezium infrared nebula in Orion although the fact that the position of the center of the source is not exactly centered on the Trapezium indicates that a complicated model for the distribution of dust would be required. In M17, the spatial coincidence of the infrared source with the maximum of the radio continuum suggests a common physical origin; Kleinmann (1970) has, therefore, argued that the infrared is thermal reradiation of the absorbed ultraviolet radiation.

The possibility that the observed flux of the infrared nebula 1' north of the Trapezium is re-radiation of the ultraviolet energy is unlikely since there is no evidence for a maximum in the radio continuum at that position (Webster & Altenhoff 1970). Kleinmann & Low (1967) have suggested that the region is a large collapsing cloud, which converts gravitational energy into infrared radiation, but Hartmann (1967) has shown that this would require too large a mass of gas to produce the observed luminosity. As an alternative explanation, he has suggested that the object is a pre-main-sequence stage of a cluster; the large amount of infrared radiation results because the cluster is still heavily embedded in dust.

The 700°K point source in Orion could be an evolved infrared star. Spectral scans of the object between 2.0 and 2.4 μ, however, show no molecular features characteristic of these objects (Hyland & Frogel 1971). Furthermore, the periodic time variations at 2.2 μ common to late-type infrared stars have not been observed (Lee 1969). Alternately, the object might be a very highly reddened early-type supergiant; however, the absolute visual magnitude of a supergiant capable of producing the observed 10 μ flux would be brighter than -10.5, much brighter than any known star. Finally, the object could be a newly forming star. Larson (1969a,b) has made model calculations which indicate that collapsing protostars will radiate strongly in the infrared and has discussed the infrared properties of such dust clouds (see Section IVA). The predicted spectral distribution and luminosity of a 1 $M_\odot$ object are similar to those observed for the point source; Larson calculated the time scale of this object to be about 10^5 years, which is consistent with the age of the Orion Nebula.

The star Herschel 36 can supply the observed energy to the infrared source in M8 if the latter has an extent of several arc seconds. At the observed angular separation between the infrared source and Herschel 36, however, radiation balance requires significantly lower temperatures than those apparently observed. If the source of the energy is a protostar within the cloud no serious difficulties arise. The angular size of the source would be about 0.01″; thus a measurement of the size of the source would provide a test of these alternatives.

The origin of the 50–300 μ radiation in H II regions is uncertain. Kleinmann (1970) and Low (1970b) have suggested that in M17, this radiation, as well as the 10 μ flux, is thermal re-emission of ultraviolet energy within the nebula. This is supported by the observations of Low (1970b) who finds an apparent correlation between the strength of the radio continuum and the 50–300 μ radiation for nine H II regions.

B. Planetary Nebulae

The energy distributions of the four best measured planetary nebulae[4] are shown in Figure 6 (Gillett et al 1967, Gillett & Stein 1970, Neugebauer & Garmire 1970); five additional planetary nebulae have been observed near 10 μ (Woolf 1969, Gillett, Knacke & Stein 1971) and eight more have been measured at 1.6 and 2.2 μ (Willner et al 1971). In those observed at 10 μ, the flux at 10 μ is about 100 times larger than that expected from free-free emission; an excess is evident at 3.5 μ only in the three shown in Figure 6 plus IC 418. The less compact planetaries generally have not shown observable excesses at 3.5 μ. It is, however, not clear whether this arises from observational problems caused by low surface brightness or is a real physical correlation.

A convincing model for the infrared emission has been discussed by Krishna Swamy & O'Dell (1968) who showed that the spectral distribution of NGC 7027 (Figure 6) can be quantitatively explained by thermal re-emission of Lyman α radiation which has been absorbed by graphite grains. Line emission as the source of the excess has been proposed by Goldberg (1968), and line radiation has, in fact, been observed in four planetary nebulae: the Ne II line at 12.8 μ in IC 418 (Gillett & Stein 1969) and the S IV at 10.5 μ in NGC 7027 (Rank et al 1970), NGC 6572, and NGC 7009 (Holtz et al 1971). The line strengths are, however, small compared to predicted levels, and high-resolution spectral searches for other lines show only continuum radiation; thus line emission apparently cannot significantly contribute to the observed excesses (Rank et al 1970).

C. Crab Nebula

The Crab Nebula is a well-known example of a nonthermal galactic source; the infrared data present no surprises. Extensive measurements of the integrated background flux of the Crab from 1 to 5 μ have been obtained

[4] It is not clear whether K3-50 is a planetary nebula or compact H II region.

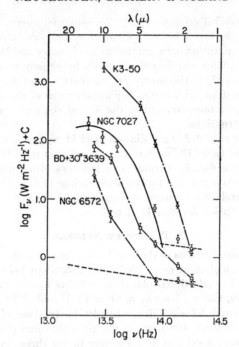

FIGURE 6. Log flux density vs log frequency for planetary nebulae K3-50 ($C=27.0$), NGC 7027 ($C=26.0$), BD+30° 3639 ($C=26.0$), and NGC 6572 ($C=26.0$). The solid curve drawn through the observations of NGC 7027 represents a 350°K blackbody. Dashed curves drawn through the 1.65 μ points of NGC 7027 and NGC 6572 are the expected free-free and bound-free emission from ionized hydrogen (Hilgeman 1970). Data on K3-50 are from Neugebauer & Garmire (1970). Observations of NGC 7027, BD +30° 3639, and NGC 6572 are from Gillett et al (1967) and Gillett & Stein (1970) for $\lambda \geq 5$ μ and from Willner et al (1971) for $\lambda \leq 3.5$ μ.

by Ney & Stein (1968) and Becklin & Kleinmann (1968); a summary of the radio, optical, and infrared energy distribution is shown in Figure 7. The infrared data are consistent with synchrotron radiation with a spectral index $\alpha = 0.65$.

Measurements of pulses from the pulsar NP 0532 located in the Crab Nebula have been extended only to 2.2 μ (Neugebauer, Becklin et al 1969). The energy distribution (Figure 7) continues smoothly from the optical pulse data and, although it falls above the extrapolation of the radio pulsed energy, clearly shows that a minimum must occur in the energy distribution of the pulsar at wavelengths between 2 μ and 20 cm. No coherent picture for this behavior has been presented.

D. Sco XR-1

Sco XR-1, the strongest known galactic X-ray source, is the only known

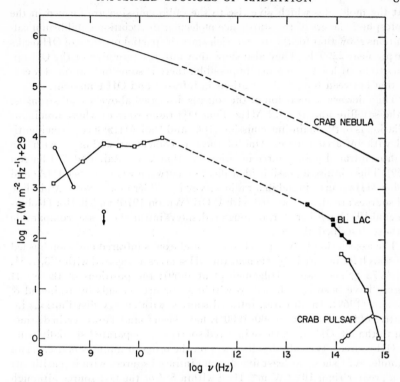

FIGURE 7. Log flux density vs log frequency for the Crab Nebula (Ney & Stein 1968, Becklin & Kleinmann 1968), the Crab Pulsar (Neugebauer, Becklin et al 1969), and BL Lac. For BL Lac, the observations of Oke et al (1969) for 19 August 1968 (□), those of Stein et al (1971) for 23 and 25 August 1970 (⊠), and unpublished California Institute of Technology data for 25 July 1970 (■) are shown. Radio flux densities are taken from the figures given by the above authors.

X-ray source for which infrared data have been published; it has not been detected at infrared wavelengths $> 2.2\ \mu$. The infrared photometry of Neugebauer, Oke et al (1969) gives results consistent with radiation from an optically thick gas with $T \sim 5 \times 10^{7}\,°K$. Since the gas apparently becomes optically thick at wavelengths of about 1 μ, the infrared can be combined with X-ray data to place very stringent limitations on the model of the gas cloud; this is discussed in a review by Oda (1971).

E. OH MICROWAVE EMISSION SOURCES

At least 30 oxygen-rich Mira stars and M supergiants, with photospheric temperatures near 2000°K and large excess radiations at 5 and 10 μ, have OH microwave emission associated with them (Wilson & Barrett 1968, Wilson et al 1970, Wilson 1970, Hyland et al 1971). Hyland et al (1971) conclude that the molecules which give rise to the radio emission are formed in the photospheric layers of the stars since molecular dissociation equilibrium calculations show that for the oxygen-rich stars the partial pressure of OH peaks sharply near 2200°K. They also show that infrared pumping of the OH appears to be at least energetically possible; there is some indication of a correlation between temporal variations in infrared and OH emissions.

The microwave emission of the sources discussed above are all strongest in the satellite line at 1612 MHz. Four OH radio sources whose dominant radiation is in the main-line emission (1665 and 1667 MHz) have been identified with infrared sources. One of these is coincident within 5″ with the brighter infrared point source in Orion (Section VA) (Raimond & Eliasson 1969). There is also a possible identification between BC Cyg (Sec IIIA) and a 1665 MHz source found in a radio survey by Elldér et al (1969). Main-line OH sources are also associated with U Ori (Wilson 1970) and R Hor (Robinson et al 1971); neither star, although red, shows infrared excesses comparable to those described above.

The association between OH sources and excess infrared emission has led to a search for infrared objects near the OH sources associated with W3, W51, and W75 (Neugebauer, Hilgeman et al 1969); the positions of these OH sources have been established to within a few arc seconds by Raimond & Eliasson (1969). In all cases, infrared sources with energy distributions indicative of temperatures $\sim$500–1000°K have been found. However, in distinction to the IR/OH stars, these infrared sources are separated spatially from the OH sources, typically by distances on the order of minutes of arc. As an example, W51 shows at least five distinct infrared sources with intensities at 2.2 μ greater than 10^{-27} W m^{-2} Hz^{-1} within 5$^{\square\prime}$ of the OH source although none is coincident with the OH position. The data are not sufficient to discover the mechanisms which relate the infrared to the OH emission.

VI. INFRARED EMISSION FROM THE GALACTIC NUCLEUS

Infrared observations of the galactic nucleus result in a complex picture which must be represented by three main components defined by both their wavelength and spatial distributions: (a) An extended source, observed between 1 and 5 μ, which has been measured over a 1° region. (b) A 15″ source, coincident with Sgr A, which has been isolated mainly at 10 and 20 μ. (c) A 2°×4° source found at 100 μ within which much fine structure is observed. A contour map of the 2.2 and 100 μ flux is shown in Figure 8.

The 1–5 μ radiation has been studied extensively by Becklin & Neugebauer (1968, 1969). Measurable flux is observed over at least a 1° diameter area, with the central region showing a very high surface brightness; the maximum brightness occurs at a position within 15″ of the center of the radio source Sgr A and within 4′ of the dynamical center of the Galaxy. In both cases coincidence is within the quoted uncertainty. Over the central 10′ of the galactic center, the 2.2 μ brightness falls off approximately as $r^{-0.8\pm0.1}$ and the radiation is distributed along the galactic plane with an axial ratio of approximately 3 to 1 (Figure 8).

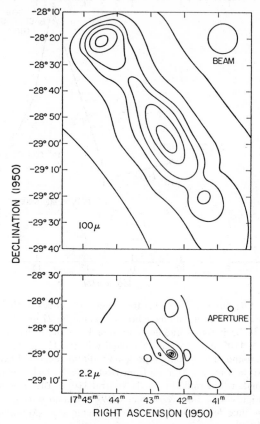

FIGURE 8. Contour maps of the galactic center region at 2.2 μ (Becklin & Neugebauer 1968) and at 100 μ (Hoffman et al 1971). Contour intervals at 2.2 μ are in units of 2.5×10^{-19} W m^{-2} Hz^{-1} ster^{-1} while those at 100 μ are 5×10^{-17} W m^{-2} Hz^{-1} ster^{-1}.

Most of the 1–5 μ radiation is believed to originate from a high spatial density of stars whose normal energy distribution is altered by about 30 mag of visual absorption (Becklin & Neugebauer 1968, Spinrad, Leibert et al 1971); the corrected energy distribution for this radiation is presented in Figure 9. Observations of the nucleus of M31 in the 1–3.5 μ region show a spatial distribution and surface brightness closely similar to that observed in the galactic nucleus (Sandage et al 1969). The density of stars in the central 1 pc diameter core of the galactic center must be at least 10^7 times larger than in the solar neighborhood if the source of this radiation is, in fact, stellar.

Within 10$''$ of the maximum surface brightness at 2.2 μ there is a point-like source whose energy distribution is very similar to that of the 2.2 μ extended background, and whose total luminosity, if it is situated at the distance of the galactic center, is 3×10^5 $L_\odot$. The energy distribution, when cor-

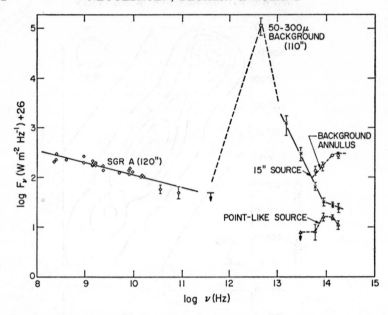

FIGURE 9. Log flux density vs log frequency for sources in the central 3′ of the galactic center. All infrared data have been corrected for 27 mag of visual absorption. Flux density in background annulus corresponds to the radiation of the extended source within a 110″ diam area less the energy in the 15″ source. The curve drawn through background annulus data represents radiation expected from M and K type stars. Infrared data for λ<20 μ are from Becklin & Neugebauer (1969). The 100 μ point corresponds to the radiation in the central 110″ of the galactic center as determined from Figure 1 of Low & Aumann (1970). Radio data of Sgr A correspond roughly to the flux density within a 120″ diameter area (Maxwell & Taylor 1968, Downs et al 1970).

rected for galactic extinction, fits a thermal spectrum, with $T > 2000°K$ (Figure 9); it is probably a bright star like α Ori at or near the galactic center.

The second major infrared component is also within 10″ of the maximum 2.2 μ background brightness, but has a finite diam ∼15″ (Becklin & Neugebauer 1969, Low et al 1969). Its luminosity between 3 and 20 μ is about 10^6 $L_⊙$; observations at wavelengths shorter than 2 μ become confused with the extended background source, while measurements at wavelengths longer than 20 μ have lacked sufficient angular resolution to isolate the source. The energy distribution of the source shown in Figure 9 can be approximated by $F_ν ∝ ν^{-2}$ over the spectral region 3 to 20 μ; this resembles the spectrum of the nuclei of several active extragalactic objects and has led to speculation that similar physical processes are involved. Possibly there is a concentration of dust and stars in the central 1 pc core and the source is thermal re-emission of absorbed starlight. In this case, the observed color temperature of 200°K implies an optical thickness of about 10^{-3}. The source may be nonthermal, but if so, it is not clear that it is phenomenologically associated with the nonthermal radio source Sgr A, since the radio emission comes from a region whose diameter is five times larger than that of the infrared source. Furthermore, the flux density at 20 μ is a factor of 10 larger than that measured at 2 cm, in disagreement with the characteristic energy distribution expected from a nonthermal source. Significant measurements of variability within the source have not been made.

The third, and most luminous, component of the galactic center region, again an extended source, was measured by Hoffman & Frederick (1969) and Hoffman et al (1971) over the central $4° \times 2°$ area. A map at 100 μ, as observed with a 15' angular resolution, is shown in Figure 8; it has three major constituents which are closely similar to the observed 11 cm radio distribution. The brightest region found corresponds approximately with the dynamical center of the Galaxy while a weaker component is coincident with the thermal H II region Sgr B. This first source is definitely extended; Low & Aumann (1970) have studied it with angular resolutions as small as 3' and have found that the average brightness within a region of diameter D varies approximately as $D^{-1.1 \pm 0.2}$. The spatial resolution at 100 μ has not been sufficient to determine if a portion of the 100 μ flux originates from the discrete 15" source. Significantly, Low & Aumann (1970), who also studied the 100 μ secondary sources, could not locate the two outermost sources at 10 μ. They also found that the peak in flux density for the source coincident with the nucleus of the Galaxy occurs near 4.5×10^{12} Hz (70 μ) (Figure 9).

Although the origin of the 100 μ flux is unknown, the similarity in the spatial distributions at 2.2 and 100 μ suggests that a significant portion of the 100 μ flux is probably thermal re-radiation of the starlight. The energy observed between 50 and 300 μ from the central 3' of the nuclear region is about 100 times greater than that observed at shorter wavelengths. Predictions based on the observed 2 μ flux and the calculated interstellar extinction show, however, that there is probably sufficient visual and ultraviolet flux to supply the infrared energy needed. The large 100 μ flux emitted from the vicinity of the H II region Sgr B is probably of the same nature as the radiation from the well-known galactic H II regions discussed in Section VA.

VII. EXTRAGALACTIC SOURCES

Observations.—The first infrared data on galaxies were published by Johnson (1966b), who measured the 0.4–3.5 μ flux in the central 35" of ten bright galaxies. Generally he found that the energy distributions could be synthesized from those of stars in the solar neighborhood although he required that some extremely red stars, such as NML Cyg, be included. An intensive study of the nuclear region of M31 by Sandage et al (1969) also showed that the 1.6–3.5 μ flux near the center ($r \leq 4"$) was not significantly greater than that farther out (Figure 10).

At 10 μ, the detection of only 13 external galaxies has been reported in the

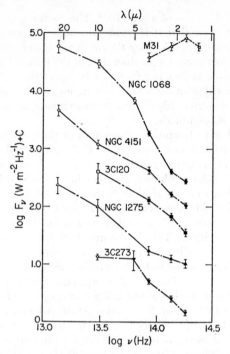

FIGURE 10. Log flux density vs log frequency for the bright Seyfert galaxies NGC 1068 ($C=29.0$), NCG 4151 ($C=29.0$), 3C120 ($C=29.0$), NGC 1275 ($C=28.0$), the bright quasistellar source 3C273 ($C=27.5$), and the central $7.6''$ of M31($C=31.15$). Filled symbols are unpublished data at California Institute of Technology; open symbols are from Kleinmann & Low (1970a,b) except for M31 (Sandage et al 1969). To eliminate background galaxy radiation at $\lambda<2.2$ μ, observations presented for NGC 1068, NGC 4151, and 3C120 were made with a $5''$ aperture, those of NGC 1275 with a $13''$ aperture.

literature (Kleinmann & Low 1970a, b); of these, eight are bright Seyfert galaxies whose nuclei show evidence of violent activity. Near-infrared data on these have been published by several authors (Pacholczyk & Wisniewski 1967, Oke et al 1967, Pacholczyk & Weymann 1968); typical data from 1 to 20 μ are shown in Figure 10. All of these Seyfert galaxies show a rise into the infrared; in general the strong infrared component dominates the stellar thermal radiation at wavelengths greater than about 3 μ. From 5 to 20 μ the continua rise with a power law close to $F_\nu \propto \nu^{-8/2}$; this slope is strikingly similar for all the Seyfert galaxies observed, although NGC 1068 may have a steeper spectrum than the others shown in Figure 10. NGC 1068 is the only Seyfert galaxy which has been observed in the infrared beyond 25 μ; the energy in its continuum continues to rise to a peak near 70 μ before dropping at the radio wavelengths (Low & Aumann 1970) (Figure 12).

In addition to the bright classical Seyfert galaxies, there is a growing list of fainter compact galaxies which, in their optical spectroscopy, share many of the properties, in particular the broad emission lines, of the well-known Seyfert galaxies. These are generally too faint to have been measured in the far infrared, but some 1.6, 2.2, and 3.5 μ data have been obtained (Neugebauer 1971). Representative energy distributions are shown in Figure 11; although all show evidence for nonthermal optical emission, these galaxies do not all have the same infrared spectrum. Some of the energy distributions are similar to those of the brighter Seyferts, and others do not show any conspicuous infrared excess.

So far, a sufficient body of data to provide tentative tests of the infrared variability has been obtained only for NGC 1068 and NGC 4151; the bulk of the data exists for NGC 4151. Fitch et al (1967) and Pacholczyk (1971) have reported changes as large as 1 mag in the 2.2 μ flux of NGC 4151 in time scales of weeks or shorter. Penston et al (1971) report similar variability in both the intensity and spatial dependence of the 2.2 and 3.5 μ flux which is consistent with a central varying nonthermal source superimposed on a

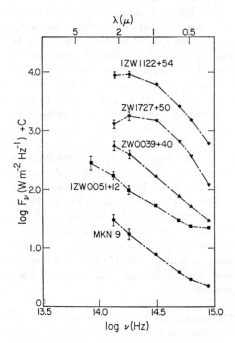

FIGURE 11. Log flux density vs log frequency for the compact galaxies 1ZW 1122 +54 ($C=32.0$), ZW 1727+50 ($C=31.5$), ZW 0039+40 ($C=31.0$), 1ZW 0051 +12 ($C=29.5$), and Markarian No. 9 ($C=29.0$). Infrared data plotted are unpublished observations at California Institute of Technology; the optical data are from Oke (1970).

quiescent outer galaxy. Kleinmann & Low (1970a) conclude that at both 10 and 20 μ a decrease by a factor of 10 occurred between 1965 and 1968 (see also Stein & Gillett (1969b)), although the conclusion relies on high values obtained in 1965 which have not since been observed. None of the authors is willing to exclude the possibility that no fluctuations actually occurred, since the changes are close to the statistical uncertainties of the measurements. There are no convincing measurements of infrared variability in NGC 1068.

Of the non-Seyfert galaxies measured at 10 μ by Kleinmann & Low (1970a, b) M82 is of special interest since the authors were able to map the galaxy at 10 μ with a 5" aperture and show that the infrared radiation comes from an ellipsoidal area roughly $25 \times 8"$ or 400×125 pc.

Of the remaining non-Seyfert-like galaxies listed by Kleinmann & Low (1970a, b), it is perhaps significant that NGC 5195, classified as an I0 irregular galaxy, has been detected at 10 μ, while NGC 5194 (M51), the visually brighter galaxy with which it interacts, has not been detected.

Only one quasistellar object, 3C273, has been measured at wavelengths beyond 3.5 μ; the first infrared measurements were obtained by Johnson (1964) and Low & Johnson (1965). The energy distribution to 10 μ of that object, the brightest known quasar, is shown in Figure 10. The increase of the flux density into the infrared is, however, apparently not an uncommon feature in quasistellar objects as is shown in Figure 12 which summarizes the energy distributions of a number of quasars. A study by Oke et al (1970) of 30 quasars visually brighter than 17th mag showed that the infrared energy distributions to 2.2 μ generally can be characterized by power-law spectra of the form $F_\nu \propto \nu^{-\alpha}$ where the exponents α fill a range of values between $+0.2$ and $+1.6$. The steepest of the continua at 2.2 μ resemble the rapidly rising continua seen at longer wavelengths in some of the brighter Seyfert galaxies; none of the quasar energy distributions shows a decrease into the infrared.

Those quasars which show large variability in their optical output (3C279, 3C345, 3C446, 3C454.3) are among those which show the steepest rise at infrared wavelengths. The 2.2 μ intensities of these four also show variability on a time scale comparable to that observed in the visual wavelengths; the simultaneous measurements necessary to study the correlations have not been made. In contrast, 3C273 and 3C48, which also have steep slopes in the infrared, have shown no variability at 2.2 μ greater than 10% since 1967 (Neugebauer 1971).

Discussion.—The outstanding feature of the extragalactic infrared observations is the large amount of energy emitted in the infrared. For example, NGC 1068 (Figure 12) has an infrared luminosity of 2×10^{12} $L_\odot$ (Low & Aumann 1970); Hoffman et al (1971) estimate that the luminosity of the nuclear region of the Galaxy is 3×10^8 $L_\odot$. The Seyfert galaxies measured by Kleinmann & Low (1970a, b) at 10 μ have total infrared luminosities which, if their energy distributions follow that of NGC 1068, range from 2×10^{10} $L_\odot$ (NGC 3077) to 2×10^{12} $L_\odot$ (NGC 1068, NGC 1275, 3C120).

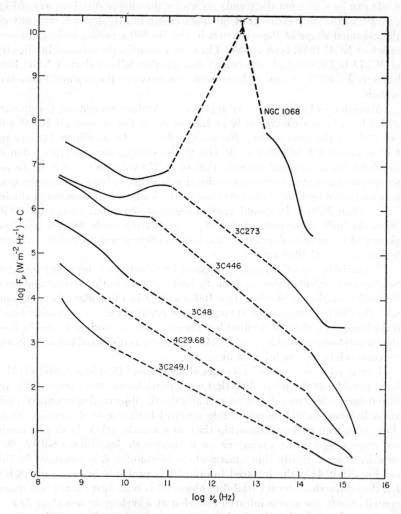

FIGURE 12. Log flux density vs log rest frequency for the quasistellar sources 3C273 (C=31.0), 3C446 (C=31.0), 3C48 (C=30.0), 4C29.68 (C=30.0), 3C249.1 (C=29.0) and the Seyfert galaxy NGC 1068 (C=32.0). Data on the quasistellar sources are from Figure 10 of Oke et al (1970) except for the infrared observations of 3C273. Infrared data at $\lambda < 22$ μ for 3C273 and NGC 1068 are as in Figure 10. The 100 μ observation of NGC 1068 is from Low & Aumann (1970). Radio data on NGC 1068 are adapted from Figure 11 of Becklin & Neugebauer (1967) and Wilson (1971).

It can be seen from the trends shown by the energy distributions of Figures 10–12 that any estimate of the infrared luminosity depends critically on the assumed shape of the spectrum in the 50–300 μ region which, in every case but NGC 1068, is unknown. Thus, for example, the infrared luminosity of 3C273 is $3 \times 10^{14}\ L_\odot$ if the energy distribution follows that of NGC 1068 but only $3 \times 10^{12}\ L_\odot$ if a smooth interpolation between the 10 μ and 1 mm data is valid.

Much larger luminosities are apparently obtained for some of the quasars which have been measured only at 2.2 and 3.5 μ. For example, if 3C446 and 3C273 have the same energy distribution beyond 3.5 μ (Figure 12), the intrinsic infrared luminosity of 3C446, if the cosmological interpretation of the redshifts is accepted, exceeds that of 3C273 by a factor of 8. If the assumption of a common energy distribution is valid, at least eight of the quasars measured by Oke et al (1967) are intrinsically more luminous in the infrared than 3C273. It should again be emphasized that, except for NGC 1068, the bulk of the estimated energy comes from unobserved spectral regions and the present data are insufficient to tell how widespread the infrared features of NGC 1068 are.

No satisfactory origin for the infrared luminosity has been put forward. Studies of particular objects seem to indicate that both thermal and nonthermal radiation is present. The 100 μ source in the Galaxy is consistent with the thermal emission from large dust concentrations. On the other hand, if the longer-wavelength infrared has the same physical origin as the 3 μ flux, the variations observed in, e.g., 3C446, preclude, as discussed below, a purely thermal origin for the infrared flux.

Low & Kleinmann (1968), Rees et al (1969), and Burbidge & Stein (1970) have considered the possibility that solid particles are the source of the infrared excess in extragalactic as well as galactic objects. Arguments of radiation balance, including reasonable spectral behavior of the emissivity of the grains, lead to the relationship that, at a wavelength λ (in μ), one would not expect significant variations on a time scale less than $\tau \sim 0.1 L^{1/2} \lambda^{5/2}$ months in an optically thin source whose luminosity L is measured in 10^{10} $L_\odot$. For NGC 4151 the infrared luminosity is probably in excess of 3×10^{10} $L_\odot$; thus if the short-term variability observed is real, dust cannot be a major contributor to the excess infrared emission at wavelengths less than 20 μ.

A nonthermal origin for the infrared excess has been discussed by several authors, often in connection with the nonthermal optical radiation. The main difficulties encountered are the need for a sharp cutoff at low frequencies and the condition that the electron scattering losses not exceed synchrotron losses (Pacholczyk & Weymann 1968). Burbidge & Stein (1970) conclude that synchrotron models are energetically possible although they require magnetic fields $\sim$1–100 G and, in some cases, nuclei made up of many discrete components. Synchrotron emission in the infrared as the primary energy source has been discussed by Cavaliere et al (1970) who conclude that the infrared emission peak and the subsequent minimum at radio wavelengths can be explained by synchrotron emission from optically thin electrons moving at small pitch angles within moderately high magnetic fields. The optical emission results from inverse Compton scattering of the infrared photons while the cyclotron turnover discussed by O'Dell & Sartori (1970) accounts for the low-frequency decrease. The energy distribution between 20 and 500 μ is, however, critical to any discussion, and discussion of detailed models is probably premature until these data become available.

Special objects.—Two further extragalactic objects should be mentioned— Maffei 1 and BL Lac. The former is visually an extremely faint, red, and extended object lying in the plane of the Galaxy, which, at 2.2 μ, is of comparable brightness to M31. A detailed analysis by Spinrad, Sargent et al (1971) has shown that this "infrared object" is most likely a normal giant elliptical galaxy which has been reddened by about 5 mag of visual absorption. Its distance and mass were estimated as roughly 1 Mpc and $2 \times 10^{11} M_\odot$; it must, therefore, be included in any consideration of the dynamics of the Local Group. Its discovery raises the question if other nearby, but obscured, galaxies exist. Maffei 2, located near Maffei 1 in the sky, is another similar object and apparently is also an external galaxy. Its faintness has precluded a definitive study up to the present time.

BL Lac is an enigmatic object with rapid variability observed at both the visual and radio wavelengths; it is not certain whether it is galactic or extragalactic. The energy distribution over a large wavelength region is characteristic of synchrotron radiation from nonthermal electrons generated in the source (Figure 7); infrared data have been published at 1.6 and 2.2 μ (Oke et al 1969) and at 3.5 and 11 μ (Stein et al 1971). The latter data show a change in the spectral index from which Stein et al (1971) concluded that a major portion of the emitted energy is probably concentrated around 3 μ. This conclusion, plus significant short-term variability observed at 2.2 μ (Neugebauer 1971), enabled those authors to set limits on the size of BL Lac which they then further used to establish an upper limit of 1 Mpc for the distance to the object.

CONCLUDING REMARKS

We have tried to describe the celestial sources of infrared radiation which are currently known; unfortunately this review is probably out of date already.

The infrared excesses of galactic sources between 1 and 20 μ are probably well understood. Thermal emission from circumstellar shells offers a consistent explanation for the excesses observed from most stars; the energy is thus ultimately derived from the normal nuclear processes inside the central star. However, while we think we understand the mechanism which produces the infrared excesses, we cannot predict with certainty which stars will have such circumstellar shells. In a few stars the infrared emission apparently originates in an ionized gas cloud.

The excesses observed in most nonstellar galactic objects are explainable as thermal emission from either dust or a gas. In some cases the infrared is re-radiated ultraviolet stellar radiation while in others the source of energy is probably the gravitational energy in a contracting protostar. The Crab Nebula is the only galactic source for which a nonthermal origin of the infrared is suspected, except perhaps for one source in the nucleus of the Galaxy. The origin of the 1–20 μ radiation in extragalactic objects is not clear; apparently both thermal and nonthermal infrared sources are observed.

At infrared wavelengths beyond 20 μ we know nothing about the emission from any star but the Sun. In some H II regions, in the galactic nucleus, and in NGC 1068, we see tantalizing evidence of luminosities comparable to or greater than that emitted in the rest of the observed spectral regions. Of the three decades of the electromagnetic spectrum covered by the infrared, only one decade has been explored; the exploration of the other two is certain to give us new and different challenges.

We thank M. Neugebauer, R. Leighton, D. Allen, M. Bessel, and G. Münch for criticism and detailed discussions about this paper; W. Stein, E. Ney, N. Woolf, and F. Low for reading the paper and, in many cases, providing preprints of their work; J. Bennett for much assistance in collating and checking the data; and K. Macdonnell and M. Katz for help in the preparation of the paper. The work was supported in part by National Aeronautics and Space Administration grant NGL 05–002–007.

LITERATURE CITED

Abt, H. A. 1955. *Ap. J.* 122:72

Ackermann, G. 1970. *Astron. Ap.* 8:315

Ackermann, G., Fugmann, G., Hermann, W., Voelcker, K. 1968. *Z. Ap.* 69:130

Bidelman, W. P. 1954. *Ap. J. Suppl.* 1:175

Becklin, E. E., Frogel, J. A., Hyland, A. R., Kristian, J., Neugebauer, G. 1969. *Ap. J. Lett.* 158:L133

Becklin, E. E., Kleinmann, D. E. 1968. *Ap. J. Lett.* 152:L25

Becklin, E. E., Neugebauer, G. 1967. *Ap. J.* 147:799

Ibid 1968. 151:145

Becklin, E. E., Neugebauer, G. 1969. *Ap. J. Lett.* 157:L31

Burbidge, G. R., Stein, W. A. 1970. *Ap. J.* 160:573

Cannon, R. D. 1966. *Observatory* 86:150

Ibid 1967. 87:231

Cavaliere, A., Morrison, P., Pacini, F. 1970. *Ap. J. Lett.* 162:L133

Chavira, E. 1967. *Bol. Observ. Tonantzintla Tacubaya* 4: No. 29, 197

Davidson, K., Harwit, M. 1967. *Ap. J.* 148:443

Donn, B., Wickramasinghe, N. C., Hudson, J. P., Stecher, T. P. 1968. *Ap. J.* 153:451

Downs, D., Maxwell, A., Rinehart, R· 1970. *Ap. J. Lett.* 161:L123

Dyck, H. M., Forrest, W. J., Gillett, F. C., Stein, W. A., Gehrz, R. D., Woolf, N. J., Shawl, S. J. 1971. In press

Eliasson, B., Bartlett, J. F. 1969. *Ap. J. Lett.* 155:L79

Elldér, J., Ronnang, B., Winnberg, A. 1969. *Nature* 222:67

Feldman, P. D., McNutt, D. P., Shivanandan, K. 1968. *Ap. J. Lett.* 154:L131

Fitch, W. S., Pacholczyk, A. G., Weymann, R. J. 1967. *Ap. J. Lett.* 150:L67

Forbes, F. F. 1967. *Ap. J.* 147:1226

Ford, W. K. Jr., Rubin, V. C. 1965. *Ap. J.* 142:1303

Friedlander, M. W., Joseph, R. D. 1970. *Ap. J. Lett.* 162:L87

Gaustad, J. E., Gillett, F. C., Knacke, R. F., Stein, W. A. 1969. *Ap. J.* 158:613

Gehrz, R. D., Ney, E. P., Strecker, D. W. 1970. *Ap. J. Lett.* 161:L219

Gehrz, R. D., Woolf, N. J. 1970. *Ap. J. Lett.* 161:L213

Ibid 1971. *Ap. J.* 165:285

Geisel, S. L. 1970. *Ap. J. Lett.* 161:L105

Geisel, S. L., Kleinmann, D. E., Low, F. J. 1970. *Ap. J. Lett.* 161:L101

Gillett, F. C., Hyland, A. R., Stein, W. A. 1970. *Ap. J. Lett.* 162:L21

Gillett, F. C., Knacke, R. F., Stein, W. A. 1971. In press

Gillett, F. C., Low, F. J., Stein, W. A. 1967. *Ap. J. Lett.* 149:L97

Gillett, F. C., Low, F. J., Stein, W. A. 1968. *Ap. J.* 154:677

Gillett, F. C., Merrill, K. M., Stein, W. A. 1971. *Ap. J.* 164:83

Gillett, F. C., Stein, W. A. 1969. *Ap. J. Lett.* 155:L97

Gillett, F. C., Stein, W. A. 1970. *Ap. J.* 159:817

Ibid 1971. *Ap. J.* 164:77

Gillett, F. C., Stein, W. A., Solomon, P. M. 1970. *Ap. J. Lett.* 160:L173

Gilman, R. C. 1969. *Ap. J. Lett.* 155:L185

Goldberg, L. 1968. *Ap. Lett.* 2:101

Grasdalen, G. L., Gaustad, J. E. 1971. *Astron. J.* 76:231

Hackwell, J. A., Gehrz, R. D., Woolf, N. J. 1970. *Nature* 227:822

Hartmann, W. K. 1967. *Ap. J. Lett.* 149:L87

Hashimoto, J., Maihara, T., Okuda, H., Sato, S. 1970. *Publ. Astron. Soc. Japan* 22:335

Herbig, G. H. 1960. *Ap. J. Suppl.* 4:337

Herbig, G. H. 1969. *Mém. Soc. Roy. Sci. Liège Collect.* In press. *Contrib. Lick Observ. No. 302.* In press

Herbig, G. H. 1970. *Ap. J.* 162:557

Herbig, G. H., Zappala, R. R. 1970. *Ap. J. Lett.* 162:L15

Hetzler, C. 1937. *Ap. J.* 86:509

Hilgeman, T. W. 1970. PhD thesis. California Inst. Tech.

Hilgeman, T., Neugebauer, G., Westphal, J. A. 1968. Unpublished data

Hoffmann, W. F., Frederick, C. L. 1969. *Ap. J. Lett.* 155:L9

Hoffmann, W. F., Frederick, C. L., Emery, R. J. 1971. *Ap. J. Lett.* 164:L23

Hoffmann, W. F., Woolf, N. J., Frederick, C. L., Low, F. J. 1967. *Science* 157:187

Holtz, J. Z., Geballe, T. R., Rank, D. M. 1971. *Ap. J. Lett.* 164:L29

Hoyle, F., Wickramasinghe, N. C. 1962. *MNRAS* 124:417

Humphreys, R. M. 1970. *PASP* 82:1158

Hyland, A. R., Becklin, E. E., Frogel, J. A., Neugebauer, G. 1971. In preparation

Hyland, A. R., Becklin, E. E., Neugebauer, G., Wallerstein, G. 1969. *Ap. J.* 158:619

Hyland, A. R., Frogel, J. A. 1971. In preparation

Hyland, A. R., Neugebauer, G. 1970. *Ap. J. Lett* 160:L177

Johnson, H. L. 1964. *Ap. J.* 139:1022

Ibid 1965. 141:923

Johnson, H. L. 1966a. *Ann. Rev. Astron. Ap.* 4:193

Johnson, H. L. 1966b. *Ap. J.* 143:187

Ibid 1967a. 149:345

Johnson, H. L. 1967b. *Ap. J. Lett.* 150:L39

Ibid 1968. 154:L125

Johnson, H. L., Low, F. J., Steinmetz, D. 1965. *Ap. J.* 142:808

Johnson, H. L., Mitchell, R. I., Iriarte, B., Wisniewski, W. Z. 1966. *Commun. Lunar Planet. Lab.* 4:99

Kleinmann, D. E. 1970. *Bull. Ann. Astron. Soc.* 2:No. 4, 325

Kleinmann, D. E., Low, F. J. 1967. *Ap. J. Lett.* 149:L1

Ibid 1970a. 159:L165

Ibid 1970b. 161:L203

Krishna Swamy, K. S., O'Dell, C. R. 1968. *Ap. J. Lett.* 151:L61

Kruszewski, A., Coyne, G. V., Gehrels, T. 1969. In *Mass Loss From Stars*, ed. M. Hack, 42. Dordrecht, Holland: Reidel

Kuhi, L. V. 1964. *Ap. J.* 140:1409

Ibid 1966. 143:991

Lambert, D. L. 1969. *Nature* 223:726

Larson, R. B. 1969a. *MNRAS* 145:271

Ibid 1969b. 145:297

Lee, O. J., Baldwin, R. J., Hamlin, D. W., Bartlett, T. J., Gore, G. D. 1947. *Ann. Dearborn Ob.* Vol. 5

Lee, T. A. 1969. *PASP* 81:878

Lee, T. A. 1970. *Ap. J.* 162:217

Lee, T. A., Feast, M. W. 1969. *Ap. J. Lett.* 157:L173

Lee, T. A., Nariai, K. 1967. *Ap. J. Lett.* 149:L93

Low, F. J. 1970a. *Semi-Ann. Tech. Rep. Contract No. F 19628-70-C-0046. Project No. 5130 (ARPA)*

Low, F. J. 1970b. *Dec. AAS Meet.* Tampa, Fla.

Low, F. J., Aumann, H. H. 1970. *Ap. J. Lett.* 162:L79

Low, F. J., Johnson, H. L. 1965. *Ap. J.* 141:336

Low, F. J., Johnson, H. L., Kleinmann, D. E., Latham, A. S., Geisel, S. L. 1970. *Ap. J.* 160:531

Low, F. J., Kleinmann, D. E. 1968. *Astron. J.* 73:868

Low, F. J., Kleinmann, D. E., Forbes, F. F., Aumann, H. H. 1969. *Ap. J. Lett.* 157:L97

Low, F. J., Krishna Swamy, K. S. 1970. *Nature* 227:1333

Low, F. J., Smith, B. J. 1966. *Nature* 212:675

Maas, R. W., Ney, E. P., Woolf, N. J. 1970. *Ap. J. Lett.* 160:L101

Maxwell, A., Taylor, J. H. 1968. *Ap. Lett.* 2:191

McCray, R. 1967. *Ap. J.* 147:544
Mendoza, E. E. 1966. *Ap. J.* 143:1010
Mendoza, E. E. 1968. *Ap. J.* 151:977
Mendoza, E. E., Johnson, H. L. 1965. *Ap. J.* 141:161
Merrill, P. W. 1928. *Ap. J.* 67:405
Miller, J. S. 1970. *Ap. J. Lett.* 161:L95
Neugebauer, G. 1971. Unpublished data
Neugebauer, G., Becklin, E. E., Kristian, J., Leighton, R. B., Snellen, G., Westphal, J. A. 1969. *Ap. J. Lett.* 156: L115
Neugebauer, G., Garmire, G. 1970. *Ap. J. Lett.* 161:L91
Neugebauer, G., Hilgeman, T., Becklin, E. E. 1969. *Bull. Am. Astron. Soc.* 1:201
Neugebauer, G., Leighton, R. B. 1969. *Two-Micron Sky Survey—a Preliminary Catalog (NASA SP-3047)*
Neugebauer, G., Martz, D. E., Leighton, R. B. 1965. *Ap. J.* 142:399
Neugebauer, G., Oke, J. B., Becklin, E., Garmire, G. 1969. *Ap. J.* 155:1
Neugebauer, G., Westphal, J. A. 1968. *Ap. J. Lett.* 152:L89
Ney, E. P., Allen, D. A. 1969. *Ap. J. Lett.* 155:L193
Ney, E. P., Stein, W. A. 1968. *Ap. J. Lett.* 152:L21
Oda, M. 1971. In press
O'Dell, S. L., Sartori, L. 1970. *Ap. J. Lett.* 162:L37
Oke, J. B. 1970. Private communication
Oke, J. B., Neugebauer, G., Becklin, E. E. 1969. *Ap. J. Lett.* 156:L41
Oke, J. B., Neugebauer, G., Becklin, E. E. 1970. *Ap. J.* 159:341
Oke, J. B., Sargent, W. L. W., Neugebauer, G., Becklin, E. E. 1967. *Ap. J. Lett.* 150:L173
Pacholczyk, A. G. 1971. *Ap. J.* 163:449
Pacholczyk, A. G., Weymann, R. J. 1968. *Astron. J.* 73:870
Pacholczyk, A. G., Wisniewski, W. Z. 1967. *Ap. J.* 147:394
Pagel, B. E. J. 1969a. *Nature* 221:325
Pagel, B. E. J. 1969b. *Ap. Lett.* 4:221
Penston, M. V., Penston, M. J., Neugebauer, G., Tritton, K. P., Becklin, E. E., Visvanathan, N. 1971. In press
Pesch, P. 1967. *Ap. J.* 147:381
Poveda, A. 1965. *Bol. Obs. Tonantzintla Tacubaya* 4:No. 26, 15
Preston, G. W. 1962. *Ap. J.* 136:866
Ibid 1964. 140:173
Preston, G. W., Krzeminski, W., Smak, J., Williams, J. R. 1963. *Ap. J.* 137:401
Racine, R., Becklin, E. E., Hyland, A. R., Neugebauer, G. 1971. In preparation
Raimond, E., Eliasson, B. 1969. *Ap. J.* 155:817
Rank, D. M., Holtz, J. Z., Geballe, T. R., Townes, C. H. 1970. *Ap. J. Lett.* 161:L185

Rees, M. J., Silk, J. I., Werner, M. W., Wickramasinghe, N. C. 1969. *Nature* 223:788
Robinson, B. J., Caswell, J. L., Goss, W. M. 1971 *Ap. Lett.* 7:163
Rodgers, A. W. 1971. *MNRAS*. In press
Rodgers, A. W., Searle, L. 1967. *MNRAS* 135:99
Sandage, A. R., Becklin, E. E., Neugebauer, G. 1969. *Ap. J.* 157:55
Schraml, J., Mezger, P. J. 1969. *Ap. J.* 156: 269
Searle, L., Rodgers, A. W., Sargent, W. L. W., Oke, J. B. 1965. *Nature* 208: 1190
Serkowski, R. 1970. *Ap. J.* 160:1107
Spinrad, H., Liebert, J., Smith, H. E., Schweizer, F., Kuhi, L. V. 1971. *Ap. J.* 165:17
Spinrad, H., Sargent, W. L. W., Oke, J. B., Neugebauer, G., Landau, R., King, I. R., Gunn, J. E., Garmire, G., Dieter, N. H. 1971. *Ap. J. Lett.* 163:L25
Spinrad, H., Wing, R. F. 1969. *Ann. Rev. Astron. Ap.* 7:249
Stein, W. A., Gaustad, J. E., Gillett, F. C., Knacke, R. F. 1969a. *Ap. J. Lett.* 155: L177
Ibid 1969b. 155:L3
Stein, W. A., Gillett, F. C. 1969a. *Ap. J. Lett.* 155:L197
Stein, W. A., Gillett, F. C. 1969b. *Nature* 224:675
Stein, W. A., Gillett, F. C., Knacke, R. F. 1971. In press
Stothers, R. 1969. *Ap. J.* 155:935
Stothers, R., Chin, C. 1968. *Ap. J.* 152:225
Ibid 1969. 158:1039
Toombs, R. 1971. Private communication
Ulrich, B. T., Neugebauer, G., McCammon, D., Leighton, R. B., Hughes, E. E., Becklin, E. 1966. *Ap. J.* 146:288
Wallerstein, G. 1958. *PASP* 70:479
Webster, W. J., Altenhoff, W. J. 1970. *Ap. Lett.* 5:233
Westerlund, B. 1961. *PASP* 73:72
Westphal, J. A., Neugebauer, G. 1969. *Ap. J. Lett.* 156:L45
Willner, S., Becklin, E. E., Visvanathan, N. 1971. In preparation
Wilson, W. J. 1970. PhD thesis. MIT
Wilson, W. J. 1971. In press
Wilson, W. J., Barrett, A. H. 1968. *Science* 161:778
Wilson, W. J., Barrett, A. H., Moran, J. M. 1970. *Ap. J.* 160:545
Wing, R. F., Spinrad, H., Kuhi, L. V. 1967. *Ap. J.* 147:117
Woolf, N. J. 1969. *Ap. J. Lett.* 157:L37
Woolf, N. J., Ney, E. P. 1969. *Ap. J. Lett.* 155:L181
Woolf, N. J., Stein, W. A., Strittmatter, P. A. 1970. *Astron. Ap.* 9:252

THE NEARBY STARS

PETER VAN DE KAMP

Sproul Observatory, Swarthmore College, Swarthmore, Pennsylvania

I. INTRODUCTION

Standard sources of information for stars nearer than 20 pc are the valuable *Catalogue of Nearby Stars* by Gliese (1969) as well as the analysis of his earlier catalogue (Gliese 1956). A limited, but more complete and more precise sample of the stellar population is obtained from the very nearest stars. Over the past several decades occasional lists of stars nearer than 5 pc have been published. The distance limit is traditional and convenient and essentially covers present knowledge of the fainter portion of the mass-luminosity relation.

In this article I shall again limit myself to the stars nearer than 5.2 pc (17 lyears). Table 1 is an updated version of Table I in *Stars Nearer than Five Parsecs* (van de Kamp 1969); the position angles of the proper motions and radial-velocity values have been added. The comments on this table will be rather more detailed than those given in the earlier article; a brief "Who's Who" of stars of particular interest is presented in Section XI. On the other hand, no attempt has been made to include the detailed data given by Gliese.

In view of the author's field, astrometric rather than astrophysical aspects will be stressed, and in this respect, as usual and once more—the present view on this sample is created in the author's image. The technique and methods of long-focus photographic astrometry are of primary significance for the study of nearby stars (van de Kamp 1967). Parallax, orbital motion, and perturbation and, hence, luminosities and masses may thus be derived. The completeness of this sample of the stellar population depends to a considerable extent on the effort of parallax observers. These, in turn, as far as the nearby stars are concerned, are determined primarily by the discovery of large proper-motion stars.

II. LONG-FOCUS PHOTOGRAPHIC ASTROMETRY

In the first decade of the 20th century, long-focus parallax techniques were developed by Frank Schlesinger (1910, 1911, 1924), who used the visual refractor of the Yerkes Observatory, which has an aperture of 102 cm (40 inches) and a focal length of 19.37 m, yielding a scale of 1 mm = 10″.65. Schlesinger derived stellar parallaxes with a probable error of about ±0″.01. The high positional accuracy is the result of three contributing factors: long focal length, i.e., large scale; stability of photographic plates, and pre-

103

TABLE 1.

No.	Gliese No.	Name	RA	Decl (1950)	Proper motion	Position angle	Radial velocity	Parallax	Distance light years
1		Sun					km/sec		
2	559,551	α Centauri[a]	14ʰ36ᵐ2	−60°38′	3″.68	281°	− 22	0″.760	4.3
3	699	Barnard's star	17 55.4	+ 4 33	10.31	356	−108	.552	5.9
4	406	Wolf 359	10 54.1	+ 7 19	4.71	235	+ 13	.431	7.6
5	411	BD +36°2147	11 00.6	+36 18	4.78	187	− 84	.402	8.1
6	244	Sirius	6 42.9	−16 39	1.33	204	− 8	.377	8.6
7	65	Luyten 726-8	1 36.4	−18 13	3.36	80	+ 30	.365	8.9
8	729	Ross 154	18 46.7	−23 53	0.72	103	− 4	.345	9.4
9	905	Ross 248	23 39.4	+43 55	1.58	176	− 81	.317	10.3
10	144	ε Eridani	3 30.6	− 9 38	0.98	271	+ 16	.305	10.7
11	866	Luyten 789-6	22 35.7	−15 36	3.26	46	− 60	.302	10.8
12	447	Ross 128	11 45.1	+ 1 06	1.37	153	− 13	.301	10.8
13	820	61 Cygni	21 04.7	+38 30	5.22	52	− 64	.292	11.2
14	845	ε Indi	21 59.6	−57 00	4.69	123	− 40	.291	11.2
15	280	Procyon	7 36.7	+ 5 21	1.25	214	− 3	.287	11.4
16	725	Σ 2398	18 42.2	+59 33	2.28	324	+ 5	.284	11.5
17	15	BD +43°44	0 15.5	+43 44	2.89	82	+ 17	.282	11.6
18	887	CD −36°15693	23 02.6	−36 09	6.90	79	+ 10	.279	11.7
19	71	τ Ceti	1 41.7	−16 12	1.92	297	− 16	.273	11.9
20	273	BD +5°1668	7 24.7	+ 5 23	3.73	171	+ 26	.266	12.2
21	825	CD −39°14192	21 14.3	−39 04	3.46	251	+ 21	.260	12.5
22	191	Kapteyn's star	5 09.7	−45 00	8.89	131	+245	.256	12.7
23	860	Krüger 60	22 26.3	+57 27	0.86	246	− 26	.254	12.8
24	234	Ross 614	6 26.8	− 2 46	0.99	134	+ 24	.249	13.1
25	628	BD −12°4523	16 27.5	−12 32	1.18	182	− 13	.249	13.1
26	35	van Maanen's star	0 46.5	+ 5 09	2.95	155	+ 54	.234	13.9
27	473	Wolf 424	12 30.9	+ 9 18	1.75	277	− 5	.229	14.2
28		G158-27	0 4.2	− 7 48	2.06	204		.226	14.4
29	1	CD −37°15492	0 02.5	−37 36	6.08	113	+ 23	.225	14.5
30	380	BD +50°1725	10 08.3	+49 42	1.45	249	− 26	.217	15.0
31	674	CD −46°11540	17 24.9	−46 51	1.13	147		.216	15.1
32	832	CD −49°13515	21 30.2	−49 13	.81	185	+ 8	.214	15.2
33	682	CD −44°11909	17 33.5	−44 17	1.16	217		.213	15.3
34	83.1	Luyten 1159-16	1 57.4	+12 51	2.08	149		.212	15.4
35	526	BD +15°2620	13 43.2	+15 10	2.30	129	+ 15	.208	15.7
36	687	BD +68°946	17 36.7	+68 23	1.33	194	− 22	.207	15.7
37	440	L145-141	11 43.0	−64 33	2.68	97		.206	15.8
38	876	BD −15°6290	22 50.6	−14 31	1.16	125	+ 9	.206	15.8
39	166	40 Eridani	4 13.0	− 7 44	4.08	213	− 43	.205	15.9
40	388	BD +20°2465	10 16.9	+20 07	0.49	264	+ 11	.202	16.1
41	768	Altair	19 48.3	+ 8 44	0.66	54	− 26	.196	16.6
42	702	70 Ophiuchi	18 02.9	+ 2 31	1.13	167	− 7	.195	16.7
43	445	AC +79°3888	11 44.6	+78 58	0.89	57	−119	.194	16.8
44	873	BD +43°4305	22 44.7	+44 05	0.83	237	− 2	.193	16.9
45	169.1	Stein 2051	4 26.8	+58 53	2.37	146		.192	17.0

[a] The position of α Centauri C ("Proxima") is 14ʰ26ᵐ3, −62°28′; 2°11′ from the center of mass of α Centauri A and B. The proper motion of C is 3″.84 in position angle 282°.

Stars nearer than five parsecs

No.	Visual apparent magnitude and spectrum			Visual absolute magnitude			Visual luminosity			No.
	A	B	C	A	B	C	A	B	C	
1	−26.8 G2			4.8			1.0			1
2	0.1 G2	1.5 K6	11 M5e	4.5	5.9	15.4	1.3	0.36	0.00006	2
3	9.5 M5	b		13.2	b		.00044	b		3
4	13.5 M8e			16.7			.00002			4
5	7.5 M2	b		10.5	b		.0052	b		5
6	−1.5 A1	8.3 DA		1.4	11.2		23.	.0028		6
7	12.5 M6e	13.0 M6e		15.3	15.8		.00006	.00004		7
8	10.6 M5e			13.3			.0004			8
9	12.2 M6e			14.7			.00011			9
10	3.7 K2			6.1			.30			10
11	12.2 M6			14.6			.00012			11
12	11.1 M5			13.5			.00033			12
13	5.2 K5	6.0 K7	b	7.5	8.3	b	.083	.040	b	13
14	4.7 K5			7.0			.13			14
15	0.3 F5	10.8		2.6	13.1		7.6	.0005		15
16	8.9 M4	9.7 M5		11.2	12.0		.0028	.0013		16
17	8.1 M1	11.0 M6		10.4	13.3		.0058	.00040		17
18	7.4 M2			9.6			.012			18
19	3.5 G8			5.7			.44			18
20	9.8 M4	b		11.9	b		.0014	b		20
21	6.7 M1			8.8			.025			21
22	8.8 M0			10.8			.0040			22
23	9.7 M4	11.2 M6		11.7	13.2		.0017	.00044		23
24	11.3 M5e	14.8		13.3	16.8		.0004	.00002		24
25	10.0 M5			12.0			.0013			25
26	12.4 DG			14.2			.00017			26
27	12.6 M6e	12.6 M6e		14.4	14.4		.00014	.00014		27
28	13.8 m			15.5			.00005			28
29	8.6 M3			10.4			.00058			29
30	6.6 K7			8.3			.040			30
31	9.4 M4			11.1			.0030			31
32	8.7 M3			10.4			.0058			32
33	11.2 M5			12.8			.00063			33
34	12.3 M8			13.9			.00023			34
35	8.5 M2			10.1			.0076			35
36	9.1 M3.5	b		10.7	b		.0044	b		36
37	11.4			12.6			.0008			37
38	10.2 M5			11.8			.0016			38
39	4.4 K0	9.5 DA	11.2 M4e	6.0	11.2	12.8	.33	.0027	.00063	39
40	9.4 M4.5	b		10.9	b		.0036	b		40
41	0.8 A7			2.3			10.			41
42	4.2 K1	6.0 K6		5.7	7.5		.44	.083		42
43	11.0 M4			12.4			.0009			43
44	10.1 M5e	b		11.5	b		.0021	b		44
45	11.1 M5	12.4 DC		12.5	13.8		.0008	.0003		45

b Unseen components.

cision measuring engine. The attainable accuracy is affected by atmospheric and optical effects, which are limited by maintaining stability of the mounting of the optical parts, objective or mirror, by reducing the spectral bandwidth and by always observing one and the same field in the same hour angle (van de Kamp 1967).

The visual refractor has continued to be proven well suited for precise photographic astrometric measures. As an illustration, the Sproul refractor (aperture 61 cm, focal length 10.93 m, scale 1 mm = 18″87) is used in conjunction with 5″ × 7″ Eastman Kodak 103aG plates and a minus blue (at present Schott OG-515) filter. A bandwidth of about 600 Å is obtained around the minimum focal length λ5607 (Strand 1946). The effective wavelength ranges from about λ5480 for spectral type A0 to λ5525 for spectral type M0. The diameter of a well-defined, sharp, well-blackened star image ranges from about 1 to 2 sec of arc. By appropriate measuring techniques the relative position of two photographic star images on a plate may be obtained with a probable error of ±0.002 mm, or even smaller, which corresponds to about ±0″02 for the Yerkes refractor, ±0″04 for the Sproul refractor.

Early parallax work was done mostly with refractors, reflectors being considered unstable because of limited field and changes in figures. The situation has changed; witness the success of the quartz astrometric reflector (scale 1 mm = 13″56) in operation at the Flagstaff station of the US Naval Observatory (Strand 1964 a, b, Worley 1966, Hoag et al 1967, Riddle 1971); other astrometric reflectors are under construction.

III. PARALLAX DETERMINATIONS

Parallax measurements are made on a background of faint stars, normally of mag 10 or fainter; generally these stars are sufficiently distant to serve as a close approximation to a fixed background, and the measured relative parallax π requires only a small correction to absolute parallax, ranging from about 0″002 to 0″007 and averaging about 0″003. This correction may be evaluated from the estimated photometric parallaxes based on magnitudes and spectra of the reference stars, or from statistical considerations based on the same parameters, taking into account also galactic latitude. Useful tables for this purpose are given by Binnendijk (1943), by Vyssotsky & Williams (1948), and by Heintz (1955).

The positions of the *parallax* star and the reference stars are measured on various types of measuring engines, at present mostly of the precision long-screw type. Considerable gain in accuracy is reached through the use of new types of measuring machines such as the machine now in operation at the US Naval Observatory (Strand 1966). The plates of any one parallax series generally are reduced by a linear transformation to a standard frame based on the reference stars. Schlesinger used three or four references stars and at the Sproul Observatory we do the same. Ideally all reference stars should have the same brightness and spectrum (color) as the central, parallax star, but in practice this can only be approximated. The brightness of the central star, if need be, is reduced by a rotating sector and thus approximate magnitude compensation between central and reference stars is obtained. Grating techniques may also be used. There is a current tendency to use more reference stars, which permits the introduction of nonlinear terms in the reduction, or terms involving magnitudes and spectra of the reference stars. Greatest parallactic shift is obtained shortly after dusk and shortly before dawn. However, at the Sproul Observatory plates are taken all through the night.

A conventional parallax determination based on some 20 or 30 plates, each with 2 or 3 exposures, extending over several years yields a probable error of $\pm 0\overset{''}{.}01$ for the relative parallax. By increasing the observational material, and by multiple parallax determinations at different observatories, higher accuracy is reached. It has been quite customary for some time to extend series of plates over several decades for the purpose of obtaining mass ratios and studying perturbations. Improved parallax results have been the obvious byproduct; internal probable errors of $\pm\overset{''}{.}005$, $\pm\overset{''}{.}002$, and even lower have been obtained.

The parallax calculations are based on the following equations:

$$X = c_x + \mu_x t + q_x t^2 + \pi P_\alpha$$
$$Y = c_y + \mu_y t + q_y t^2 + \pi P_\delta$$

1.

where X, Y is the position of the central star measured and reduced to the background of reference stars; the first three terms on the right-hand side represent the heliocentric path, the last term the parallactic displacement. The quadratic (acceleration) term may be required for series of plates covering several decades. While most parallax information is obtained from right ascension (X) measures it is customary to measure the declination (Y) coordinate also (van de Kamp 1967).

The absolute magnitude M is obtained from the absolute parallax p and apparent magnitude m by the relation:

$$M = m + 5 + 5 \log p$$

2.

The bulk of several of the major parallax programs consists of faint large-proper-motion stars and in my opinion this is sound. It has been argued that this leads to selective information. But, then, what doesn't? Parallax observations of the naked-eye stars, although of value, have led to large numbers of parallaxes too small to be of much individual significance. Selecting stars on the basis of brightness is only mildly productive—and remains a selection also. Statistical parallax methods often seem to be the preferred approach for stars which, as a group, have small parallaxes.

In no case, therefore, should we neglect parallax observations of faint large-proper-motion stars; they will lead to a significant improvement of knowledge of individual stars in our neighborhood, particularly regarding the masses and luminosities of faint stars. We recognize, therefore, the efforts of such past observers as Max Wolf and Frank Ross, and of the current observers, Henry Giclas (1970) and, particularly, Willem J. Luyten (1970), who is pushing his Palomar Schmidt search for faint proper-motion stars to the limit and thereby provides potential employment for parallax observers for decades to come.

IV. GENERAL SURVEY

Table 1 includes all stars known as of January 1, 1971 to have parallaxes of $0\overset{"}{.}192$ or over, i.e., up to a distance of 5.2 pc, or 17 lyears. Generally, the parallaxes are well determined and based on results obtained at several observatories. The average probable error is slightly below $\pm 0\overset{"}{.}006$; the individual values, however, range from $\pm 0\overset{"}{.}001$ (Barnard's star) to $\pm 0\overset{"}{.}016$ (CD-36°15693). All these are internal errors and the true probable error of the parallax for Barnard's star may well be larger, say $\pm 0\overset{"}{.}003$. In any case, further parallax determinations should be stressed and welcomed to refine our knowledge of the nearby stars.

The parallax values given in Table 1 are reduced to absolute; the majority are based on the Jenkins *Catalogue* (1952, 1963). On the average these differ less than $0\overset{"}{.}0005$ from the values given in Gliese's *Catalogue* (1969); for a study of nearby stars, this difference is negligible. Some unpublished Sproul material is included, as well as the recent newcomer G158-27 (Riddle et al 1971). The visual luminosities L are based on a visual absolute magnitude of 4.8 for the Sun, i.e., $\log L = 0.4 (4.8 - M)$. The list contains 60 separate visible stars, including the Sun. In addition, there are several unseen companion objects. It is natural to remain skeptical about the reality of these objects, but some encouragement and comfort may be derived from the fact that at one time, and for many years, the faint visible companions, Sirius B, Procyon B, and Ross 614B were classified as unseen companions.

Thirty-one stars, including the Sun, are single, though half a dozen of these appear to have unseen companions, astrometric or spectroscopic. Twenty-two stars are grouped two by two in 11 binary systems; 2 of these, 61 Cygni (A) and BD+43°44 A, appear to have unseen companions. Six stars are grouped three by three in 2 triple systems, which consist of an *inner* binary system and a distant companion. Altogether there are 45 stellar systems: 32 single, 11 double, and 2 triple. Slightly over one half of the stars in this nearby sample appear to be single, slightly under one half are components of double and triple systems.

From Gliese's *Catalogue* (1957), Heintz (1969) finds a lower incidence of duplicity among M dwarfs, which was also found by Worley (1962), but not confirmed by him later (1969).

V. DENSITY FUNCTION

The 60 visible stars fill a volume of 589 pc³, corresponding to a density of 0.098 stars per pc³, or about 1 star per 350 lyear³. Or, if we consider systems rather than separate stars we find a spatial density of 0.077 object per pc³, or about 1 object per 450 lyear³. This corresponds to an average separation of somewhat over 2 pc, or 8 lyear, between the systems (the distance from the Sun to α Centauri is 4.3 lyear).

Incompleteness of the sample is indicated through the decreasing value of the density function with increasing distance. This is seen if we divide the sample into two equal volumes: an inner sphere with a radius of 13.5 lyear (4.14 pc) and an outer shell between radii 13.5 and 17.0 lyear:

	Number of	
	Systems	Stars
Within 13.1 lyear	25	35
Between 13.9 and 17.0 lyear	20	25

The decrease in density becomes, obviously, still more pronounced if we extend the distance limits. Gliese's (1969) *Catalogue* contains 1049 stars (systems) nearer than 20 pc, yielding an average density of only 0.03 stars per pc³. Making allowance for incompleteness and taking into account possible radio stars and unseen companions, Gliese (1956) estimates a true density of 0.15 to 0.20 stars per pc³.

In contemplating the average spacing between stellar systems, one must keep in mind the spacing of stars within these systems which range from several astronomical units to the extreme value of over 12,000 au for the separation of α Centauri C from α Centauri A, B. Even this considerable separation is only 0.06 pc, or 0.2 lyear, a small value compared with the average spacing between systems.

VI. Velocity Distribution

Radial velocities are known for 38 systems of the present sample. Space velocities may be computed for these stars by converting the $\mu\alpha''$ and $\mu\delta''$ as given, for example, in Gliese's *Catalogue* to galactic components through the relation

$$\mu_l = \mu_\alpha \cos \phi + \mu_\delta \sin \phi$$
$$\mu_b = -\mu_\alpha \sin \phi + \mu_\delta \cos \phi$$

3.

where ϕ is the galactic parallactic position angle counted from the galactic latitude circle to the declination circle (Ohlsson 1932).

Linear values in km/sec of the galactic components are obtained through

$$L = \frac{4.74}{p} \mu_l, \qquad B = \frac{4.74}{p} \mu_b$$

4.

and are thus compatible with the radial-velocity component R. Using the current "new" galactic coordinates l and b, referred to the galactic center at RA 17^h39^m3, Decl $-28°54'$ (1900), we obtain the following expressions for the rectangular components of the space velocity

$$U = - L \sin l + (-B \sin b + R \cos b) \cos l$$
$$V = + L \cos l + (-B \sin b + R \cos b) \sin l \qquad 5.$$
$$W = + B \cos b + R \sin b$$

The coordinates U, V are in the galactic plane and point to the galactic center ($l=0°$, $b=0°$) and the direction of galactic rotation ($l=90°$, $b=0°$), respectively; W points to the galactic north pole ($b=+90°$).

In the present sample, Kapteyn's star (Gliese No. 191) has a strikingly high radial, and hence space velocity. According to Eggen (1965) there is evidence that this subdwarf belongs to a moving group of stars.

With the space velocities given by Gliese (1969), the following values are found for the galactic components of the group motion:

		Group motion		Galactic circular velocity
		All 38 stars	Excluding Kapteyn's star	
Toward galactic center	U	$-$ 8 km/sec	$-$ 9 km/sec	$-$ 9 km/sec
Direction galactic rotation	V	-28 km/sec	-21 km/sec	-12 km/sec
Toward galactic north pole	W	-12 km/sec	-11 km/sec	$-$ 7 km/sec

For comparison are given the components of the galactic circular velocity (Delhaye 1965). The asymmetry in stellar motions is clearly exhibited in the negative excess in $\overline{V}$ for the nearby stars. The preferential *stream motion* is also clearly exhibited in the large value of the dispersion in the component U, as compared with components V and W (excluding Kapteyn's star):

$$\sigma_U = \pm 49 \text{ km/sec}$$
$$\sigma_V = \pm 26 \text{ km/sec}$$
$$\sigma_W = \pm 25 \text{ km/sec}$$

The following two objects show both large deviations from group motion and sufficient similarity in space motion to be mentioned:

	U	V	W
BD$+36°2147$	$+46$ km/sec	-53 km/sec	-74 km/sec
AC$+79°3888$	$+71$ km/sec	-60 km/sec	-77 km/sec

In view of the uncertainties of the data, particularly of the radial velocities, an actual closer agreement is not excluded for these stars.

VII. Luminosity Function

Only four stars are more luminous than the Sun. Sirius is the brightest with a visual luminosity (23), larger than the combined visual luminosity (21.8) of all the other stars in this sample. Next comes Altair (10) and Procyon (7.6) while α Centauri A (1.3) appears to be slightly more luminous than the Sun. The faintest four stars in the list are Ross 614 B (0.00002), Wolf 359 (0.00002), L726-8 B (0.00004) and G158-27 (0.00005).

Still fainter stars have been found beyond the limits of the present sample; there appear to be two well-established cases of stars intrinsically fainter than visual mag 16.8. Both stars are common proper-motion companions of brighter stars whose trigonometric parallaxes have been measured. These stars are the following:

1. L745–46 B ($7^h38.0$, $-17°17.$, $\mu = 1''26$ in 117°), which is 21'' distant in position angle 276° from L745–46 A (12.9, spectrum DF, parallax $+0''142$ $\pm''012$). L745–46 B has an apparent photographic mag 17.6, corresponding to an absolute photographic mag 18.4.

2. Van Biesbroeck No. 10 ($19^h14^m6, +5°7'$, $\mu = 1''46$ in 203°), the original *Van Biesbroeck's star*, is the star of lowest known luminosity. It is a flare star. This star is the proper-motion component of BD$+4°4048$ (visual mag 9.1, spectrum M3.5 V, parallax$+0''173 \pm''004$) at a distance 74'' in position angle 150°. Its photovisual magnitude is 17.4, corresponding to a visual absolute magnitude of 18.6; the spectrum appears to be dM5.

It is not unreasonable to expect and be prepared for the discovery of still fainter stars within the limits of the present sample.

Table 2 gives the luminosity function, i.e., frequency distribution of absolute visual magnitudes. The sharp drop beyond $+15$ must be partly due to incompleteness of the data, and may well be reduced by parallax measures of the numerous faint proper-motion stars awaiting study.

TABLE 2. Luminosity function

Vis abs mag		Number of stars
$+ 1.3$ to	$+2.4$	2
2.5	4.9	3
5.0	7.4	6
7.5	9.9	6
10.0	12.4	19
12.5	14.9	18
$+15.0$ to	$+16.7$	6
		—
	Total	60

And then, of course, the unseen companions belong in the entry $+15$ and fainter; one or more at some future time may be seen and thus reduce the incompleteness of the visible stars of low luminosity.

Luyten (1968) finds a maximum of the luminosity function at $M = +15.7$ pg, only slightly fainter than his value $M = +14.5$ pg found in 1938; he feels inclined to believe that the maximum has now been reached, and that it is unlikely that it could be shifted more than 1 mag.

VIII. SPECTRUM-LUMINOSITY RELATION

Except for the six white dwarfs (Sirius B, Procyon B, van Maanen's star, L145-141, 40 Eridani B, and Stein 2051 B), all stars appear to be main-sequence stars, predominantly red dwarfs, and two subdwarfs (Barnard's star and Kapteyn's star). This is illustrated in Table 3. The spectra for G158-27, Luyten 1159-16 are inferred from their color indices. Low-dispersion spectra of Stein 2051, as well as color indices, indicate spectra of M5 for the A (Treanor 1966), DC for the B component (Greenstein, private communication). The spectrum of Procyon B is assumed to be DF. No spectrum, or color index, is as yet known for Ross 614 B; we have assumed it to be a M6 dwarf. An H-R diagram of the sample appears elsewhere (van de Kamp 1967).

TABLE 3. Frequency of spectra

Spectrum	Number of stars
A	2
F	1
G	3
K	8
M	40
White dwarfs	6
Total	60

While this is the most complete spatial sample for stars of low luminosity, the much less abundant high-luminosity stars are woefully lacking; to reach a blue supergiant like Deneb (α Cygni) or a red supergiant like Betelgeuse (α Orionis), one has to increase the volume of the present sample over a hundred thousand fold.

Flare-ups have been observed for α Centauri C, Wolf 359, Luyten 726-8 B (UV Ceti), Ross 154, Groombridge 34 B, Krüger 60 B, Luyten 1159-16, BD+20°2465 and BD+43°4305.

IX. DOUBLE STARS. MASS RATIO AND MASSES

The technique and methods of long-focus photographic astrometry are ideally suited for the study of orbital motions of visual binaries. Precise measures for pairs wider than about 3″ are obtained by the multiple-exposure technique developed by Hertzsprung (1920) and continued by Strand (1946) and others. Mass ratios are obtained by the same technique and reduction methods as originated by Schlesinger for parallax determinations and described in Sections II and III (van de Kamp 1967).

Mass ratio calculations are based on the following equations:

$$X = c_x + \mu_x t + q_x t^2 + \pi P_\alpha - B\Delta x$$
$$Y = c_y + \mu_y t + q_y t^2 + \pi P_\delta - B\Delta y$$

6.

where B is the fractional mass of the secondary and Δx, Δy the separation of primary and secondary. In case of photographic blending of the components the following equations may be used:

$$X = c_x + \mu_x t + q_x t^2 + \pi P_\alpha - (B - \beta)\Delta x$$
$$Y = c_y + \mu_y t + q_y t^2 + \pi P_\sigma - (B - \beta)\Delta y$$

7.

where β is the fractional luminosity of the secondary.

Or the following equations may be used:

$$X = c_x + \mu_x t + q_x t^2 + \pi P_\alpha + \alpha Q_\alpha$$
$$Y = c_y + \mu_y t + q_y t^2 + \pi P_\delta + \alpha Q_\delta$$

8.

where Q_α, Q_δ are the orbital factors, analogous to the parallax factors P_α, P_δ. The semiaxis major α of the photocentric orbit is the $(B-\beta)$ fraction of the semiaxis major a of the relative orbit of the two components of the binary. Hence

$$B = \frac{\alpha}{a} + \beta$$

9.

is the significant relation from which the fractional mass is derived. Formulae 8 and 9 have limited validity in case the blending effect changes with changing separation of the components (Feierman 1971).

The combined mass of a visual binary is found from

$$\mathfrak{M}_A + \mathfrak{M}_B = \frac{a^3}{P^2}$$

10.

expressed in astronomical units of mass (relative to Sun), space (au) and time (year). The semimajor axis in linear measure is obtained from

$$a(\text{au}) = \frac{a''}{p''}$$

11.

where a'' is the angular value of the (unforeshortened) semiaxis major of the relative orbit and p'' the parallax. Figure 1, kindly prepared by W. D.

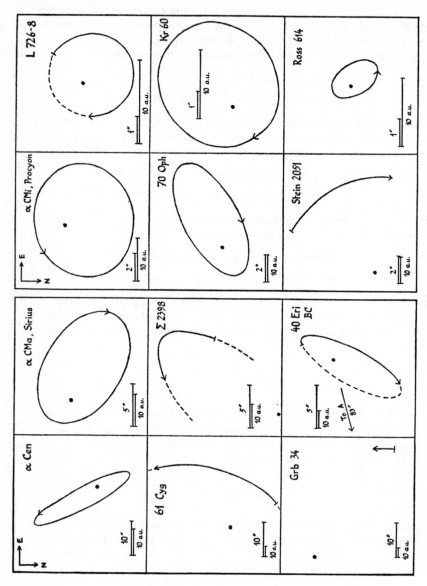

FIGURE 1. Apparent orbital motion of 12 visual binaries within 5.2 pc. The *full drawn lines* represent the observed motion; the *arrow* indicates the direction of motion and the present position.

TABLE 4. Masses of binary components

	a (au)	P (years)	Total mass ($\odot$)	B	$\mathfrak{M}_A$ ($\odot$)	$\mathfrak{M}_B$
α Centauri A, B	23.1	79.92	1.93	.449	1.06	.87
Sirius A, B	19.9	50.09	3.14	.30	2.20	.94
Luyten 726-8 A, B	5.8	25	0.30	(.5)	.15	.15
61 Cygni A, B	84.2	720	1.15	(.5)	.58	.57
Procyon A, B	15.9	40.65	2.43	.268	1.78	.65
Σ 2398 A, B	55.1	453	0.82	(.5)	.41	.41
Krüger 60 A, B	9.5	44.6	0.43	.377	.27	.16
Ross 614 A, B	3.9	16.5	0.22	.32	.15	.07
40 Eridani B, C	33.6	247.9	0.62	.32	.42	.20
70 Ophiuchi A, B	23.3	87.85	1.64	.42	.95	.69

Heintz, illustrates the apparent orbits of 12 visual binaries nearer than 5 pc. The full drawn lines represent the observed motion; the arrow indicates the direction of motion and the present position (1971). Ten of the 12 systems have sufficient orbital data to permit, with varying degrees of accuracy, the derivation of the masses of their components. The results for mass ratio, and masses are given in Table 4. References to individual orbits as well as mass ratios are given in Section XI. Improved orbital and mass ratio data may soon be obtained for 40 Eridani B, C, for Wolf 424, and eventually for 61 Cygni, Σ2398, BD+43°44, and Stein 2051. A value for the mass ratio of Luyten 726-8 should be expected in the near future. More precise photometric measurements are desirable for the components of Ross 614.

While a decade and a half ago the lower end of the mass-luminosity relation was not known beyond Kriiger 60 B, this relation has now been extended through the data for Ross 614 and Luyten 726-8. We note that the lowest value for the mass of a star is $0.07\mathfrak{M}_\odot$ (Ross 614 B). Whether lower masses exist for luminous stars is a challenge to theoreticians and, we trust, will be settled by observers. The mass-luminosity relation permits us to estimate the masses of the other single main-sequence stars; the masses of the remaining three white dwarfs (van Maanen's star, L145-141 and Stein 2051 B) are estimated at $0.6 \mathfrak{M}_\odot$ each. In this way a total mass of 25 $\mathfrak{M}_\odot$ is found for the (visible) stars in the present sample. This leads to an average density of 0.042 $\mathfrak{M}_\odot$ per pc³, or a mass density of nearly 3×10^{-24} g/cm³.

From his study of stars nearer than 20 pc Gliese (1956) estimated a total density of 0.057 $\mathfrak{M}_\odot$ per pc³, or 3.9×10^{-24} g/cm³. Luyten (1968) derives a density of 0.064 $\mathfrak{M}_\odot$ per pc³, or 4.4×10^{-24} g/cm³ based on an evaluation of the masses of stars within 10 pc. A study by Sandage & Luyten (1967) yields a mass density of 2×10^{-24} g/cm³ for white dwarfs alone (Weidemann 1969).

These values are below the star density 1.0×10^{-23} g/cm³ in our neighborhood based on dynamical considerations (J. H. Oort 1962).

X. PERTURBATIONS

Astrometric perturbations have been established for several objects in Table 1. The dynamical results are presented below under the discussion of individual objects. Information about the mass of the unseen component is contained in the mass function

$$\frac{\alpha^3}{P^2} = (B - \beta)^3 (\mathfrak{M}_A + \mathfrak{M}_B) \qquad 12.$$

or

$$\mathfrak{M}_B - \beta(\mathfrak{M}_A + \mathfrak{M}_B) = \alpha P^{-2/3} (\mathfrak{M}_A + \mathfrak{M}_B)^{2/3} \qquad 13.$$

where α is the semiaxis major of the photocentric orbit expressed in astronomical units (van de Kamp 1967). The observed quantities α and P have to be supplemented by an assumption of the total mass of the system and, if necessary, by an assumption of the fractional luminosity of the unseen companion.

Apart from the planetlike companion indicated for Barnard's star, we note a range from 0.008 to 0.026 $\mathfrak{M}_\odot$ for the masses of the unseen companions. Higher values have been found for the masses of unseen companions beyond the present sample, for spectroscopic binaries particularly. The possible influence of systematic errors on the interpretation of perturbations of small amplitude must always be kept in mind.

XI. INDIVIDUAL OBJECTS

We now turn to a more complete description of individual objects, numbered according to their listing in Table 1.

1. *Sun.*—In reviewing those nearby objects, which for one reason or another are of special interest, we must not forget the nearest star, our Sun. In the context of the present survey we shall stress only those properties and aspects of the Sun which are comparable to those obtainable for the other nearby stars. In other words, we shall take a bird's eye view of the Sun, or rather a view from a nearby star.

The Sun (visual mag -26.8, spectrum G2), is a main-sequence star. The total energy output or luminosity of the Sun amounts to 3.79×10^{33} ergs/sec. Solar flares are frequent but have no measurable effect on the Sun's luminosity. Variation in the Sun's brightness has been suspected, and may possibly be related to the sunspot cycle (Blanco & McCuskey 1961). The Sun's diameter is 13.914×10^{10} cm, or 0.0093 au, its volume 1.41×10^{33} cm³. The mass of the Sun amounts to 1.987×10^{33} g; its mean density is 1.41 g/cm³.

The Sun appears to be a single star, although from an observational view-point, the existence of a distant dwarf star companion is not excluded (van de Kamp 1961). On the other hand, the Sun is generously provided with relatively close companions—a *disk* of planets (giant, terrestrial, and minor), their satellites,—and a *halo* of comets and meteors.

With present terrestrial equipment, even Jupiter, the largest planet, could hardly be seen from an observing station at the distance of α Centauri. At best, Jupiter would appear as a pointlike object of the twenty-third visual magnitude (van de Kamp 1944); also, its angular proximity of less than 4″ from the Sun would render its visual detection well nigh impossible. Similar considerations would exclude visual detection of any of the other planets.

It is of interest to consider the gravitational effect of the planets on the Sun. The Sun describes a complicated path around the center of mass of Sun and planets. The perturbing effects by the giant planets have been tabulated by G. M. Clemence (1953). The effects depend on the masses of the planets and the location in their orbit. Their effects are essentially in the plane of the ecliptic; the maximum values (total amplitude) are listed below:

| | Perturbing effects of giant planets on Sun | | | | | | |
| | Orbit | | Mass | | | Maximum effect at distance of | |
Planet	Semiaxis major	Period	$(\odot)$	$\jupiter$	au	α Centauri	Barnard's star
	au	years					
Jupiter	5.20	11.86	1/1047	1	.099	″0076	″0055
Saturn	9.54	29.46	1/3500	.30	.055	.0042	.0030
Uranus	19.19	84.01	1/22,900	.036	.017	.0013	.0009
Neptune	30.07	164.78	1/18,890	.055	.032	.0024	.0018

Over an interval covering a few decades, the combined effect of Jupiter and Saturn may, therefore, yield total amplitudes of 0″.0118 and 0″.0085 at the distance of α Centauri and Barnard's star, respectively. And, over long time intervals, the effects of Uranus and Neptune play a role.

Quantities of this order can be measured by the techniques and methods of long-focus photographic astrometry. Average values based on measurements of, say, 100 plates yield a probable error of something like ±″.003; only a slight increase in accuracy, such as may be expected from improved measuring techniques, will push the predicted perturbations well above the threshold of accidental and, it is to be hoped, systematic errors. It is obvious that these effects are reflected in any measured positions referred to the Sun as origin.

In the recent Sproul study of the long-range series of plates on Barnard's star (1916–1919, 1938–1969), the perturbing effects of Jupiter and Saturn have been taken into account (van de Kamp 1969b). Their total range was

only 0".008, appreciably below the total range (0".04) of the observed fluctuations in the measured positions of Barnard's star. The applied corrections, however, did reduce some of the large residuals. We do not wish to
say that, thus, we could have rediscovered Jupiter and Saturn, but clearly
their perturbing effects had to be removed by referring the observations to
the barycenter of the solar system.

Conventional *annual parallax factors* are referred to the Sun. The question arises whether they should be referred to the barycenter of the solar
system. This matter is only of academic interest for the vast majority of
stars. However, with increased, or increasing accuracy and time interval,
the perturbing effects of the giant planets, particularly Jupiter and Saturn,
must be taken into account for the nearer stars.

The cometary companions of the Sun form a cloud, extending not beyond
10^5 au, or $\frac{1}{2}$pc (Oort 1950). This cloud has been moulded into an approximately spherical shape by the perturbing action of nearby stars. Oort estimates a total population of 10^{11} comets with a total mass of some 10^{27} g, or
$\sim 10^{-6}$ the Sun's mass. Jupiter plays a prominent role in capturing comets
and causing them to have perihelia near Jupiter's orbit.

As we leave the solar system and study the nearby stars, it is at once
obvious that the direct visual approach cannot reveal even the giant planets,
not to speak of lesser dependents.

Our lack of knowledge and the difficulties in finding unseen planetary
companions of stars should certainly be no reason for accepting the absence
of such objects. We shall review below (Barnard's star) current observational astrometric studies. No matter how inadequate these attempts may
seem, the results obtained thus far are encouraging. Any concern because of
the minuteness of the observed perturbations must be countered with the
reflection: "Naturally, this is what you would expect from planetary companions. Larger perturbations, though of greater relative accuracy, would
eliminate the interpretation of planetary masses."

In reviewing briefly some stars of individual interest we point out the
usefulness of the Gliese (1969) *Catalogue*, of Worley's list of double and
multiple stars nearer than 10 pc (1969), and of the *Third Catalogue of Orbits
of Visual Binary Orbits* (Finsen & Worley 1970), all of which contain a
wealth of information for the stars in our sample and beyond.

2. *Alpha Centauri.*—The position of α Centauri C (sometimes named
Proxima) is $14^h26^m3, -62°28', 2°11'$ from the center of mass of α Centauri A
and B. The period of A, B is 79^y92, the semiaxis major is $17".58$, the fractional mass of α Centauri B is 0.449. The current separation is $19"$ and is
increasing slowly till a maximum of $21".8$ is reached in a decade (Heintz
1960). Gasteyer (1966) finds a value for the parallax of C which exceeds
that of A by $+".021 \pm ".015$, in agreement with earlier results (Jenkins 1952).
This difference need not be taken seriously but it does exclude an appreciable
difference in parallax for C, relative to A, B. Because of the relatively small

radial velocity of A, B, −22.7 km/sec (Wesselink 1953), secular changes in parallax, proper-motion, and orbital elements are comparatively small, but in future astrometric studies will have to be taken into account.

The relative proper, i.e. orbital, motion of C and A, B is appreciable, and in agreement with what would be expected. The projected astrometric separation of C to A, B on the sky is 12,000 au, still a small fraction of the average spacing between stellar systems (Section V) and, therefore, an additional confirmation of the physical relationship between A, B, and C.

α Centauri C is a flare star (V645 Centauri).

3. *Barnard's star* (BD+4°3561).—This is the star of largest known proper motion as found by Barnard (1916). The parallax is known with high accuracy, primarily from the extended series of observations at the Sproul Observatory. The formal uncertainty in the distance is less than 1 light week (probable error); systematic errors, of course, may vitiate this result.

The tangential velocity of Barnard's star is 90 km/sec, the radial velocity −108 km/sec, and the space velocity is 140 km/sec. The distance to Barnard's star will gradually diminish to a minimum of 3.75 lyear (less than the present distance of the α Centauri system) in about AD 11,800. At that time its brightness will have increased 2.5 times (1 mag) and Barnard's star will appear as a star of visual mag 8.5.

Barnard's star is the example par excellence of observed secular perspective acceleration (van de Kamp 1970). This quantity $\Delta\mu$ is directly proportional to the parallax p, proper motion μ, and radial velocity V (van de Kamp 1967).

$$\Delta\mu = -2\rlap{.}{''}05 \times 10^6 \mu V p/\text{year} \qquad\qquad 14.$$

For Barnard's star

$$p = +\quad 0\rlap{.}{''}552 \pm 0\rlap{.}{''}001$$

$$\left.\begin{array}{l} \mu_x = -\quad 0.8038 \pm 0.0001 \\ \mu_y = +\quad 10.2745 \pm 0.0001 \end{array}\right\} \text{total } \mu = 10\rlap{.}{''}31 \text{ in } 355\rlap{.}{°}5$$

$$V = -108 \pm 2.5 \text{ km/sec}$$

The predicted acceleration is:

$$\left.\begin{array}{l} \Delta\mu_x = -0\rlap{.}{''}00010 \pm 0\rlap{.}{''}00001 \\ \Delta\mu_y = +0.00125 \pm 0.00003 \end{array}\right\} \text{total } \Delta\mu = +0\rlap{.}{''}00125 \pm 0\rlap{.}{''}00003$$

A recent determination based on some 3500 plates taken with the Sproul refractor over the intervals 1916–1919 and 1938–1969 yields

$$\left.\begin{array}{l} \Delta\mu_x = -0\rlap{.}{''}00011 \pm 0\rlap{.}{''}00003 \\ \Delta\mu_y = +0.00120 \pm 0.00003 \end{array}\right\} \text{total } \Delta\mu = +0\rlap{.}{''}00121 \pm 0\rlap{.}{''}00003$$

in excellent agreement with the predicted acceleration (van de Kamp 1970).

To obtain the above values for the acceleration components, allowance was made for the spurious acceleration caused by the changing influence of the proper motions of the reference stars, with time (van de Kamp 1967).

If we like we may use the observed acceleration, in combination with the very well-known values of parallax and proper motion, to derive the radial velocity. Thus

$$V = -4.88 \times 10^5 \frac{\Delta\mu}{\mu p} \qquad\qquad 15.$$

yields

$$V = -104 \pm 2.6 \text{ km/sec}$$

While for Barnard's star it may prove feasible to improve the observed value of the Doppler shift, the procedure just carried out illustrates the value of deriving the radial velocity by a purely geometric method. True there are few stars for which the product μp is sufficiently large to render this method effective, but with increasing time intervals the accuracy of any observed value of $\Delta\mu$ increases rapidly, provided adequate allowance can be made for the spurious acceleration. This astrometric problem of the future has special significance for stars which may have an appreciable gravitational redshift, or other shifts, which would thus lead to erroneous radial velocities. This problem will be taken up again for No. 26, van Maanen's star.

Barnard's star appears to have a perturbation with a total amplitude of 0″.04, which may be attributed to one or two planetlike companions. The one-planet hypothesis yields an orbit with a period of 25 years, a semi-major axis of 4.5 au, an eccentricity of 0.75, and a companion mass of 1.5 times Jupiter (van de Kamp 1969a). The alternate hypothesis yields two circular corevolving orbits with periods of 26 and 12 years, radii of 4.7 and 2.8 au, and companion masses, each slightly less than Jupiter (van de Kamp 1969b). Future observations may decide which hypothesis is to be preferred.

Brief comments on further objects in Table 1 include:

4. *Wolf 359.*—The faintest star in this sample, one of the faintest stars known (Section VII), with frequent flare-ups.

5. *BD+36°2147* (Lalande 21185).—Indications of a perturbation.

6. *Sirius.*—Sirius is a member of the Ursa Major group. The companion, discovered by Bessel through perturbation (1844), was first seen by Alvan Clark in 1862 with the 48 cm refractor of Dearborn Observatory.

Both Sirius A and B have been suspected of duplicity, A on theoretical grounds (Heintze 1968), B from observations (Eggen 1956). Perturbations have been suspected by Volet (1932) and Zagar (1932) from visual micro-metric observations but these results should be considered doubtful. Perhaps during the current decade new evidence may be obtained. The separation is now (1971.0) 11″31, increases to 11″35 in 1973, then slowly decreases to 10″30 in 1980.0, after which it begins to decrease rapidly, reaching a minimum of 2″6 in 1993 (Heintz 1960). Sirius B is one of the few white dwarfs for which the mass is known. The orbit of Sirius A, B is well known (van den Bos 1960): $a = 7″50$, $P = 50.09$ years. The fractional mass of the companion is 0.30 (van de Kamp 1954).

Lindenblad (1970) has made a new photographic determination of the magnitude of Sirius B, using a wire grating with hexagonal diaphragm on the 66 cm refractor of the US Naval Observatory. He finds a visual magnitude difference $(B - A)$ of $+9.83 \pm .03$ (pe), thus yielding an apparent visual magnitude of 8.37 for Sirius B, in good agreement with Hardie's photoelectric determination (1969) of $+8.29$.

7. *Luyten 726-8.*—The B component is the well-known flare star UV Ceti. Several orbit determinations were made, all leading to total mass of about $0.08 \, \mathfrak{M}_\odot$ for the system. The most recent orbit by Luyten (1971) yields $a = 2″13$, $P = 25$ years. The corresponding total mass is $0.30 \, \mathfrak{M}_\odot$; assuming equal masses, the component masses are $0.15 \, \mathfrak{M}_\odot$ each. The components of L726-8 would, therefore, no longer be the visible stars of smallest known mass; this record would seem to revert to Ross 614 B (see below).

L726-8 remains an important object for establishing the lower end of the mass-luminosity relation. Further astrometric data are needed to furnish more precise information on parallax, mass ratio, and orbital motion.

A perturbation has been suspected from astrometric measures at the McCormick Observatory (Fredrick & Shelus 1969).

8. *Ross 154* (V126 Sgr).—Flare star.

9. *Ross 248.*—Emission-line subdwarf with a light variation of 0.06 mag in a period of 115 days, which may be attributed to a patchy, or spotted surface (Kron 1950b). Sproul astrometric positions do not reveal this cycle.

10. ϵ *Eridani.*—This is one of the two single stars in this sample (the other is No. 19, τ Ceti), sometimes considered as candidates for a sunlike star with possibly a life-supporting companion. The measured positions on 740 plates taken over the interval 1938 to 1969 with the Sproul refractor do not exclude the possibility of a perturbation. The period might be something like 25 years, the total amplitude not over 0″04, corresponding to a companion with a mass of approximately 6 times Jupiter. This tentative interpretation must be regarded with considerable reserve.

12. *Ross 128.*—Slightly variable (Kron 1950a).

13. *61 Cygni.*—The classical binary, at one time referred to as Bessel's star. The first object for which an accurate parallax was measured by Bessel

in 1838. According to Strand (1942) the semiaxis major is 24".59, the period 720 years. There is no accurate value for the mass ratio (van de Kamp & Damkoehler 1953); for the time being, the component masses are assumed to be equal. This is the first binary star for which an accurate photographic perturbation was found, leading to the hypothesis of a companion of very low mass (Strand 1957). The period is 4.8 years, the apparent semiaxis major is 0".0102; the companion mass $0.008\mathfrak{M}_\odot$. Strand's data are based on relative positions of A and B, obtained by the multiple-exposure technique. According to Deutsch (1951), the companion is attached to 61 Cygni A.

15. *Procyon.*—Companion discovered by Bessel through perturbation (1844) and first seen by Schaeberle with the 91 cm Lick refractor in 1896. There is a good orbit by Strand (1951), who gives $a = 4".55$, $P = 40.65$ years, and the accurate fractional mass of 0.268 for the companion. The minimum separation of this difficult object was 2".22 in 1968, is now 2".96 (1971.0) and increases to 5".17 between 1987 and 1990 (Heintz 1960). The spectrum of Procyon B is unknown, the mass is well determined, the star is probably a white dwarf.

16. *Σ 2398.*—Long-period binary for which an astrometric study was made by van de Kamp, Gökkaya & Heintz (1968), yielding $a = 15".66$ and $P = 453$ years. Since over the interval 1913–1966 the maximum deviation from linearity in the relative motion of A and B amounts to only about ".08 and ".17 in RA and Decl, respectively, no precise determination of the mass ratio is, as yet, possible. Intensive observations over the next decade may lead to substantial improvement in the orbital data and, hence, the masses. Near equality of the mass is indicated and, for the time being, adopted; the likelihood of unseen companions is not excluded.

17. *BD+43°44* (Groombridge 34).—Assuming a circular orbit, Lippincott finds a period of 2600 years and a semiaxis major of 41".15. These values are very provisional as is the corresponding value $0.46\ \mathfrak{M}_\odot$ for the combined mass. A range in radial velocity of $+2$ to $+28$ km/sec indicates that Groombridge A is a spectroscopic binary; the H and K lines appear in emission (Joy 1947). BD+43°44 B is a flare star.

19. *τ Ceti* (see *ε* Eridani).—Difficult to observe in Northern Hemisphere; insufficient material to draw any conclusions regarding a possible perturbation.

20. *BD+5°1668.*—Indication of perturbation with period of the order of several decades and a total amplitude of about 0".03. Promising candidate for unseen companion of low (but not of planetary) mass.

22. *Kapteyn's star.*—High-velocity subdwarf; star of largest known proper motion in the Southern Hemisphere.

23. *Krüger 60.*—Orbit well determined: $a = 2\rlap{.}''41$, $P = 44.60$, and the accurate fractional mass of 0.377 for the B component (Lippincott 1953, Wanner 1967). Krüger 60 B = DO Cep is a flare star (van de Kamp & Lippincott 1951, Lippincott 1953, 1956).

24. *Ross 614.*—This is the classical example of a perturbation in a "single" star discovered by the technique and methods of long-focus photographic astrometry (Reuyl 1936). The companion was seen and photographed in 1955 by Baade and has also been observed by Worley & van Biesbroeck. An orbital analysis made by Lippincott (1951, 1955) yields $a = 0\rlap{.}''98$, $P = 16.5$ years. The fractional mass of the companion is 0.32, allowing for Feierman's correction (1971), the corresponding mass of the companion is 0.07 $\mathfrak{M}_\odot$. Ross 614 B appears to be the visible star of lowest known mass (Lippincott 1955), a record which, for a while, seemed to be held by the components of L726-8 (see Luyten 726-8).

The spectrum of the secondary is not known (Section IX).

26. *van Maanen's star,* also named van Maanen 2, Wolf 28.—This star is of particular interest because it is a nearby white dwarf for which eventually the gravitational redshift may be determined by the method described under No. 3. For van Maanen's star

$$p = +0\rlap{.}''234 \pm 0\rlap{.}''004$$

$$\mu_x = +1.2214 \pm 0\rlap{.}''0009$$

$$\mu_y = -2.6843 \pm 0.0009$$

The observed acceleration, based on 281 plates taken at the Sproul Observatory over the interval 1937.0–1970.0, is found to be

$$\Delta\mu_x = -0\rlap{.}''00013 \pm 0\rlap{.}''00006$$

$$\Delta\mu_y = -0.00019 \pm 0.00006$$

Allowance was made for spurious acceleration. Using formula 12 given above, we find values of $+220 \pm 110$ km/sec, and 150 ± 50 km/sec for the radial velocity from measures in RA and Decl, respectively, or combined with relative weights of 1 and 5, a final value of $+160 \pm$ km/sec. The observed spectral shift, interpreted as Doppler shift, is something like $+40$ km/sec (recent letter by Greenstein).

The present astrometric result for the secular perspective acceleration of van Maanen's star is not anywhere near as accurate as the corresponding determination for Barnard's star. In that study the accuracy was limited by the obtainable accuracy ($\pm 0\rlap{.}''00003$) of the spurious acceleration, due to the large annual changes in the dependences of the reference star. For van Maanen's star the spurious acceleration is very accurately known (pe

$\pm 0\rlap{.}''000007$ in x, $\pm 0\rlap{.}''000005$ in y); however, **the** accuracy of the measured acceleration is relatively low. Observational material, distributed uniformly in time, yields an acceleration whose weight increases with the fifth power of the time interval. In the present case the material has a far from uniform, poor distribution in time, observations between 1953 and 1962 being scarce. With current short exposure times, adequate future material appears to be assured and a vast improvement of the acceleration and, hence, the radial velocity is expected within a decade.

27. *Wolf 424.*—Duplicity first noticed by Reuyl (1941). A close ($\sim 1''$) pair with almost equal magnitudes. A minimum period of 16 years is indicated (Lippincott 1958).

28. *G158-27.*—The recently measured parallax (Riddle et al 1971) puts this star well within 5 pc. Photoelectric measurements indicate spectral type M.

34. *Luyten 1159-16.*—Flare star.

36. *BD+68°946.*—Sproul measures (Lippincott 1967) show a perturbation with period of 24.5 years, and the high eccentricity of 0.9. The mass of the companion evaluated to be $0.026\ \mathfrak{M}_\odot$, with a visual absolute magnitude fainter than $+16$.

37. *L145-141.*—No known spectrum; color estimates indicate white dwarf.

39. *40 Eridani.*—The B companion is one of the few white dwarfs for which a value of the mass and of the gravitational redshift can be measured (Popper 1954). The B, C system is $82\rlap{.}''7$ from 40 Eridani A; orbital motion of B, C relative to A is evident; a long period, with tentative value of 8610 years, is indicated. The orbit of B, C was determined by van den Bos (1926), who finds $a = 6\rlap{.}''89$, $P = 247.9$ years, yielding a combined mass of $0.62\ \mathfrak{M}_\odot$. The mass ratio of 0.32, given by van den Bos, yields masses of $0.42\ \mathfrak{M}_\odot$ for the white dwarf companion and $0.20\ \mathfrak{M}_\odot$ for the mass of the M-type companion. The present separation (1971.0) is $8\rlap{.}''2$, increases slowly, reaching $8\rlap{.}''9$ in 1980.

40. *BD+20°2465*, AD Leonis.—A flare star. A perturbation is indicated but there is considerable uncertainty about the elements.

41. *70 Ophiuchi.*—The orbital elements are well established (Strand 1952): $a = 4\rlap{.}''55$, $P = 87.85$ years as well as the fractional mass 0.42 of the B component (van de Kamp 1954); the photographic measures do not confirm the frequently quoted perturbation.

44. $BD+43°4305$, EV Lac.—Strong indication of perturbation with period of the order of two decades and amplitude of about $0''.04$. Promising candidate for unseen companion of low (not planetary) mass. Flare star.

45. *Stein 2051*.—Interesting pair with one red and one white dwarf component (Hardie & Heizer 1966, Treanor 1966). Appreciable orbital motion, tentative period of 350 years. Accurate masses may be expected after several decades.

For helpful advice and criticism, thanks are due to J. L. Greenstein, W. D. Heintz, S. L. Lippincott, W. J. Luyten.

The support of the National Science Foundation for much of the work done on the nearby stars at the Sproul and several other observatories is gratefully acknowledged.

LITERATURE CITED

Barnard, E. E. 1916. *Astron. J.* 29:181

Binnendijk, L. 1943. *Bull. Astron. Inst. Neth.* 10:15, table 8

Blanco, V. M., McCuskey, S. W. 1961. *Basic Physics of the Solar System.* Addison-Wesley

Clemence, C. E. 1953. *Astron. Papers Am. Ephemeris* 13:Part 4

Delhaye, J. 1965. *Stars and Stellar Systems.* 5. *Galactic Structure*, Chap. 4:61

Deutsch, A. J. 1951. *Izv. Pulkova*, No. 146:1

Eggen, O. 1956. *Astron. J.* 61:416

Eggen, O. 1965. *Stars and Stellar Systems.* 5. *Galactic Structure*, Chap. 6:111

Feierman, B. H. 1971. *Astron. J.* 76:89

Finsen, W. S., Worley, C. E. 1970. *Republic Obs. Circ.* 7, No. 129

Fredrick, L. W., Shelus, P. J. 1969. *Bull. Am. Astron. Soc.* 1:No. 3, 241

Gasteyer, C. 1966. *Astron. J.* 71:1017

Giclas, H. L. 1970. *Int. Astron. Union Colloq. No. 7*, 64. Univ. Minnesota

Gliese, W. 1956. *Z. Ap.* 39:1. *Astron. Rechen-Inst., Heidelberg, Mitteil. Ser. A:* No. 3, 1

Gliese, W. 1957. *Mitteil. Astron. Rechen-Inst., Heidelberg*, No. 8

Gliese, W. 1969. *Veröff. Astron. Rechen-Inst., Heidelberg*, No. 22

Hardie, R. H. 1969. *Bull. Am. Astron. Soc.* 1:1, 11

Hardie, R. H., Heizer, A. M. 1966. *PASP* 78:171

Heintz, W. D. 1955. *Astron. Nachr.* 282:221

Heintz, W. D. 1960. *Veröff. München* 5:No. 10

Heintz, W. D. 1969. *J. Roy. Astron. Soc. Can.* 63:283

Heintze, J. R. W. 1968. *Bull. Astron. Inst. Neth.* 20:1

Hertzsprung, R., 1920. *Publ. Ap. Obs. Potsdam* 24:No. 75

Hoag, A. A., Priser, J. B., Riddle, R. K., Christy, J. W. 1967. *Publ. US Naval Obs.* 20:Part 2

Jenkins, L. F. 1952. *General Catalogue of Trigonometric Stellar Parallaxes*

Jenkins, L. F. 1963. *Supplement*

Joy, A. H. 1947, *Ap. J.* 105:96

Kron, G. E. 1950a. *Astron. J.* 55:69

Kron, G. E. 1950b. *Sky Telescope* 9:No. 7, 161

Lindenblad, I. W. 1970. *Astron. J.* 75:841

Lippincott, S. L. 1951. *Astron. J.* 55:236

Lippincott, S. L. 1953. *Astron. J.* 58:135

Lippincott, S. L. 1953. *PASP* 65:248

Lippincott, S. L. 1955. *Astron. J.* 60:379

Lippincott, S. L. 1956. *Hemel Dampkring* 54:58

Lippincott, S. L. 1967. *Astron. J.* 63:322

Lippincott, S. L. 1967. *Astron. J.* 72:1349

Luyten, W. J. 1938. *MNRAS* 98:677

Luyten, W. J. 1968. *MNRAS* 139:221

Luyten, W. J. 1970. *The Stars of Low Luminosity.* Minneapolis, Minn: Univ. Minnesota

Luyten, W. J. 1971. Private communication

Ohlsson, J. 1932. *Ann. Obs. Lund*, No. 3, Oort, J. H. 1950. *Bull. Astron. Inst. Neth.* 11:91

Oort, J. H. 1962. *The Distribution and Motion of Interstellar Matter in Galaxies*, 10. New York: Benjamin

Popper, D. 1954. *Ap. J.* 120:316

126   VAN DE KAMP

Reuyl, D. 1936. *Astron. J.* 45:133
Reuyl, D. 1941. *PASP* 53:336. 1942. 54:52
Riddle, R. K. 1971. *Publ. US Naval Obs.*
20:Part 3a
Riddle, R. K. et al 1971. *PASP* 83: No. 492
Sandage, H., Luyten, W. J. 1967. *Ap. J.*
148:767
Schlesinger, F. 1910, 1911. *Ap. J.* 32:372–
87; 33:8–27
Schlesinger, F. 1924. *Seeliger Festschr.* 422.
Berlin: Springer
Strand, K. Aa. 1942. *PASP* 55:29
Strand, K. Aa. 1946. *Astron. J.* 52:3
Strand, K. Aa. 1951. *Ap. J.* 113:1
Strand, K. Aa. 1952. *Astron. J.* 57:97
Strand, K. Aa. 1957. *Astron. J.* 62:35
Strand, K. Aa. 1964a. *Sky Telescope* 27:
204. 1964b. *Science* 144:1299
Strand, K. Aa. 1966. *Astron. J.* 71:873
Treanor, P. T. 1966. *Observatory* 86:152
van de Kamp, P. 1944. *Sky Telescope* 4:No.
2, 5
van de Kamp, P. 1954. *Astron. J.* 59:447
van de Kamp, P. 1961. *PASP* 73:404
van de Kamp, P. 1967. *Principles of Astrometry.* Freeman
van de Kamp, P. 1969. *PASP* 81:5
van de Kamp, P. 1969a. *Astron. J.* 74:238

van de Kamp, P. 1969b. *Astron. J.* 74:757
van de Kamp, P. 1970. *Int. Astron. Union
Colloq. No. 7*, 77. Univ. Minnesota
van de Kamp, P., Damkoehler, J. E. 1953.
Astron. J. 58:21
van de Kamp, P., Gökkaya, N. G., Heintz,
W. D. 1968. *Astron. J.* 73:361
van de Kamp, P., Lippincott, S. L. 1951.
PASP 63:141
van den Bos, W. H. 1926. *Bull. Astron.
Inst. Neth.* 3:128
van den Bos, W. H. 1960. *J. Obs.* 43:145
Volet, Ch. 1932. *Bull. Astron.* 8:51
Vyssotsky, A. N., Williams, E. T. R. 1948.
Publ. Leander McCormick Obs. 10:33,
table 8.I; p. 36, table 8.VII
Wanner, J. F. 1967. *Sky Telescope* 33:No.
1, 16
Weidemann, V. 1969. *Low-Luminosity
Stars*, 311. Gordon & Breach
Wesselink, A. J. 1953. *MNRAS* 113:505
Worley, C. E. 1962. *Astron. J.* 67:396
Worley, C. E. 1966. *Vistas Astron.* 8:33
Worley, C. E. 1969. *Low-Luminosity Stars*,
122. Gordon & Breach
Zagar, F. 1932. *Publ. Roy. Oss. Astron.
Padova No. 23*

CENTRAL STARS OF PLANETARY NEBULAE 2018

E. E. SALPETER[1]

Cornell University, Ithaca, New York

1. INTRODUCTION

An earlier volume in this series contains an excellent general review of planetary nebulae by Osterbrock (1). All aspects of planetary nebulae were discussed at the IAU Symposium No. 34 in September 1967 and the symposium proceedings are now published. This review treats mainly a restricted aspect of the central stars of planetary nebulae, the present-day theoretical ideas on the evolution of these objects. The usual excuse for writing another review so soon after a previous one—that the subject has been clarified in the meantime—is only partially valid in this case: A large amount of theoretical work has been carried out in the last few years and some aspects have indeed been clarified, but the present dilemma is that there are too many rival theoretical ideas on the structure and previous history of these stars. For these reasons I will concentrate on (*a*) which possibilities can be excluded, (*b*) general theoretical considerations on which there is substantial agreement, and (*c*) my own nonexpert view of where the controversy stands at the moment.

A planetary nebula is a gaseous region of roughly stellar mass, ionized by the far-ultraviolet radiation from a very hot central star contained inside the nebula. It is clear that the nebula was ejected from the central star only $\sim 10^5$ years ago; the star must be several billion years old and yet it evolves noticeably during the expansion time of the nebula (the *Harman-Seaton sequence*). We shall face three major theoretical questions: 1. the nature and history of the star before the ejection of the nebula, 2. the cause for the ejection of the nebula, and 3. the reason for the apparent rapid evolution of the central stellar remnant after the explosion.

Section 2 summarizes some of the general observational data which have indirect but important bearing on the history of the central star, e.g. frequency of occurrence, age, and chemical composition. Section 3 discusses the sequence in the luminosity-color diagram for the central stars themselves. Section 4 gives a review of elementary stellar evolution theory and discusses unrealistically simplified models which nevertheless illustrate some of the basic features. Section 5 discusses advanced evolution in globular cluster stars. These have somewhat different masses and chemical composition from

[1] Work supported in part by National Science Foundation grant GP-26068.

the progenitors of planetary nebulae, but are nevertheless relevant and are reasonably well understood. Section 6 gives conjectures on the progenitor star just before the ejection of the planetary nebula and Section 7 discusses the possible causes and mechanism for this ejection. Finally, Section 8 gives conjectures on the presence and future of the central star after the ejection of the planetary nebula.

2. GENERAL OBSERVATIONAL DATA

For stars in general in our Galaxy it is easy to distinguish the extreme Population I (stars of age $\lesssim 10^9$ years) from extreme Population II (the very earliest stars, $\sim 10^{10}$ years old, found particularly in globular clusters) from their spatial and velocity distribution. However, most of the stars fall in an intermediate category, the disk population, which can also be distinguished from the two extreme populations and inside of which there is a mild correlation between stellar mass and the spatial and velocity distribution. The distribution of planetary nebulae (2, 3) is quite typically that of the disk population and the mean height above the galactic plane of planetary nebulae is the same as that of stars of mass about 1.3 $M_\odot$. It is most likely that a large fraction of planetary nebulae came from stars whose original mass was somewhere between 1.0 and 1.5 $M_\odot$. It is much harder to determine how pure a population they represent and the observational data are quite compatible with the hypothesis that stars of all masses are equally likely to become planetary nebulae (after they have evolved sufficiently far away from the main sequence). The fact that one planetary nebula has been found in a globular cluster so far is also compatible with this hypothesis. However, we shall restrict ourselves to the least controversial statement that some stars between 1 and 1.5 $M_\odot$, typically $\sim 5 \times 10^9$ years old, can become planetary nebulae.

In contrast with drastic variations of heavy-element abundances amongst extreme Population II stars, stars in the disk population and in Population I (as well as interstellar H II regions) seem to have similar cosmic abundances. Although there are some individual variations (including a low oxygen abundance for the one nebula in a globular cluster), the average abundances in the ionized gas of planetary nebulae are quite close to the cosmic abundances (4, 5). In particular this is true for the hydrogen-helium abundance ratio, which is well measured in a number of cases and gives a mean helium abundance (by mass) of $Y \approx 0.30$ to 0.35.

For each planetary nebula it is easy to measure its angular size ϕ and the surface brightness in Hβ which is proportional to $\langle n_e^2 \rangle R$ where n_e is the electron density in the gaseous nebula and $R \sim \phi \times$ (distance) is its radius. If the distance is known (and if the *filling factor* $\epsilon = \langle n_e^2 \rangle \langle n_e \rangle^{-2}$ is known or assumed) one can then obtain luminosity L, n_e, R, and the mass M_{ig} of the ionized gas. Conversely, if M_{ig} (and ϵ) is known one can determine the distance. For a few planetary nebulae in our Galaxy their individual distances have been measured directly and for a few with bright forbidden lines from [O II] the mean densities are known. For most of the nebulae, however, one can only give statistical arguments which connect average values of M_{ig} (or of n_e) with a distance scale factor k. We have

$$\langle L \rangle \propto k^2, \quad \langle n_e^2 \rangle \propto k^{-1}, \quad \langle M_{ig} \rangle \propto k^{2.5} \qquad 1.$$

and for the value of k adopted by O'Dell (6) one finds $\langle M_{ig} \rangle \sim 0.17\ M_\odot$.

The determination of the distance scale for galactic planetary nebulae is not very accurate; another similar determination (7), for instance, gave a value of k about 1.5 times larger than O'Dell's. Fortunately, Webster (8) has recently carried out measurements on planetary nebulae in the Magellanic Clouds, where the distance is known accurately, and found values of M_{ig} in the range of 0.1 to 0.25 $M_\odot$. The presently adopted values (9) are close to $\langle M_{ig} \rangle \sim 0.17\ M_\odot$ and to O'Dell's galactic distance scale. Apart from remaining uncertainty in the exact numerical value of k, two cautionary remarks are in order about the mass of the gaseous nebula: 1. Even for optically thin nebulae, where most the *volume* contains ionized gas, dense filaments of neutral gas are found to be imbedded in the ionized medium. It is hard to estimate the total mass of these neutral substructures, but it could well be appreciably larger than M_{ig}, the mass of the ionized (lower density) gas. 2. Even for the ionized component, $\langle M_{ig} \rangle$ only represents a statistical mean (or median) and there is *no* reason to believe that all nebulae have even approximately the same value of M_{ig}. In fact, the few cases with well-documented measured distances (including those in the Magellanic Clouds) at least suggest a considerable spread in values of M_{ig}. To summarize the situation on M_g, the total gas mass of a planetary nebula: It is safe to say that $M_g \gtrsim 0.2\ M_\odot$ but the upper limit of this mass range is not known.

From the spectroscopically measured radial expansion velocity of a planetary nebula and from its angular size one can obtain its age t (since its ejection from the central star), if its distance is known. Estimates have also been made (3, 6, 9, 10) of the spatial density of planetary nebulae (density $\propto k^{-3}$) and one can therefore estimate the rate χ at which stars are processed through a planetary nebula phase. The numerical value depends on the distance scalefactor,

$$t \propto k, \quad \chi_{PN} \propto k^{-4} \qquad 2.$$

and density estimates vary somewhat. However, adopting the distance scale of O'Dell leads to most estimates (9) giving $\chi_{PN} \sim 2 \times 10^{-12}$ per pc^3 per year. Quite independently of planetary nebulae, estimates have been made for the rate χ_{WD} at which white dwarf stars are created (11). These two rates are very close to each other and also to a third rate, the theoretical estimates for the rate at which deaths of stars originally on the main sequence now occur.

There are of course large numerical inaccuracies in the estimates for those three rates, but the simplest hypothesis is that an appreciable fraction (or even the majority) of all main-sequence stars evolve through a planetary nebula phase at some stage before settling down to become white dwarfs. The more massive stars, $M \gtrsim 3$ $M_\odot$, contribute rather little to the overall rate of star deaths and we have little direct evidence about their probability of forming planetary nebulae. However, both the spatial-velocity distribution and the large value of χ_{PN} indicate that the most common producers of star deaths, stars of original main-sequence mass 1.0 to 1.5 $M_\odot$, must have a good chance of forming planetary nebulae.

Since white dwarfs are the likely endproduct of central stars of planetary nebulae, the mass distribution of white dwarfs is of interest. If one uses the theoretical mass-radius relation (see Section 4) and photometric data to determine radii, one finds that most white dwarfs lie in the mass range 0.3 to 1.0 $M_\odot$ and the latest estimate (11) for the mean mass is about 0.65 $M_\odot$. A different method uses mass-radius ratios as determined from measurements of the gravitational redshift. The latest measurements (12) indicate somewhat larger mean values (~ 0.9 $M_\odot$) than obtained from the photometric data, but it seems well established that a large fraction of white dwarfs have masses somewhere between 0.5 and 1.0 $M_\odot$.

3. The L-T_e Sequence

As suggested some time ago by Zanstra (13), one can use the hydrogen emission-line intensities in the visible from a planetary nebula to calculate the total luminosity beyond the Lyman edge of the central star (as long as the nebula is optically thick to these photons in the far uv). If one assumes a blackbody spectrum for the surface emission from the central star, this calculation plus a measurement of the direct visual magnitude of the central star gives the bolometric correction and thus the bolometric luminosity L and the effective surface temperature T_e. To obtain L in absolute units one of course needs to know (or assume) the distance, whereas T_e is independent of distance (1, 4).

For a planetary nebula of known distance, besides obtaining L and T_e the measurements of radial expansion velocity and angular size also give the age t of the nebula since its ejection from the central star. These ages t range from a fraction of 10^4 years for small, dense nebulae to about 3×10^4 years for highly expanded nebulae (older nebulae exist but cannot be analyzed reliably). Besides plotting points for individual central stars of planetary nebulae on an L-T_e diagram, we can now ask if these points form some kind of track or sequence as a function of the age t. Some time ago, Shklovsky (14) suggested that T_e increases appreciably (i.e. the radius decreases) with increasing age t and this was confirmed later by O'Dell and others (15, 17). Later work by Harman & Seaton (16, 7) also indicated that L first increased slightly and later decreased (more appreciably) as t increased.

A number of uncertainties remain in the details of the L-T_e plot. If the nebula does not absorb all of the star's radiation in the far uv, one cannot obtain the total luminosity L, but criteria (18) exist to suggest when one can determine L and when one only obtains a lower limit for L. One is handi-

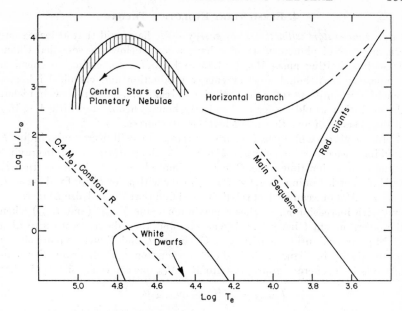

FIGURE 1. Observational luminosity-T_e diagram (with T_e in °K). The main sequence, the red giant branch, and the horizontal branch (with the *asymptotic branch* indicated by the dashed extension) for globular clusters are also shown, as is the cooling curve at a constant radius (the zero-temperature radius of a 0.4 $M_\odot$ star).

capped because in most cases we have only statistical information on distances; Webster's (19) investigation of planetary nebulae in the Magellanic Clouds is of help and has also confirmed that the luminosity L does indeed increase with age t at early times before reaching a maximum. The mean L-T_e sequence (for galactic nebulae), based on Webster's work and on Seaton's (7), adjusted to O'Dell's distance scale, is shown in Figure 1. There is considerable scatter (2 or 3 mag) in the data for the L-T_e diagram—more than can be attributed to observational inaccuracies. For instance, the sequence in the Small Magellanic Cloud (19) seems to be about a magnitude brighter than elsewhere. It is not yet known whether this variation in the L-T_e diagram is due to variation in stellar mass, chemical composition, or other causes.

For the analyses leading to Figure 1 a blackbody spectrum was assumed for the stellar emission. In the meantime, progress has been made (20–24) towards detailed model atmospheres for the central stars. When fully reliable atmosphere models become available, the quantitative aspects of Figure 1 will have to be altered. However, these changes are likely to affect L very little and, although the numerical values of T_e might be altered appreciably, it is still safe to state that T_e increases strongly and monotinocally with increasing age t.

4. ELEMENTARY EVOLUTION THEORY

Low-mass stars without nuclear energy.—We have said that at least some central stars of planetary nebulae have masses below the so-called Chandrasekhar limiting mass M_{Ch} ($\sim$1.25 to 1.43 $M_\odot$, depending on chemical composition). Although nuclear energy production and chemical inhomogeneities are important for the evolution of real stars, let us review first what the evolution would be for a chemically homogeneous star with $M < M_{\mathrm{Ch}}$ in the absence of any thermonuclear energy sources.

In the absence of nuclear energy sources a star will decrease its radius R (and increase central density $\rho_c \propto MR^{-3}$) appreciably during a Kelvin gravitational contraction time scale. One can estimate how central temperature T_c and bolometric luminosity L ($\propto R^2 T_{\mathrm{max}}^4$) vary with ρ_c (or R) for fixed mass M. For $M \ll M_{\mathrm{Ch}}$ one can neglect relativity and find that T_c (and also L) increases first with increasing ρ_c, reaches a maximum value $T_{c\ \mathrm{max}}$ (and L_{max}) when the electrons start becoming degenerate, and then decreases to zero at a slightly larger limiting density ρ_{max}. Curves of L and T_c versus ρ are shown schematically in Figure 2 as is the corresponding L-T_e diagram. For nonrelativistic degeneracy appropriate to $M \ll M_{\mathrm{Ch}}$ one can easily show (25) that

$$T_{c,\mathrm{max}} \propto M^{4/3} \qquad \rho_{c,\mathrm{max}} \propto M^2 \qquad\qquad 3.$$

As M approaches M_{Ch}, Equation 3 breaks down because of effects of special relativity (in particular, $T_{c,\mathrm{max}}$ and $\rho_{c,\mathrm{max}}$ approach infinity as $M \to M_{\mathrm{Ch}}$, as shown by Chandrasekhar). Nevertheless, for any $M < M_{\mathrm{Ch}}$

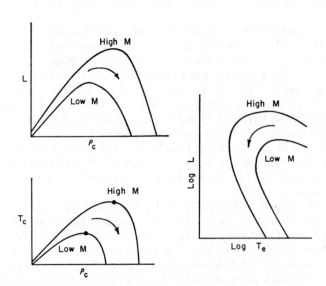

FIGURE 2. Schematic evolution of luminosity L and central temperature T_c as a function of increasing central density ρ_c (the schematic L-T_e diagram is also shown).

there is a finite $\rho_{c,\max}$ and limiting radius $R_{\min}$; once T_c has decreased well below $T_{c,\,\max}$ we are dealing with cooling white dwarf stars with radii only slightly larger than $R_{\min}$. The numerical value of $R_{\min}$ depends on chemical composition (atomic charge Z, weight A) through the combination (MA^2/Z^2) and only weakly on Z itself because of some small Coulomb corrections (26). For ^{12}C or ^{16}O, for instance, $R_{\min} \approx 0.015\ R_\odot$ for $M = 0.4\ M_\odot$ (the cooling curve at this constant radius is also shown as a dashed line in Figure 1) and $R_{\min} \approx 0.006\ R_\odot$ for $M = 1.05\ M_\odot$.

L-T_e diagrams have also been constructed from actual evolutionary calculations but, even for homogeneous models without nuclear energy sources, the results depend on whether neutrino energy losses are included. Although there are still no directly relevant laboratory experiments, present theories of weak interactions suggest a current-current interaction which leads to the production of neutrino-antineutrino pairs. There is also no conclusive single verification of this interaction from astrophysics, but a number of independent astrophysical considerations (27) (including a weak indication from planetary nebulae themselves) make it very plausible. Henceforth we shall assume the presence of this current-current interaction with the coupling constant given by the Universal Fermi Coupling theory. Accepting these assumptions, the neutrino energy-loss rates as a function of density and temperature are now known quite accurately (28). A different kind of neutrino interaction which would involve a direct photo-neutrino coupling has also been tentatively suggested recently and its rate calculated (29). However, this interaction can already be ruled out conclusively by astrophysical evidence alone (30), e.g. by the fact that it would predict numbers of hot, low-mass white dwarfs which are thousands of times smaller than the observed numbers.

The inclusion of the current-current neutrino interaction has three main effects on stellar models (for central temperatures $T_c \gtrsim 10^8 °$K): 1. Evolutionary lifetimes are decreased because of the extra energy loss, 2. the optical luminosity is increased slightly, and 3. if the stellar core is degenerate, the maximum temperature occurs not at the center of the star but slightly further out. Figure 3 gives the L-T_e curves and the time scales for evolution along these curves for homogeneous stars consisting of a carbon-oxygen mixture for masses of 0.75 and 1.02 $M_\odot$, based on some evolutionary calculations (31, 32) with neutrino energy losses included.

Simplified models with nuclear energy sources.—We have to ask next under what circumstances our neglect of thermonuclear reactions would be justified for a star of originally homogeneous composition. The thermonuclear reaction rate for a given composition depends very strongly on temperature and only fairly weakly on density. There is then a reasonably well-defined temperature below which the thermonuclear energy production is much smaller than the optical luminosity of the star. The maximum temperature $T_{\max}$ reached by a star of mass M increases with M, according to Equation 3.

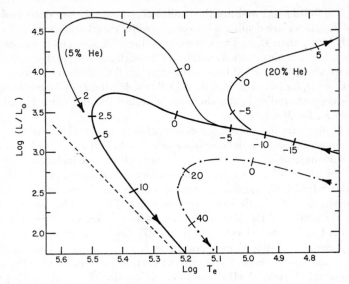

FIGURE 3. The theoretical L-T_e diagram for stars consisting mainly of carbon and oxygen. The solid curves are for total mass 1.02 $M_\odot$: the thick curve with no helium envelope, the two thin curves for stars with a helium envelope containing 5% and 20% of the mass, respectively. The dash-dot curve is for a 0.75 $M_\odot$ star without helium envelope. The evolutionary age (from an arbitrary zero point) in units of 10^4 years is shown at a few places along each track.

For given chemical composition there is then a characteristic mass M_{min}, so that for M appreciably less than M_{min} nuclear reactions are unimportant whereas a star appreciably more massive than M_{min} can settle down to a state where the nuclear energy release feeds most of the star's luminosity. For stars of "ordinary composition" the energy source is hydrogen burning and this state of course refers to main-sequence stars. The value of M_{min} obtained from actual evolutionary calculations (33), is quite low for this "ordinary main sequence" (see Table 1). For material in which all the hydrogen has been converted into helium, an appreciably higher temperature is required for helium to burn rapidly enough (through the triple-alpha reaction, mainly giving ^{12}C, ^{16}O, and a little ^{20}Ne and ^{24}Mg). It is of interest to consider hypothetical homogeneous stars consisting of such hydrogen-depleted material, even though realistic evolved stars are expected to have a more complex composition. Models for such *homogeneous helium stars* have been calculated (34–36) and the value of M_{min} is given in Table 1 as well as the approximate maximum temperature which stars near this mass can reach. Similarly, models have been calculated for material in which helium is also depleted and ^{12}C is the main nuclear fuel (37–43) and for material in which ^{12}C is also depleted and ^{16}O is the main nuclear fuel (32, 44). Values of M_{min} for such materials are also given in Table 1.

TABLE 1. Minimum masses and temperatures for thermonuclear reactions

	H→He	He→C, O	C→Ne, Mg	O→Si, S
$M_{min}/M_\odot$	0.08	0.35	1.04	1.4
T_{max} (°K)	4×10^6	1.2×10^8	6×10^8	1.3×10^9

Thermonuclear reaction rates are continuous (although steep) functions of temperature and one might expect the importance of nuclear reactions to change only gradually as a function of stellar mass. However, the onset of nuclear burning is slightly more dramatic, due to the following stellar structure detail: In a gravitationally contracting star the energy source is rather extended and the temperature distribution rather flat in the star's interior. If the star switches to a concentrated nuclear energy source in the star's inner core (if the maximum temperature occurred in the center), the temperature gradient in the deep interior is increased, i.e. the temperature in the center increases but the temperature further out (and the optical luminosity L) decreases slightly. For stellar masses slightly above M_{min} the electrons are somewhat degenerate in the center before nuclear ignition; after ignition the degeneracy is relieved partially and the structure of the star changes appreciably. We shall encounter such rearrangements in more dramatic form for inhomogeneous stars (e.g. the *helium flash*, see Section 5), but it also accentuates the onset of nuclear burning for homogeneous stars: For 0.9 M_{min}, say, nuclear reactions are unimportant whereas for 1.1 M_{min}, say, nuclear burning becomes very pronounced.

Central stars of planetary nebulae must previously have converted helium into C, O, and Ne in their interior and have lost much of their hydrogen-rich envelope. They most likely possess a helium envelope surrounding the carbon-rich core and this envelope can affect the star's properties appreciably, but we first compare the observations (Figure 1) with homogeneous models which do not burn carbon. As Figure 3 shows, the homogeneous stars of lower mass (e.g. the 0.75 $M_\odot$ sequence) have too low a maximum value of the optical luminosity L and too long evolutionary time scales compared with the observed evolution of central stars of planetary nebulae. For masses approaching M_{min}, however (e.g. the 1.02 $M_\odot$ sequence), the evolutionary timescales are close to the observed ones and the discrepancy in maximum luminosity is small [this small discrepancy could be removed altogether (45) by considering stars consisting mainly of material in the iron region of elements, but it does not seem feasible to process matter into such elements in stars of low mass, $M \lesssim 1.4 \ M_\odot$, and we shall disregard this possibility]. Although the luminosities are of the right order, the homogeneous models give much too large values for surface temperature T_e (too small radii R). This is not necessarily very embarrassing, since even small amounts of hydrogen in an outer envelope can increase R greatly [e.g. an evenlope (31)

of fractional mass 10^{-4} increases R by about 30%]. Such a hydrogen envelope is indeed likely to be present, but so is an intermediate layer where
helium burns in a shell (into C and O), which can have additional effects
on the models.

Evolutionary sequences have been calculated (46, 48–51) for stars containing a carbon-oxygen core (52) with a helium envelope of low mass (but
no hydrogen). If the envelope mass exceeds a critical value ∼0.01 $M_\odot$ (value
depending on stellar mass), helium burns in a thin shell and the luminosity
is increased greatly. Figure 3 contains curves for one 1.02 $M_\odot$ star with a
helium envelope comprising 20% of the mass, in which case the radius also
increases greatly (51), and one with an envelope of 5% of the mass, in which
case only the luminosity is increased for a while (some variability due to
thermal relaxation oscillations is smoothed over in these curves but will be
discussed later). Fortunately the helium layer is likely to be less than 5% of
the mass in the remnant star after the planetary nebula has been ejected,
but an additional hydrogen envelope can cause further complications.

5. Evolution of Globular Cluster Stars

Although globular clusters belong to the "extreme Population II"
whereas most planetary nebulae belong to the disk population, we briefly
review globular cluster stars before returning to planetary nebulae. This will
be useful because the evolution away from the main sequence of stars in
globular clusters has been studied extensively, both theoretically and observationally, and the evolving stars have masses comparable with one solar
mass. Figure 1 gives a schematic view of the red giant branch and horizontal
branch in globular clusters and a detailed discussion of the observational
data is given by Sandage (57).

The basic features of globular cluster evolution have been clear since the
early theoretical work by Schwarzschild and others: When a main-sequence
star (in the mass range 0.5 to 2 $M_\odot$, say) has converted hydrogen into helium
in a core of 10 to 20% of the stellar mass, the core contracts, the envelope
expands, and the luminosity rises (53–55). This evolution upwards along
the red giant branch is characterized by an outer convective zone, a hydrogen-burning shell source of energy (at the boundary between the helium
core and the hydrogen-rich envelope), and a core where density increases
rapidly and temperature increases slowly, resulting in pronounced electron
degeneracy. At the top of the red giant branch the central temperature
finally becomes high enough for the helium to ignite in the core. The initia-
tion of the conversion of ^{4}He into ^{12}C, ^{16}O, ^{20}Ne, and ^{24}Mg alters the structure
of the star greatly (expanding the core sufficiently to remove electron de-
generacy) and the rearrangement, the so-called *helium flash* (56), takes
place in a small fraction of a Kelvin contraction time scale. The helium flash,
although violent, is believed *not* to result in complete mixing of the star.
Stars on the horizontal branch presumably have two nuclear energy sources,
an outer hydrogen-burning shell and an inner helium-burning zone (in a
core at first and then in a shell surrounding the carbon-rich core).

For accurate theoretical work on a cluster diagram one must assume an age for the cluster and initial main-sequence values for the helium abundance Y (by mass) and the abundance Z of the heavier elements. Once these assumptions are made, the calculations predict the mass of stars at the *turn off* near the main sequence, the position of this turnoff point on an L-T_e diagram, the shape of the giant and horizontal branches, and the luminosity function along these branches. This wealth of theoretical predictions based on a few parameters gives some check on implicit assumptions in the theory. One particularly interesting question is how much evidence one can find for mass loss along the giant or horizontal branches. The light curves of the RR Lyrae variable stars, which are situated on the horizontal branch, are of great help since their theoretical analysis (58–60) leads to values of mass and helium abundance for these stars, $M \approx 0.55 \ M_\odot$ and $Y \approx 0.3$. The latest L-T_e calculations (discussed below) together with the observations (57) indicate cluster ages of about 10×10^9 years and $Y \approx 0.3$ to 0.35 for the initial main sequence. Considerations of the luminosity function (61) independently lead to similar values of Y. These values are close to those of RR Lyrae stars and of Population I stars (Z varies from cluster to cluster but is known). The calculations also give the mass of stars near the turnoff from the main sequence, $M \approx 0.85 \ M_\odot$. We thus have some indirect evidence for mass loss between the main sequence and the horizontal branch, but not a very severe loss.

The evolution from the main sequence along the red giant branch towards the helium flash is now well understood (62, 63). There is some observational evidence (64) for continuous, mild mass loss during the red giant phase but this does not have a drastic effect on the evolution. The details of the first helium flash at the tip of the red giant branch are affected by the inclusion of neutrino energy losses, even though the loss rates are not yet very high at temperatures of $\sim 10^{8\circ}$K. Comparison of these calculations (65–68) with globular cluster observations argues in favor of including the neutrino processes, corroborating other (individually weak) arguments for the inclusion. The mass M_c of the hydrogen-free helium core at the onset of the flash (65, 68) is 0.45 to 0.50 $M_\odot$. The flash not only leads to a rearrangement of the structure of the star, but temporarily gives an extended convective core (driven by the helium burning) whose outer edge comes close to the hydrogen-burning shell. Detailed calculations (68, 69) indicate that the flash does *not* give substantial mixing between the helium core and hydrogen envelope, and do *not* suggest the violent ejection of an appreciable fraction of the star's mass. Some hydrodynamic effects may have been overlooked (70) in the calculations to date, but indirect evidence from the observations (57, 71) allows only a slight helium enrichment of the envelope.

For calculations of horizontal-branch stars it is then reasonable to assume the presence of the neutrino energy-loss rate, the absence of any mixing of helium into the hydrogen-rich envelope, and an initial mass M_c of the helium core of 0.45 to 0.5 $M_\odot$. However, the detailed results depend rather sensitively on the total mass M_{HB} of a horizontal-branch star and M_{HB} has to be considered as a parameter in the calculations (72–74). To fit the observations (57, 71) the latest calculations (74) require masses in the range $M_{HB} \approx 0.5$ to 0.65 $M_\odot$. Furthermore, the observations probably cannot be fitted by a single evolutionary sequence of unique mass and chemical composition and a spread of values of M_{HB} in the above range is probably indicated. It is gratifying that the analysis of RR Lyrae variable light curves also indicates masses within this range. We can be fairly confident (but not certain) that some modest (and variable) mass loss has occurred before the star enters the horizontal branch, but that it still has an appreciable hydrogen-rich envelope.

At the start of evolution on the horizontal branch we have helium burning in the interior of a helium core and hydrogen burning in a shell. The ratio of the two energy production rates, and the initial evolution, depend rather sensitively on M_{HB}. In a typical case the initial evolution is probably towards the blue, the importance of helium burning increases slowly, and the evolutionary track turns towards the red again. As ^{12}C and ^{16}O build up and helium becomes exhausted in the innermost core, the luminosity and radius both start to increase and T_e decreases. In the observational L-T_e diagram the horizontal branch is continued "upward and to the right" almost to the tip of the ordinary red giant branch (this region is sometimes called the *asymptotic branch*). Evolution along this asymptotic branch presumably corresponds to the exhaustion of helium in the innermost core and the subsequent evolution when an inert carbon-oxygen core is surrounded by a helium-burning shell source and a hydrogen-burning shell further out. However, as soon as helium burning in a shell commences we have to deal with a new and complex phenomenon of thermal instability (75).

This thermal instability in intermediate regions of a star requires an energy source which is temperature sensitive and occurs in a region which is thin compared with layers above it (so that pressure changes little in an expansion) and yet thick enough so that radiative flux divergence does not eliminate excess heat too rapidly. One is then dealing with thermal relaxation oscillations whose time scale is of the order of the time that would be required to radiate away the heat content of the intermediate region. This *Kelvin time* for the intermediate region is much shorter than the Kelvin time for the whole star, but much longer than dynamic time scales. Similar phenomena can be encountered in even more evolved, more massive stars (e.g. carbon-shell burning) but in globular clusters, models with helium-shell burning (75–77) are required to drive these thermal relaxation oscillations. The calculations (76, 77) indicate that the amplitude of the oscillations increases slowly with time until it takes on the character of a succession of helium-shell flashes which can eventually lead to some hydrogen mixing. The complexity of evolutionary calculations increases enormously when thermal instability is encountered and we do not yet have a quantitative understanding of the asymptotic branch of globular clusters, when the star ascends into the red giant region for the second time. Thus our present insight into globular cluster evolution starts to peter out just as the story becomes most relevant for planetary nebulae!

6. THE STAR BEFORE THE EJECTION OF THE NEBULA

We have discussed in detail the red giant and horizontal branches of globular clusters with low metal abundances ($Z \approx 0.001$), typical of Population II stars. Most planetary nebulae occur in Population I, or in the intermediate disk population. Both these populations have $Y \sim 0.3$ as does Population II but their metal abundances are higher, $Z \sim 0.01$ to 0.03. Fortunately, the cluster diagrams for old galactic clusters are qualitatively very similar to those for globular clusters, and the horizontal-branch stars in old galactic clusters should be typical of progenitors of planetary nebulae. However, the stellar masses are somewhat different. Under the assumption of Population I abundances for the oldest galactic clusters ($Y \sim 0.3$, $Z \sim 0.01$ to 0.03), the mass of stars near the turn off from the main sequence is $M \approx 1.1$ $M_\odot$, slightly larger than for globular clusters of the same age. The mass at the turnoff is greater for a younger cluster and the most common initial masses of stars in the galactic disk which have evolved far away from the main sequence are in the range of about 1.1 to 1.5 $M_\odot$. We should also bear in mind the one planetary nebula in a globular cluster, presumably with initial mass ≈ 0.85 $M_\odot$, and the open question whether the rarer stars of initial mass 1.5 to 4 $M_\odot$, say, can also evolve into planetary nebulae.

The early red giant evolution leading up to the first helium flash at the tip of the red giant branch has been studied in detail for various values of the total mass M (68, 78–80). One remarkable result of these calculations is the fact that the mass M_c of the hydrogen-free helium core at the onset of its flash is almost independent of total mass, $M_c \sim 0.4$ to 0.5 $M_\odot$ over a wide range of masses. The rearrangement of the core at the top of the red giant branch and the helium-core burning plus hydrogen-shell burning on the horizontal branch also does not depend very strongly on initial mass. In particular, the star is likely to have survived these evolutionary stages without thorough mixing and with only mild mass loss from the hydrogen-rich envelope.

We are most interested in the next stage of evolution where most of the helium in the core has already been converted into carbon, oxygen, neon, and magnesium. The helium burning in this stage takes place in a shell immediately outside this inert core (mass M_{He} interior to this shell) and hydrogen continues to burn in a shell (with mass M_H interior to this shell) outside of the intermediate hydrogen-free helium zone. The onset of this stage occurs when $M_H \sim 0.55$ $M_\odot$ and $M_{He} \sim 0.45$ $M_\odot$ and is already marked by 1. the presence of thermal relaxation oscillations, 2. a luminosity L considerably greater than ~ 100 $L_\odot$, the luminosity during the previous horizontal-branch stage, and 3. a pronounced "red giant structure" with the photospheric radius R orders of magnitude greater than the radius of the hydrogen-burning shell.

The subsequent evolution during this double-shell-source stage again

depends remarkably little on the mass $(M - M_H)$ of the hydrogen envelope:
Considering M_H, instead of time, as the independent variable, values of
M_{He}, luminosity L, central temperature, and density are all of the same
order of magnitude in Paczynski's calculation (81) for $M = 3$ $M_\odot$ and Rose &
Smith's calculations (77) for $M = 0.85$ $M_\odot$, for instance. The thermal relaxation oscillations complicate the computations greatly and at present
uncertainties in the results due to computational difficulties mask the dependence on mass M. The general trend of the evolution is clear, however,
including the increasing importance with time of the effects mentioned above.

When the hydrogen-burning shell has progressed to $M_H \sim 0.6$ $M_\odot$, the
mass $M_H - M_{He}$ of the intermediate helium zone is almost 0.1 $M_\odot$, the radius
at M_H is about 3×10^9 cm, and the photospheric radius R is almost $\sim 10^{13}$
cm. At this epoch the two shell sources produce comparable amounts of
energy; the total luminosity L does not vary much during a relaxation
oscillation (period $\sim 2 \times 10^4$ years) but its value is still uncertain with estimates (77, 81) varying from 1500 to 4000 $L_\odot$. Since the energy content per
gram is much greater for hydrogen burning than for helium burning, the
mass of the intermediate helium zone soon shrinks to $M_H - M_{He} \sim 0.01$ to
0.04 $M_\odot$ and hydrogen burning accounts for $\sim 85\%$ of the *average* luminosity
from then on (with the helium energy production rate high only during the
short shell-flash phase of each relaxation oscillation). M_H advances from
0.6 to 0.75 $M_\odot$ in about 1 to 2×10^6 years; the radius has risen above 10^{13}
cm and the luminosity to $L \sim (1$ to $1.5) \times 10^4$ $L_\odot$.

It will require future calculations to tell us accurately how the three
quantities $(M_H - M_{He})$, R, and L depend on evolutionary time (or M_H).
We don't even know the *sign* of the dependence of these quantities on the
envelope mass $(M - M_H)$ but this dependence is weak as long as $M - M_H$ is
appreciably larger than 0.05 $M_\odot$. Qualitatively, we know that R and L increase with increasing M_H until either $M - M_H$ drops below ~ 0.05 $M_\odot$ (which
does not yet extinguish the shell sources but at least alters their structure)
or some catastrophe happens. We discuss next the possible causes for such a
catastrophe, the ejection of the nebula.

7. The Ejection of the Planetary Nebula

Before invoking the detailed stellar models we have just discussed, we
note some conjectures (82–83) which can be drawn directly from observations on planetary nebulae. The remnant star has a luminosity $\sim 10^4$ $L_\odot$
and a radius not much larger than that of a typical white dwarf. It is then
likely that the star before the ejection of the nebula possessed a core of mass
and radius typical of white dwarfs (e.g. for $M_c \sim 0.8$ $M_\odot$ the radius is $r_c \sim 0.02$
$R_\odot \approx 1.4 \times 10^9$ cm, about twice the radius of a white dwarf of the same mass)
and a luminosity somewhat larger than that of the remnant, $\sim 10^4$ $L_\odot$. The
present expansion velocities of the nebulae are typically 30 km/sec and it
seems reasonable to conjecture that the material was ejected from radial
distances r where the escape velocity is comparable. This escape velocity

u_r (against gravity of mass 0.8 $M_\odot$) is

$$u_r \approx (200 \, R_\odot/r)^{1/2} \, 39 \text{ km/sec} \qquad\qquad 4.$$

indicating escape from radii $\sim$200 $R_\odot$ $\approx 1.4 \times 10^{13}$ cm, typical of red giants. We have a further clue from chemical composition considerations: The remnant star can have little hydrogen in it, but the ejected nebula's composition is close to that of Population I ($X \sim 0.7$). We must then look for a mechanism where much of a hydrogen-rich envelope (surrounding a compact core of He, C, etc) continues to be released from radii very much larger than the core radius r_c until the mass $M - M_c$ in this extended envelope has become quite small.

At least three different mechanisms have been suggested for the ejection of the nebula, each of which is at least qualitatively plausible in light of the evolutionary features discussed in Section 6: (a) violent thermal relaxations could lead to dynamic effects (47); (b) in very extended envelopes the ionization-recombination equilibrium can lead to instabilities (84–85); and (c) because of the large luminosities generated in the stellar core and the large radii of the envelope, radiation pressure can lead to mass ejection (86–88). All three effects may contribute concurrently in varying degrees and no comprehensive computations are available as yet. We shall first explain the basic ideas of the radiation pressure ejection.

We define a quantity L_{crit} by the relation

$$L_{\text{crit}} = 4\pi cGM_c/\kappa_r \qquad\qquad 5.$$

where κ_r is the mean opacity coefficient at distance r from the star's center. If the total rate of outward heat flow at r is L_r, photon scattering transfers momentum to the particles in the gas so that the effective force of gravity is reduced by a multiplying factor of $(1 - L_r L_{\text{crit}}^{-1})$. In reality the ionization-recombination equilibrium affects the thermodynamics and the opacity coefficient κ_r at least in the outermost layers of the star. For the sake of simplicity, we pretend at the moment that the material is fully ionized and that κ_r is simply the constant κ_0 for Thomson scattering by free electrons. We consider the radius r_c of the core (of mass $M_c \sim 0.8 \, M_\odot$) to be that of the hydrogen-burning shell source and assume that r_c and the luminosity L_c immediately outside the shell are known. In the hydrogen-rich ($X = 0.7$, $\mu \approx 0.6$) envelope between r_c and the photospheric radius R we have $\kappa_0 \approx 0.38$ and $L_{\text{crit}} = 3.8 \times 10^4 \, L_\odot$. The luminosity L at the photosphere must be less than L_{crit}, but L_r need not be constant in the envelope and the luminosity L_c generated by the core could exceed L_{crit}.

Consider a case where $L_c - L$, the rate at which heat is deposited in the envelope, is $\sim L_{\text{crit}}$. Without detailed evolutionary plus hydrodynamic calculations one cannot be sure whether the envelope merely heats up and expands without mass loss or whether mass loss is accompanied by a cooling and shrinking envelope or some other combination. For simplicity let us assume the case of stationary flow where the outer radius R remains constant and the slow decrease of density in the envelope is replaced by a constant mass flow ϕ,

$$\phi = 4\pi\rho_r r^2 v_r \qquad\qquad 6.$$

with varying radial velocity v_r. The velocity v_R at the photosphere must exceed the escape velocity u_R there but is very small compared with the escape velocity u_{rb} at the inner boundary of the envelope. If we (unrealistically) assume the temperature T_r (and hence the radiation energy density) not to change at all with time, the energy $L_c - L$ goes into $\frac{1}{2}u_{rb}^2\phi$, the energy required to lift the material. This gives an order of magnitude for the expected flow rate,

$$\phi \sim \frac{L_c - L}{3.8 \times 10^4\, L_\odot}\, \frac{r_b}{0.02\, R_\odot} \times 1.3 \times 10^{21}\ \text{g/sec} \qquad\qquad 7.$$

At this flow rate of $\phi \sim 10^{21}$ g/sec the ejection of 0.5 $M_\odot$, say, from the envelope would require about 2×10^4 years (comparable with the period of a relaxation oscillation). The remarkable result quoted in Section 6, that r_c and L_c (and density and temperature there) depend only weakly on the envelope mass $M - M_c$, provides one precondition for the assumption of *stationary flow*. However, in reality the matter density in the envelope must decrease with time and the uncertainty in the corresponding temperature decrease results in considerable uncertainty in the flow rate ϕ: In the extended envelopes with a large heat flux, radiation pressure dominates gas pressure and the total binding energy of the envelope (gravitational minus thermal energy) is much less than gravitational energy alone. Much of the required *lifting energy* $\frac{1}{2}u_{rb}^2\phi$ could then come from the release of stored radiation energy and ϕ could be increased (87) by a factor f of order 10, say, over the expression in Equation (7).

In Equation 7 we only estimated the rate at which the heat production from the stellar core can lift material through the envelope. We still need criteria as to whether the material will merely increase the photospheric radius R or will flow out of the (more or less stationary) photosphere with some velocity v_R which exceeds the escape velocity u_R there, given by Equation 4. The typical thermal velocity U_{th} in the photosphere is given by

$$\left.\begin{aligned}
U_{th} &\approx \left(\frac{T_e}{5700^\circ \text{K}}\right)^{1/2} 12\ \text{km/sec} \\
&\approx \left(\frac{L}{4 \times 10^4\, L_\odot}\right)^{1/8}\left(\frac{200\, R_\odot}{R}\right)^{1/4} 12\ \text{km/sec}
\end{aligned}\right\} \qquad 8.$$

and we can construct a dimensionless ratio

$$\eta \equiv \left(\frac{U_{th}}{u_R}\right)^2 \approx 0.1\left(\frac{L}{4 \times 10^4\, L_\odot}\right)^{1/4}\left(\frac{R}{200\, R_\odot}\right)^{1/2} \qquad 9.$$

For main-sequence stars η is extremely small; for our highly evolved red giants η is still small compared with unity but not by very much. We shall see that the fact of η approaching unity helps lift off the material. Nevertheless, since $\eta < 1$ and if indeed $v_R > u_R$, there must be a region between the photosphere R and the *sonic point* r_s some distance below where the flow is supersonic and in which most of the momentum is imparted to the material (acceleration from $\frac{1}{2}v_R$, say, to v_R). If this acceleration is to be done purely by radiation pressure (neglecting recombination), the criteria are qualitatively as follows.

One can show (88–89) that the energy transferred from the radiation flux into kinetic energy in the supersonic region is of order

$$\Delta L \sim L(v_R/2c)\tau \qquad\qquad 10.$$

where τ is the optical depth between the sonic point and the photosphere. The flow velocity near the sonic point is usually of order $\frac{1}{2}v_R$ and τ of order $(v_R/2U_{\text{th}})^8$. For a self-consistent stationary flow the kinetic energy transferred must also be of order $v_R^2\phi$ where ϕ is some factor f (between 1 and 100, say) times the expression in Equation 7. These two expressions for the kinetic energy transfer then give a consistency requirement involving v_R and other quantities. For instance, for v_R to be about twice the escape velocity u_R one requires

$$f\frac{L_c - L}{L}\eta^4 \sim \frac{2GM/r_b}{cu_R} = \frac{(R \times 3.4 \times 10^{-6}\,R_\odot)^{1/2}}{r_b} \qquad 11.$$

The quantity on the right-hand side is of order unity for our stars. If the left-hand side exceeds the right side, then $\frac{1}{2}v_R$ will be larger than the escape velocity; if the left-hand side is much smaller, then stationary-flow mass loss is not possible. Some solutions of the actual hydrodynamic equations (using correct opacity expressions but omitting recombination) for stationary flow have already been reported (87). The relationship of these calculations to the evolving models discussed in Section 6 is not completely clear, but it seems likely that mass loss can already proceed at rates exceeding 10^{21} g/sec when the luminosity L_c produced by shell burning in the core reaches about $\frac{1}{2}L_{\text{crit}}\sim 2\times 10^4\,L_\odot$. At this stage the mass in the core is about 0.8 to 1 $M_\odot$, the outer radius $\gtrsim 100\,R_\odot$, and the terminal flow velocity $\lesssim 100$ km/sec.

The ionization potential of hydrogen in temperature units is $I/k\sim 1.5 \times 10^{5\circ}$K. In the outer layers of the extended envelopes of our red giant stars the actual temperature is very much smaller and the gravitational potential energy per hydrogen atom at $R\sim 300\,R_\odot$, say, is only about a third of the ionization potential. The total energy (relative to neutral atoms at infinity) of the outermost layers can thus be positive and instabilities can result. The main energy required to lift material from the lower part of the envelope towards the sonic point still comes from the absorption of radiation, but the kinetic energy $\sim\phi v_R^2$ could now come from the energy released by chemical recombination. Actual stability calculations (84) indeed indicate that mass loss could occur when R becomes sufficiently large even if L is appreciably

less than L_{crit}. The only calculations to date (89) including both recombination and radiation pressure mainly correspond to very large values of L_c,
but recombination effects certainly help initiate mass loss at a slightly
earlier stage than does radiation pressure alone.

As mentioned earlier, the thermal relaxation oscillations become more
pronounced (76–77) as the two shell sources slowly burn their way outward.
These oscillations eventually involve a series of helium-shell flashes and it is
possible that an explosive event occurs which ejects part of the envelope. It
is more likely that the oscillations have an indirect effect by introducing
changing conditions over time periods $\sim 10^4$ years, compared with the overall evolution times $\sim 10^6$ years. Fortunately, all three mechanisms can act
cooperatively and it is fairly clear that much of the hydrogen-rich envelope
mass $(M - M_c)$ can be ejected in a time period less than 10^4 years when the
core mass M_c has reached some critical value between 0.6 and ~ 1 $M_\odot$.

8. THE REMNANT STAR AFTER THE EJECTION

We finally return to the question of the central star after the planetary
nebula has been ejected and its evolution in about 3×10^4 years along the
Harman-Seaton sequence. We are fairly confident about its most basic
features: Most of its mass probably consists of a core of ^{12}C and ^{16}O of mass
between 0.5 and 1 $M_\odot$, surrounded by a helium shell of small mass and
a hydrogen envelope of very much smaller mass still. We are inclined to
accept the validity of the neutrino energy-loss rates obtained from the current-current interaction. With neutrinos included, the core is not massive
enough to burn carbon and is undergoing gravitational contraction and
cooling from the stage just before its maximum temperature towards the
white dwarf stage. The luminosity-time history of such a bare core contracting is of roughly the right order of magnitude, but the radius of the
bare core is at each stage appreciably smaller than the observed radius.
Small amounts of helium and hydrogen can greatly increase the star's
radius and presumably light envelopes have to be invoked to remove the
radius discrepancy. However, we are still quite in the dark about the details
of these hydrogen and helium layers.

The present uncertainty in M_c, the mass of the carbon-oxygen core plus
helium shell of the star (denoted by M_H in Section 6), stems from three
sources: 1. the dependence on the total mass before ejection of the nebula,
2. variability from case to case, and 3. inaccuracies in present-day calculations for luminosity as a function of core mass M_c and for the minimum
luminosity required for ejection of the envelope. At the moment 3 is the
dominant cause, but the most likely values of M_c are probably between
0.6 and 0.9 $M_\odot$. The existence of a planetary nebula in a globular cluster is
on the verge of being an embarrassment, since the total mass before ejection
must be at least $M_c + 0.05$ $M_\odot$ for the ejection to work, whereas the largest
estimates for masses M_{HB} of horizontal-branch stars in globular clusters is
about 0.65 $M_\odot$. Presumably this suggests values of M_c closer to 0.6 $M_\odot$ than
to 0.9 $M_\odot$.

The mass of the intermediate helium layer was probably between 0.01 and 0.05 $M_\odot$ before the ejection. The ejection of the hydrogen-rich envelope stops when the remaining mass of this envelope has been reduced to some value ΔM_H. Our present theoretical knowledge suggests that the value of ΔM_H should be less than 0.05 $M_\odot$, but we don't know how much less. Presumably the mass of the hydrogen envelope of the remnant star of a *young* planetary nebula must be close to ΔM_H and we have another embarrassment unless (a) ΔM_H turns out to be $\lesssim 0.005$ $M_\odot$ or (b) mild mass loss continues after the ejection of the main nebula: Hydrogen-shell burning can continue as long as the envelope mass exceeds ~ 0.001 $M_\odot$ and the peak luminosity could be maintained (77, 80) for about an extra 10^4 years for each extra 0.002 $M_\odot$ burned.

For the reasons just stated we cannot estimate the masses of the helium and hydrogen layers accurately from considerations before and during the ejection of the nebula. One might hope to obtain these masses empirically by fitting theoretical models to the observed L-T_e track. Besides requiring the fitting of two parameters, one has to contend with an extra difficulty: Because of the violent ejection mechanism, the remaining envelope is left with a thermal structure which we do not know at present. Although hydrostatic equilibrium is restored quickly, the thermal structure relaxes (90–91) only with a time scale $\sim 10^4$ years, comparable with the observed stretch on the L-T_e track, and the L-T_e track thus depends on the previous history of the envelope as well as on its mass. Although we seem to end on a pessimistic note, it is good to remember that we have shifted our ignorance to what is merely the outer few percent of the central stars of planetary nebulae.

I am grateful to Drs. P. Demarque, D. G. Hummer, B. Paczyński, W. K. Rose, M. Schwarzschild, and M. J. Seaton for interesting discussions and communications.

LITERATURE CITED

1. Osterbrock, D. E. 1964. *Ann. Rev. Astron. Ap.* 2:95
2. Seaton, M. J. 1968. *Planetary Nebulae*, ed. C. R. O'Dell, D. E. Osterbrock, 1. Dordrecht-Holland: Reidel
3. Perek, L. 1968. *Planetary Nebulae*, ed. C. R. O'Dell, D. E. Osterbrock, 9. Dordrecht-Holland: Reidel
4. Aller, L., Liller, W. 1968. *Nebulae and Interstellar Matter*, ed. B. Middlehurst, Chap. 9. Univ. Chicago Press
5. Osterbrock, D. E. 1970. *Quart. J. Roy. Astron. Soc.* 11:199
6. O'Dell, C. R. 1962. *Ap. J.* 135:371
7. Seaton, M. J. 1966. *MNRAS* 132:113
8. Webster, B. L. 1969. *MNRAS* 143:79
9. Seaton, M. J. 1968. *Ap. Lett.* 2:55
10. Cahn, J H., Kaler, J. B. 1971. *Ap. J. Suppl. No. 189*
11. Weidemann, V. 1968. *Ann. Rev. Astron. Ap.* 6:351
12. Greenstein, J., Trimble, V. L. 1967. *Ap. J.* 149:283
13. Zanstra, H. 1931. *Publ. Dom. Ap. Obs.* 4:209
14. Shklovskii, I. S. 1956. *Astron. J. Sov. Union* 33:315
15. O'Dell, C. R. 1963. *Ap J.* 138:67
16. Harman, R. J., Seaton, M. J. 1964. *Ap. J.* 140:824
17. Abell, G. O. 1966. *Ap. J.* 144:259
18. Harman, R. J., Seaton, M. J. 1966. *MNRAS* 132:15
19. Webster, B. L. 1969. *MNRAS* 143:113
20. Gebbie, K. B., Seaton, M. J. 1963. *Nature* 199:580
21. Böhm, K. H., Deinzer, W. 1966. *Z. Ap.* 63:177
22. Böhm, K. H. 1969. *Astron. Ap.* 1:180

23. Hummer, D. G., Mihalas, D. 1970. *MNRAS* 147:339
24. Cassinelli, J. P., Hummer, D. G. 1971. Unpublished rep., JILA, Univ. Colorado
25. Salpeter, E. E. 1968. *Planetary Nebulae*, ed. C. R. O'Dell, D. E. Osterbrock, 409. Dordrecht, Holland: Reidel
26. Hamada, T., Salpeter, E. E. 1961. *Ap. J.* 134:683
27. Stothers, R. B. 1966. *Astron. J.* 71:943
28. Beaudet, G., Petrosian, V., Salpeter, E. E. 1968. *Ap. J.* 150:411
29. Chaudhury, P. R. 1970. *Can. J. Phys.* 48:935
30. Stothers, R. B. 1970. *Phys. Rev. 2:* D1417
31. Beaudet, G., Salpeter, E. E. 1969. *Ap. J.* 155:203
32. Boozer, A., Joss, P., Salpeter, E. *Ap. J.* To be published
33. Kumar, S. S. 1963. *Ap. J.* 137:1126
34. Cox, J. P., Salpeter, E. E. 1964. *Ap. J.* 140:485
35. Divine, N. 1965. *Ap. J.* 142:824
36. L'Ecuyer, J. 1966. *Ap. J.* 146:845
37. Deinzer, W., Salpeter, E. E. 1965. *Ap. J.* 142:813
38. Vila, S. 1966. *Ap. J.* 146:437
39. Vila, S. 1967. *Ap. J.* 149:613
40. Hayashi, C., Hōshi, R., Sugimoto, D. 1962. *Progr. Theor. Phys. Suppl. 22*
41. Murai, T., Sugimoto, D., Hōshi, R., Hayashi, C. 1968. *Progr. Theor. Phys.* 39:619
42. Sugimoto, D., Yamamoto, Y., Hōshi, R., Hayashi, C. 1968. *Progr. Theor. Phys.* 39:1432
43. Kutter, G. S., Savedoff, M. P. 1969. *Ap. J.* 156:1021
44. Rakavy, G., Shaviv, G., Zinamon, Z. 1967. *Ap. J.* 150:131
45. Savedoff, M. P., Van Horn, H. M., Vila, S. C., 1969. *Ap. J.* 155:221
46. Rose, W. K. 1966. *Ap. J.* 146:838
47. Rose, W. K. 1967. *Ap. J.* 150:193
48. Rose, W. K. 1969. *Ap. J.* 155:491
49. Paczyński, B. 1971. *Acta Astron.* 21: In press
50. Vila, S. C. 1970. *Ap. J.* 162:605
51. Joss, P., Katz, J., Salpeter, E. *Ap. J.* To be published
52. Arnett, W. D., Truran, J. W. 1969. *Ap. J.* 157:339
53. Sandage, A., Schwarzschild, M. 1952. *Ap. J.* 116:463
54. Hoyle, F., Schwarzschild, M. 1955. *Ap. J. Suppl.* 2:1
55. Haselgrove, C. B., Hoyle, F. 1956. *MNRAS* 116:515
56. Schwarzschild, M., Härm, R. 1962. *Ap. J.* 136:158

57. Sandage, A. 1970. *Ap. J.* 162:841
58. Christy, R. F. 1966. *Ap. J.* 144:108
59. Cox, J. P., Giuli, R. T. 1968. *Principles of Stellar Structure.* New York: Gordon & Breach
60. Christy, R. F. 1969. *J. Roy. Astron. Soc. Can.* 63:299
61. Hartwick, F. D. 1970. *Ap. J.* 161:845
62. Iben, I., Rood, R. T. 1970. *Ap. J.* 159:605
63. Demarque, P., Mengel, J., Aizenman, M. 1971. *Ap. J.* 163:37
64. Weymann, R. 1963. *Ann. Rev. Astron. Ap.* 1:97
65. Rood, R. T., Iben, I. 1968. *Ap. J.* 154:215
66. Eggleton, P. 1968. *MNRAS* 140:387
67. Thomas, H. C. 1970. *Ap. Space Sci.* 6: 400
68. Demarque, P., Mengel, J. G. 1971. *Ap. J.* 164:317
69. Härm, R., Schwarzschild, M. 1966. *Ap. J.* 145:496
70. Edwards, A. C. 1969. *MNRAS* 146: 445
71. Newell, E. G. 1970. *Ap. J.* 159:443
72. Iben, I., Faulkner, J. 1968. *Ap. J.* 153:101
73. Castellani, V., Giannone, P., Renzini, A. 1969. *Ap. Space Sci.* 3:518
74. Demarque, P., Mengel, J. G. 1971. *Ap. J.* 164:469
75. Schwarzschild, M., Härm, R. 1965. *Ap. J.* 142:855
76. Schwarzschild, M., Härm, R. 1967. *Ap. J.* 150:961
77. Rose, W. K., Smith, R. L. 1970. *Ap. J.* 159:903
78. Refsdal, S., Weigert, A. 1970. *Astron. Ap.* 6:426
79. Iben, I. 1968. *Ap. J.* 154:557
80. Paczyński, B. 1970. *Acta Astron.* 20:47
81. Paczyński, B. 1970. *Acta Astron.* 20: In press
82. Shklovskii, I. S. 1956. *Astron. Zh.* 33:315
83. Abell, G., Goldreich, P. 1966. *PASP* 78:232
84. Roxburgh, I. W. 1967. *Nature* 215:838
85. Paczyński, B., Ziolkowski, J. 1968. *Acta Astron.* 18:255
86. Kutter, G., Savedoff, M., Schuerman, D. 1969. *Ap. Space Sci.* 3:182
87. Finzi, A., Wolf, R. A. 1970. *Ap. J. Lett,* 5:63
88. Faulkner, D. J. 1970. *Ap. J.* 162:513
89. Zytkow, A. 1971. *Acta Astron.* 21: In press
90. Deinzer, W., Hansen, C. J. 1969. *Astron. Ap.* 3:214
91. Deinzer, W., Sengbush, K. v. 1970. *Ap. J.* 160:671

THE ATMOSPHERES OF MARS AND VENUS   2019

ANDREW P. INGERSOLL

*Division of Geological and Planetary Sciences, California
Institute of Technology, Pasadena, California*

CONWAY B. LEOVY

*Atmospheric Sciences and Geophysics, University of
Washington, Seattle, Washington*

INTRODUCTION

Of all the planets which may exist in the Universe, only nine have been studied by man. As a result, one cannot classify planets with the same confidence that one has in classifying stars; there is no theory of planetary evolution comparable in development to the theory of stellar evolution. Nevertheless, many of the goals of planetary science and stellar astronomy are the same: to classify objects according to their most fundamental properties in order to understand their present physical state and their evolution. From this point of view, the terrestrial planets comprise a group which can usefully be considered together. By comparing the similarities and differences between them, we may hope to gain insight into the evolution of the entire group.

Perhaps the most fundamental basis for identifying the terrestrial planets as a group is the similarity of their densities. The mean densities of Mercury, Venus, Earth, and Mars are 5.4, 5.1, 5.5, and 4.0 g/cm^3, respectively, and their masses are 0.054, 0.815, 1.000, and 0.108 in units of the Earth's mass. From these data, it is believed that the terrestrial planets consist mainly of the cosmically abundant metals Fe, Si, Mg, Ca, Al, and their oxides. Their atmospheres consist mainly of the cosmically abundant elements H, C, O, N, in compounds such as CO_2, H_2O, N_2, O_2. Thus the terrestrial planets are further distinguished by the high degree of oxidation of their atmospheres. It is generally assumed that the compounds CO_2, H_2O, etc, observed at the surfaces of the terrestrial planets, represent the volatile fraction released from the interiors after planetary accretion.

This review will focus on the atmospheres of Mars and Venus. These planets' atmospheres consist mainly of CO_2, and so it is convenient to consider them together. In contrast, Mercury appears to have no atmosphere. This may be due either to a different internal history or to an inability to hold an atmosphere. The Earth's atmosphere is beyond the scope of this review, but we shall compare the abundances of CO_2, H_2O, N_2, etc, on the Earth,

Mars, and Venus in order to better understand the terrestrial planets as a group.

Several published conference proceedings (Brancazio & Cameron 1964, Brandt & McElroy 1968, Jastrow & Rasool 1969, Sagan, Owen & Smith 1971), and a number of excellent review articles (Sagan & Kellogg 1963, Jastrow 1968, Hunten & Goody 1969, Goody 1969, Eshleman 1970), dealing with planetary atmospheres, have appeared in recent years. Progress in this field has been so rapid that a further survey of the present state of knowledge of the atmospheres of Mars and Venus seems warranted. In this review, we shall compare the compositions, thermal structures, and dynamics of the lower and upper atmospheres of Mars and Venus, and shall consider some of the problems of their evolution. The remainder of the Introduction consists of brief reviews of attempts to detect an atmosphere on Mercury, and of the abundances of volatile constituents of the Earth's atmosphere and oceans.

Observations of Mercury.—The search for an atmosphere on Mercury has thus far yielded negative results. The published upper limit to the CO_2 abundance is 0.58 m-atm, or 0.04 mb surface pressure (Belton, Hunten & McElroy 1967, Bergstralh, Gray & Smith 1967). An upper limit to the H_2O abundance is 30 μ precipitable water (equivalent to a layer of liquid water 30 μ thick), or 10^{-3} mb surface pressure (Spinrad, Field & Hodge 1965). The search for evidence of other gases in Mercury spectra has also yielded negative results.

Gases whose spectra are unobservable from the Earth can be detected in other ways. An upper limit to total pressure at the surface of Mercury of 0.4 mb is inferred from comparison of polarization curves for Mercury, the Moon, and powdered silicate samples observed in the laboratory (O'Leary & Rea 1967, Ingersoll 1971). The low value of surface thermal conductivity inferred from infrared and radio temperature measurements implies surface pressures less than about 0.1 mb (Morrison & Sagan 1967, Murdock & Ney 1970, Morrison 1970); for $P>0.1$ mb, heat transport by gas within the soil material would raise the thermal conductivity above the observed value. A more stringent, but a speculative upper limit of 10^{-5} mb for the surface pressure follows from the assumption that the similarity between the photometric functions of Mercury and the Moon is a consequence of bombardment by the solar wind (Sagan 1966, O'Leary & Rea 1967). In any case, we adopt the prevailing view that Mercury, like the Moon, has essentially no atmosphere.

Volatiles on the Earth.—The Earth's atmosphere is about 75% N_2 and 23% O_2 by volume, with smaller amounts of H_2O, CO_2, and other gases. However, a complete inventory of the Earth's volatiles includes the water in the Earth's oceans and the CO_2 buried in sedimentary rocks (Rubey 1951). The mass of the oceans averaged over the surface of the Earth is 300 kg/

cm², corresponding to a mean pressure at the bottom of the oceans of 300 atm. An additional 150 kg/cm² of water is thought to be buried in sedimentary rocks (Ronov & Yaroshevsky 1967). The amount of CO_2 in sedimentary rocks (as carbonate and organic carbon) is about 40 kg/cm², and an equal amount in igneous rocks may represent recycled sedimentary CO_2 (Ronov & Yaroshevsky 1967). It is believed that carbonate deposition has taken place continuously for most of the Earth's lifetime, and that the volatiles buried in this way represent part of the volatile fraction which has been released from the Earth's interior (Rubey 1951).

There is substantial evidence that these volatiles were released from the interior after the Earth formed (Brown 1952). In the first place, the low abundance of rare gases suggests that the Earth never had a primitive atmosphere of roughly solar composition, or that the primitive atmosphere was lost. Second, the fact that the volatile compounds H_2O, CO_2, N_2, O_2, etc are relatively more abundant than the rare gases indicates that these compounds were retained chemically while the rare gases were being lost. Thus the constituents of the present atmosphere and oceans of the Earth have been released from the interior. One unsolved problem of planetary science is whether the differences we find among the terrestrial planets' atmospheres are due to differences in their internal histories or to differences in conditions which have prevailed at their surfaces.

Atmospheric Abundances

Carbon dioxide.—The only direct sampling of a planetary atmosphere other than the Earth's was carried out on the Soviet entry probes Venera 4, 5, and 6 (Vinogradov, Surkov & Florensky 1968, Avduevsky, Marov & Rozhdestvensky 1970). Data on the composition of the Venus atmosphere were obtained at two levels, $P = 0.6$ and 2.0 atm, respectively, where P is total atmospheric pressure. The CO_2 gas analyzer indicated that the molar fraction of CO_2 is in the range 93–97%.

Estimates of the total amount of CO_2 in the atmosphere are obtained indirectly. The Mariner 5 radio occultation experiments give atmospheric refractivity above the $P = 7$ atm level as a function of distance to the center of the planet (Kliore & Cain 1968, Eshleman et al 1968). Pressure and temperature profiles above this level are obtained with the aid of the hydrostatic relation. Surface pressure is estimated by extrapolating to the surface, using the value of the radius of the solid planet obtained from ground-based radar data (Ash et al 1968, Melbourne, Muhleman & O'Handley 1968). The estimated value of the surface pressure is in the range 70–120 atm, corresponding to 70–120 kg/cm² of CO_2 averaged over the surface of Venus. This estimate is in accord with estimates based on the observed opacity of the atmosphere from ground-based radar data, if it is assumed that CO_2 is the principal absorber of radio waves (Wood, Watson & Pollack 1968, Muhleman 1969, Slade & Shapiro 1970).

The CO_2 abundance on Mars has been measured spectroscopically. The

Mars atmosphere, unlike that of Venus, is optically thin, and therefore the spectrum of sunlight reflected by the surface of Mars contains information about the total abundance of CO_2. This abundance is in the range 60–85 m-atm, corresponding to a partial pressure 4.5–6.5 mb (Kaplan, Munch & Spinrad 1964, Giver et al 1968, Young 1969, Carleton et al 1969). From these studies it is also possible to determine the total pressure from the broadening of spectral lines, and it appears that CO_2 is the major constituent of the Mars atmosphere as well. Total pressure is also determined from the surface refractivity obtained during the Mariner 4, 6, and 7 flybys (Kliore et al 1969a, Rasool et al 1970), and these results fall in the range 4.9–7.6 mb. Here the spread of values is real, and is probably due to topographic differences at the entrance and exit points for the different spacecraft (Pettengill et al 1969, Goldstein et al 1970). It has been established that the polar caps contain solid CO_2 (Leighton & Murray 1966, Herr & Pimentel 1969, Neugebauer et al 1969), so the total amount of CO_2 on Mars may be much greater than that observed in the atmosphere.

The CO_2 abundances for Venus, Mars, and the Earth are summarized in Table 1. For the Earth, the atmospheric abundance and the total abundance, including all forms of CO_2 known to have passed through the atmosphere, are given separately. For Venus and Mars, the total abundance is largely unknown. The Mars polar caps may contain large amounts of solid CO_2, and it is possible that the Venus surface is covered by carbonate deposits which contain an appreciable fraction of that planet's CO_2 (Mueller 1964, Lewis 1968). Clearly, CO_2 is an abundant volatile compound at the surfaces of Venus, Earth, and Mars, but it is not yet possible to compare the total amounts of CO_2 at or near the surfaces of these planets.

Water.—The Venera entry probes detected water on Venus by indicating an increase in electrical conductivity of P_2O_5 when it was exposed to the atmosphere. At the two levels $P = 0.6$ and $P = 2.0$ atm, the molar fraction of

TABLE 1. Abundance of volatiles[a]

Volatile	Earth atmosphere	Earth total	Venus atmosphere	Mars atmosphere
CO_2	0.5×10^{-3}	70 ± 30	95 ± 25	$(1.4 \pm 0.2) \times 10^{-2}$
H_2O	$(1-10) \times 10^{-3}$	375 ± 75	$10^{-2 \pm 2}$	$(0.5-2.5) \times 10^{-6}$
O_2	0.23	~0.23	$<6 \times 10^{-3}$	$<2.5 \times 10^{-5}$
N_2	0.75	~0.75	<5	$<0.8 \times 10^{-3}$
Ar	1.3×10^{-2}	$\sim 1.3 \times 10^{-2}$	<5	$<0.3 \times 10^{-2}$
CO	$(1-10) \times 10^{-7}$	$\sim 10^{-6}$	3×10^{-3}	0.7×10^{-5}
Cl	—	5.7	5×10^{-5}	—
F	—	0.3×10^{-3}	2×10^{-7}	—

[a] Units: kg/cm².

water was found to be in the range 0.1–1.0% (Vinogradov, Surkov & Florensky 1968, Avduevsky, Marov & Rozhdestvensky 1970). However, Lewis (1969) has argued that the change in electrical conductivity is due to the presence of concentrated aqueous solutions of HCl in the Venus clouds, and that the mole fraction of water in the atmosphere is only 10^{-4}. Kuiper (1969b) has argued that the change in conductivity may be due to the presence of partially hydrated $FeCl_2$ in the clouds, and that the mole fraction of water is 10^{-6}. There is a regrettable lack of agreement on the interpretation of these important measurements.

Pollack & Morrison (1970) have pointed out that the anomalously low brightness temperatures observed for Venus in the vicinity of the 1.35 cm water vapor line can be accounted for if the water vapor mixing ratio of the lower atmosphere is in the range $(0.3–1.0) \times 10^{-2}$ (Figure 1). Others (Avduevsky et al 1970, Ohring 1969, Pollack 1969) observe that this amount of water is sufficient to supply the high infrared opacity needed to maintain the high surface temperatures of Venus. However, neither of these arguments positively identifies water as the unknown absorber.

Spectroscopic studies of sunlight reflected in the Venus atmosphere imply

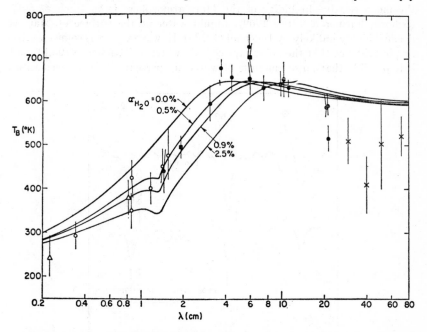

FIGURE 1. The average brightness temperature of Venus at wavelengths from 2 mm to 70 cm. The points are from Dickel (1967) and Pollack & Morrison (1970), and represent the work of many observers. The curves were computed by Pollack & Morrison (1970) assuming an adiabatic lower atmosphere, and are labeled by the assumed percentage molar abundance of water vapor.

water vapor mixing ratios $\lesssim 10^{-4}$ (Dollfus 1963, Bottema, Plummer & Strong 1965, Belton & Hunten 1966, Spinrad & Shawl 1966, Belton, Hunten & Goody 1968), that is, $<100\ \mu$ precipitable water above the level of spectral line formation (Figure 2). Studies at longer wavelengths in the stronger bands of water vapor indicate even less water, e.g. $<20\ \mu$ (Connes et al 1967) and $\sim 5\ \mu$ (Kuiper 1969a) precipitable water.

Schorn et al (1969) have detected water in a number of lines near 8200 Å, and they report that the amount of water varies from day to day between zero and 30–40 μ. Thus the spectroscopic studies can only be reconciled with the Venera results if the former refer to a higher, colder level in the atmosphere, where water vapor is effectively condensed out. The question is therefore whether ice clouds are consistent with the spectroscopic observations, and this will be considered in our general discussion of clouds. The debate over the nature of the Venus clouds is intimately tied to the question of the abundance of water on Venus.

Water has also been detected spectroscopically on Mars (Kaplan, Munch & Spinrad 1964). The amount of precipitable water in a vertical column above the surface varies with the seasons from <5 to $\sim 25\ \mu$, which indicates that the mean relative humidity of the Mars atmosphere is as much as 50% (Schorn, Farmer & Little 1969). Liquid water at the surface of Mars is deemed highly unlikely by Ingersoll (1970a). However, there is some spectroscopic evidence that the polar caps contain water ice (Kieffer 1970a,b), and it is possible that large amounts of water are present either as permafrost

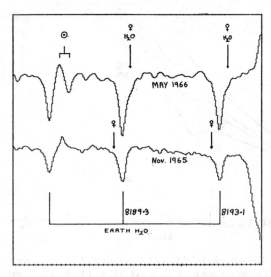

FIGURE 2. Ratio of Venus and Sun spectra near the 8189 and 8193 Å lines of water vapor, showing weak Doppler-shifted Venus lines on the wings of strong unshifted telluric lines. From Belton & Hunten (1966); copyright, the University of Chicago.

(Leighton & Murray 1966) or as water of hydration in surface minerals (Pollack et al 1970). Thus, from atmospheric abundances alone, it is impossible to estimate the total abundance of water on the surface of Mars. Water abundances are also summarized in Table 1. There are no oceans or ice caps on the surface of Venus (Gale & Sinclair 1969), but it is remotely possible that large amounts of water are bound chemically at the surface. It is more likely that all the water is in the atmosphere, in which case Venus has no more than 10^{-3} of the amount of water that the Earth has.

Other gases.—The Venera gas analyzer measured the partial pressure of gases other than CO_2 at the $P = 0.6$ and 2.0 atm level on Venus. From this, the experimenters concluded that the mole fraction of N_2 and other inert gases is in the range 2–5% (Avduevsky, Marov & Rozhdestvensky 1970). These figures, taken at face value, are significant. They imply that the partial pressure of N_2 at the surface of Venus may be as great as 5 atm, and that the partial pressure of N_2 and other inert gases is no less than 2 atm.

Other gases that have been positively identified on Venus are HCl (Figure 3), HF, and CO (Connes et al 1967, 1968). Their molar abundances relative to CO_2 are 6×10^{-7}, 5×10^{-9}, and 4.5×10^{-5}, respectively. These figures refer to conditions in the clouds, and presumably to conditions in the lower atmosphere as well. It is interesting to compare the CO abundance with the O_2 upper limit obtained by Belton & Hunten (1968, 1969a). For the mole fraction of O_2 relative to CO_2, they obtained $O_2/CO_2 < (2$ or $8) \times 10^{-5}$, the value depending on whether absorption takes place within or above the clouds, respectively. Thus the elemental abundance ratio of oxygen to carbon in the Venus atmosphere may be less than 2 by a small but significant amount. The amount of CO above the level of line formation is about 13 cm-atm, approximately 50–100 times that in the Earth's atmosphere.

Values of the total surface pressure on Mars inferred from the broadening of spectral lines fall in the range 5–7 mb (Kaplan, Connes & Connes 1969, Young 1969), in agreement with the Mariner occultation results (Rasool et al 1970). Since these values are not significantly greater than the CO_2 partial pressure, the partial pressure of other gases including N_2 is less than 1–2 mb on Mars.

Still more stringent upper limits on N_2 were obtained with the Mariner 6 and 7 ultraviolet spectrometer (Barth et al 1969). The spectrum of the Mars upper atmosphere from 1100–4300 Å showed emission features of CO, O, and CO_2^+, but no evidence of N_2 (Figure 4). The maximum mole fraction of N_2 relative to CO_2 at the level of observation (200 km) was estimated to be about 5% (Dalgarno & McElroy 1970). However, the ratio of N_2 to CO_2 is likely to increase with height as a result of diffusive separation of the lighter gas in the upper atmosphere, so the actual mole fraction of N_2 in the lower atmosphere is probably less than 5%. These results imply that the partial pressure of N_2 at the surface is less than 0.3 mb.

The only gas other than CO_2 and H_2O which has been positively identified

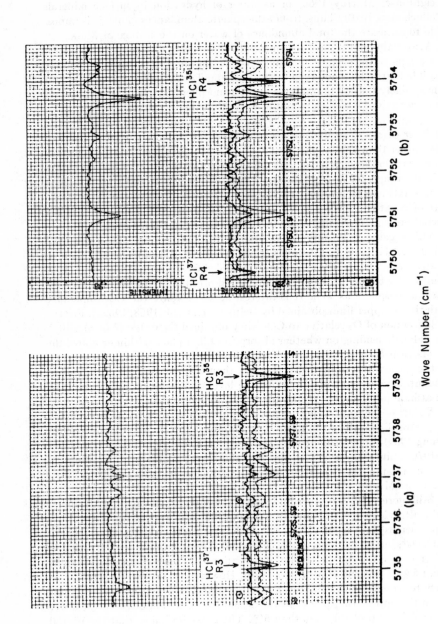

FIGURE 3. Interferometer spectra at frequencies 5733–5755 cm^{-1}, showing absorption due to HCl in the Venus atmosphere. The upper trace is an average of two solar spectra, and the lower two traces are averages of four and five Venus spectra, respectively. From Connes et al (1967); copyright, the University of Chicago.

in the lower atmosphere of Mars is CO (Kaplan, Connes & Connes 1969). They report $CO/CO_2 = 0.8 \times 10^{-3}$ by volume, or 5.6 cm-atm CO above the surface. This is about 25–50 times the amount of CO in the Earth's atmosphere.

The abundances of N_2, O_2, and CO at or near the surfaces of the Earth, Mars, and Venus are given in Table 1. The data do not exclude the possibility that the ratio N_2/CO_2 is the same on all three planets. The elements Cl and F are also listed in Table 1. For Venus, the amounts refer to the mass of Cl and F observed in the atmosphere as HCl and HF. For the Earth, only total abundances are listed. These refer to the Cl and F in solution in the oceans (Rubey 1951).

Cloud composition.—Here we consider observations which indicate the existence of clouds or haze in the Venus and Mars atmospheres, and observations which pertain directly to the chemical composition of the cloud particles. Observations of temperatures and pressures, vapor pressure relations, and models of cloud structure will be discussed later.

The reflection spectrum of Venus contains several broadband absorption features which are characteristic of liquid or solid material, as well as numerous narrowband features due to CO_2 and other gases. The gaseous features contain little or no evidence of high temperatures and pressures, so the broadband features are presumably characteristic of the cloud particles and not of the planet's surface. Figure 5 shows low-resolution spectra of Venus at wavelengths 0.2–4.0 μ (Moroz 1965, Bottema et al 1965, Irvine 1968, Kuiper 1969b). Absorptions at 1.5, 2.0, and beyond 3.0 μ are characteristic of ice (Bottema et al 1965, Pollack & Sagan 1968, Plummer 1970), but CO_2 gaseous absorptions may account for the features observed in the Venus spectrum at 1.5 and 2.0 μ (Rea & O'Leary 1968, Hansen & Cheyney 1968). On the other hand, ice is not a strong absorber below 0.4 μ, although many other solids are. Kuiper (1969b) has compared these spectra with spectra of many substances, and he concludes that the cloud particles are partially hydrated $FeCl_2$. There is no agreement as to the interpretation of these data.

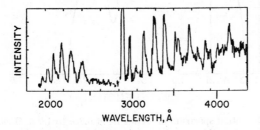

FIGURE 4. Ultraviolet spectrum of the Mars upper atmosphere showing the Cameron bands of CO below 2600 Å, the Fox-Duffendack-Barker bands of CO_2^+ above 3000 Å, the CO_2^+ ultraviolet doublet band system at 2890 Å, and the 2972 Å line of atomic oxygen. From Barth et al (1969); copyright, the American Association for the Advancement of Science.

In principle, it should be possible to derive the refractive index of the cloud particles from measurements of intensity and polarization of reflected sunlight. However, many other effects enter, including the effects of particle size, shape, and orientation with respect to the vertical, and the effects of spatial inhomogeneities within the clouds. At best, one can conclude from existing data that the cloud particles, if they are dielectric spheres, have a refractive index greater than that of water (Arking & Potter 1968, Coffeen 1969, Hansen & Arking 1971).

On Mars, several different cloud types can be distinguished, including several kinds of condensation cloud, and dust clouds. Mariners 6 and 7 observed two reflection features near 4.3 μ which are highly characteristic of solid CO_2 (Herr & Pimentel 1970). These features were observed on the limb at a height of 25 ± 7 km. Thin hazes on the limb were also observed in the Mariner 6 and 7 television pictures (Leovy et al 1971), and evidence for a uniform haze was provided by data from the Mariner uv spectrometer (Barth & Hord 1971). Ground-based polarization data have also been cited as evidence of a uniform haze in the Mars atmosphere (Dollfus & Focas 1969, Morozhenko 1970), but Ingersoll (1971) has shown that surface polarization and Rayleigh scattering by CO_2 can account for all the features cited in these studies.

Ground-based observers frequently report blue-white clouds near the limb and terminator of Mars (Dollfus 1961, Slipher 1962). These clouds often occur repeatedly in fixed locations and at fixed times, usually in the afternoon. Several of these repeatedly occurring bright objects were seen by the Mariner 6 and 7 TV cameras (Leovy et al 1971). Their unique diurnal

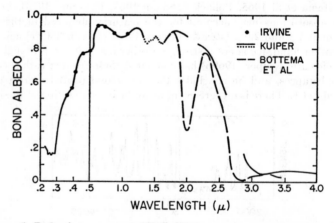

FIGURE 5. Reflection spectrum of Venus from 0.2 μ to 4.0 μ. Bond albedo is the ratio of total light reflected by the planet to total light incident. Irvine's (1968) points and Kuiper's (1969b) curve are replotted from Kuiper (1969b). The un-normalized data of Bottema et al (1965) are replotted so as to agree with Kuiper's data at 1.6–1.7 μ.

brightening behavior suggests that these are also a condensation phenomenon.

Bright, yellow clouds are also seen by ground-based observers. They are generally thought to be dust clouds which arise in the bright, yellow areas. Although infrequent, they occasionally obscure the entire planet, and may last for several weeks (Dollfus 1961, Pollack & Sagan 1970).

Some observers feel that there is evidence of a "blue haze" in the atmosphere of Mars. The dark maria are normally invisible in ground-based photographs taken through a blue filter, but there are sometimes "blue clearings" during which these features become faintly visible. However, the Mariner 6 and 7 TV pictures showed that crater visibility was the same in pictures taken with a blue filter as in pictures taken with red or green filters (Leighton et al 1969). Moreover, the normal invisibility of the maria in blue light can be explained as a surface phenomenon (McCord 1969, Pollack & Sagan 1969, McCord & Westphal 1971), and the blue clearings, which occur mostly near opposition, can be explained as an effect of the surface phase function (Slipher 1962, O'Leary 1967). The only difficulty is that blue clearings do occasionally occur far from opposition (de Vaucouleurs 1968, Capen 1970), so the possibility remains that a thin blue haze does affect the visibility of surface phenomena.

STRUCTURE OF THE LOWER ATMOSPHERES

Several convenient properties distinguish planetary lower atmospheres from upper atmospheres. First, temperature is well defined in the lower atmospheres; the population of molecular vibration and rotation states follows a Boltzmann distribution. Second, lower atmospheres are well mixed; the proportion of noncondensable gaseous constituents is the same at all levels. And third, recombination is swift; the effects of photodissociation and photoionization can generally be ignored. In this section we shall consider the characteristics of the Venus and Mars lower atmospheres: their thermal structures, the Venus cloud structure, and theories of the thermal structures. This subject was reviewed by Goody (1969), and the following discussion will therefore focus on the more recent observational and theoretical developments.

Thermal structure.—Vertical profiles of temperature and pressure in the lower atmospheres have been determined from direct measurements made by Veneras 4, 5, and 6, and from radio occultation measurements made by Mariners 4, 6, and 7 at Mars and by Mariner 5 at Venus. The derived profiles depend on the assumed composition of the lower atmosphere, but this is not a major source of error. These methods are extremely powerful, and a wealth of detailed information has been obtained.

Figure 6 shows the most recent analysis of temperature vs distance to the center of Venus, based on Mariner 5 refractivity data (Fjeldbo, Kliore & Eshleman 1971). The Venera 4, 5, and 6 data agree with these results, although absolute height is uncertain in the Venera data (Avduevsky, Marov & Rozhdestvensky 1970). The Venera data extend deeper into the atmosphere (to $P \approx 27$ atm), but the Mariner data extend farther up, and show considerably more fine structure.

The lapse rate of temperature is within a few percent of the adiabatic rate, at least from 20 km altitude ($P \approx 27$ atm) up to 60 km. If the adiabatic lapse rate is extrapolated down to the surface, temperatures are obtained which are in reasonable agreement with surface temperatures determined from microwave data (Figure 1), and with the $747 \pm 20°K$ surface temperature measured by the Soviet space probe Venera 7 (Advuevsky et al 1971). If there is an isothermal layer next to the ground (Gale, Liwshitz & Sinclair 1969, Muhleman 1969), its thickness must be less than about 3 km.

The fine structure shown in Figure 6 is also significant. The abrupt changes in lapse rate at altitudes of 50 and 60 km may indicate a transition from dry to wet adiabatic conditions, implying that clouds are forming at

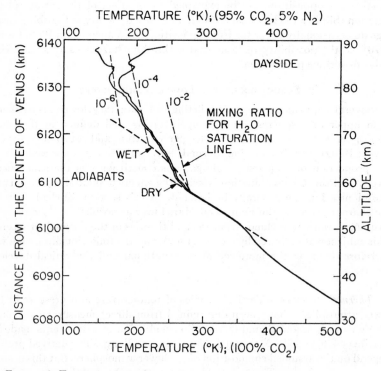

FIGURE 6. Temperature profiles of Venus from refractivity data by assuming a mixed atmosphere in hydrostatic equilibrium. Also shown are saturation lines for water vapor and adiabatic profiles for a dry and a wet (mass mixing ratio of 10^{-2}) atmosphere. From Fjeldbo, Kliore & Eshleman (1971); copyright, the American Astronomical Society.

these two levels. The region below 50 km is also a source of attenuation at 2297 MHz (Fjeldbo, Kliore & Eshleman 1971). Possible sources of attenuation at these levels are dust, turbulence (de Wolf 1970), and clouds of mercury halides (Rasool 1970).

The radio occultation experiments on Mariners 4, 6, and 7 provide data about the thermal structure of the Mars atmosphere (Kliore et al 1969b, Rasool et al 1970). Figure 7 shows the temperature profiles at the entry and exit points for Mariners 6 and 7. The profiles are uncertain at high altitudes, because of uncertainties in the motion of the spacecraft and in the refractivity of the ionosphere. Nevertheless, all published analyses of these data show an extremely cold region in the middle atmosphere, with derived temperatures falling below the saturation curve of CO_2. The point at which each profile crosses the saturation curve is indicated by a shaded line in Figure 7. Other features of importance in the Figure 7 profiles are: the subadiabatic

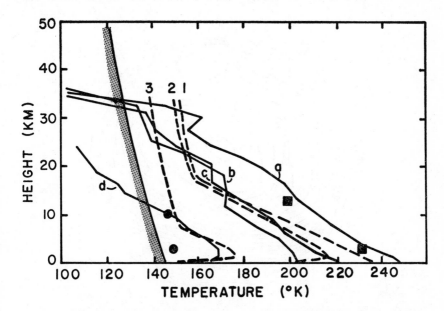

FIGURE 7. Thermal structure of the lower atmosphere of Mars. The solid lines are from Mariner 6 and 7 radio occultation data, as analyzed by Rasool et al (1970): a. latitude 3.7°N, local time 1545; b. latitude 38.1°N, local time 0310; c. latitude 58.2°S, local time 1430; d. latitude 79.3°N, local time 2210 (all for northern hemisphere early fall). The dashed lines are from the static model calculations of Gierasch & Goody (1968): 1. equator equinox, local time 1600; 2. equator equinox, local time 0400; 3. 45°N winter, local time 0800. The points are from the dynamic model calculations of Leovy & Mintz (1969): squares—equator, northern winter, local time 1600; circles—75°N winter, local time 2200. The shaded line is the CO_2 saturation vapor pressure curve.

lapse rate (about 3.5°K/km, or 70% of the adiabatic rate), observed in all
the profiles, the strong inversion at the autumn pole (profile d), and the
highly unstable temperature discontinuity at the equator during the day
(profile a). The later discontinuity is indicated by comparing the atmospheric
temperature near the ground, $T \approx 250°K$, with the surface temperature,
$T \approx 275$, measured with the Mariner infrared radiometer (Neugebauer et al
1969).

   Venus cloud structure.—Venus exhibits limb darkening in the thermal infrared, as shown in Figure 8 (Murray, Wildey & Westphal 1963, Westphal
1966). This suggests either that the visibility within the cloud is large—
about one scale height (Samuelson 1968)—or that the cloud-top surface is
bumpy and irregular. These alternatives are also suggested by studies of
optical phenomena near inferior conjunction (Goody 1967, Schilling &
Moore 1967, Abhyankar 1968). The near uniformity of integrated disk

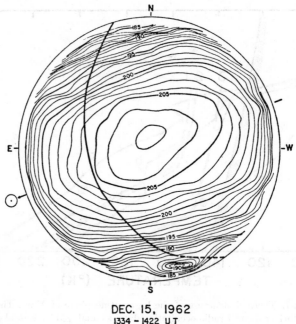

DEC. 15, 1962
1334 - 1422 U T

FIGURE 8. Brightness temperature map of Venus (8–14 μ) for the morning of
December 15, 1962. The terminator and the projected direction to the Sun are indicated, respectively, by the heavy line and by the symbol. The slightly crenelated
pattern running in the east-west direction results from uncertainties in positioning
the scans. The brightness temperatures shown are systematically too low because of
uncertain telescope transmission losses. From Murray, Wildey & Westphal (1963);
copyright, the American Geophysical Union.

temperature at wavelengths from 4 to 14 μ serves to define a mean emission temperature for the cloud. Measured values of the emission temperature fall in the range 215–250°K (Sinton 1963, Gillett, Low & Stein 1968, Hanel et al 1968).

Spectroscopic studies define an effective cloud temperature which is comparable to the emission temperature. They also provide evidence that the optical path of reflected sunlight varies by at least one scale height. The observed increase of gaseous absorption toward superior conjunction indicates that sunlight travels at least one scale height farther in the atmosphere at normal incidence than it does at grazing incidence (Chamberlain 1965, Moroz 1968). In addition the strengths of gaseous absorptions vary erratically with time and with position on the disk (Kuiper 1969b, Young, Schorn & Smith 1970).

Temperatures derived from rotational analyses of CO_2 bands fall in the range 220–280°K (Belton, Hunten & Goody 1968, Belton 1968, Young, Schorn & Smith 1970), although temperatures as high as 450°K have been inferred (Spinrad 1966). The effective pressures generally fall in the range 100–200 mb, although the effective pressure may be as low as 50 mb for strong bands, and as high as 500 mb for weak bands (Young 1970). It is tempting to associate these effective temperatures and pressures with a specific level in Figure 6. The level $T = 240°K$, $P \approx 100$ mb occurs at about 66 km, approximately 6 km above the level of transition from adiabatic to subadiabatic lapse rate.

If the clouds are water clouds with bases at 60 km, then the water vapor mixing ratio must fall from 10^{-2} at the cloud base to 10^{-3} at the level of spectral line formation. This means that the partial pressure of water vapor must be about 0.1 mb at the level of line formation, and in order that this amount be consistent with spectroscopic abundances (less than 100 μ precipitable water above the level of line formation), the scale height of water vapor must be less than 1 km. In fact, this is about one-half the scale height one obtains using the observed temperature gradient to define a vapor pressure scale height. Given the uncertainties in defining the level of line formation, these numbers are not inconsistent. The fact that temperature falls to 180°K, according to Figure 6, suggests that the clouds of Venus may be water clouds, provided the level of line formation is sufficiently high in the atmosphere. It must be remembered, however, that these arguments do not establish that the clouds are water, and they do not exclude the possibility of cloud constituents other than water.

Theories of the thermal structure.—There are many theories of the high surface temperatures on Venus, and it is still not possible to choose between them. We do not know the profiles of solar heating or of infrared emission below about one optical depth in the atmosphere. Such information could be obtained by means of an entry probe on the sunlit side of the planet, and such a probe is well within the United States' and Soviets' technological capacities

(Hunten & Goody 1969). We also do not know the opacities of CO_2 and other atmospheric constituents under conditions which prevail in the lower Venus atmosphere. Existing laboratory data cannot settle the important question whether there is significant absorption in the wings of gaseous absorption lines and in the windows between vibration-rotation bands (Plass & Stull 1963).

Theories of the thermal structure of the Venus atmosphere differ as to how the surface is heated, and how the heat is trapped near the surface. Trapping may be due to reflection of upward-propagating radiation at the base of the cloud (Avduevsky et al 1970), absorption by dust in the lower atmosphere (Hansen & Matsushima 1967), absorption and reflection by dust in the lower atmosphere (Samuelson 1967), or absorption by atmospheric gases (Sagan 1960, Ohring 1969, Pollack 1969). Most workers feel that absorption by CO_2 alone does not provide the opacity necessary to maintain high surface temperatures with an adiabatic lower atmosphere. Unless the heat source at the surface were many times greater than the solar constant, a pure CO_2 atmosphere would cool to temperatures below the present value, and an extensive isothermal layer would develop. Water vapor in concentrations 10^{-3} to 10^{-2}, together with CO_2, could provide the necessary opacity, but the extrapolation to high temperatures and pressures is uncertain.

Theories of the mechanism of heat deposition at the Venus surface also differ. In the greenhouse models, heat is deposited directly by sunlight, either at the surface or in the lower atmosphere (Sagan 1960, Pollack 1969, Ohring 1969). In the internal heating models, no solar heat is deposited, but the infrared opacity of the atmosphere is so great that planetary heat can maintain the high surface temperatures (Hansen & Matsushima 1967). In the dynamical models, energy is convected downward by large-scale motions driven by solar heating at the top of the atmosphere (Goody & Robinson 1966). Gierasch & Goody (1970) argue that convection of heat by large-scale motions is the most effective means of heating the deepest levels. They point out that unless convection extends to the ground, cloud particles of dust or low-vapor-pressure condensate will fall out, and the cloud will collapse. Gierasch & Goody do not consider clouds of water or other high-vapor-pressure condensates, mainly because they feel such clouds are excluded by optical and spectroscopic data. Without direct entry probes into the Venus atmosphere, it is unlikely that any of these questions will be settled.

Theories of the thermal structure of the Mars atmosphere differ from those of Venus and the Earth in that the Mars atmosphere has only a small effect on temperatures at the ground. This is due partly to the small heat capacity of the Mars atmosphere and partly to the small optical thickness of the atmosphere at most visible and infrared wavelengths. With these assumptions, Leighton & Murray (1966) showed that the partial pressure of CO_2 in the atmosphere might be controlled by the vapor pressure of solid CO_2 at the pole. Subsequent observations have shown that the polar caps are indeed CO_2 (Herr & Pimentel 1969) and are at the temperature necessary for equilibrium with the atmosphere (Neugebauer et al 1969).

Because the radiative adjustment time for atmospheric temperature is short (Goody & Belton 1967), the atmospheric temperatures tend to "follow" the temperature of the ground during the diurnal cycle. Gierasch & Goody (1968) have computed temperature profiles vs time of day for a radiating, convecting model of Mars (Figure 7). The significant features of these profiles are the absence of CO_2 condensation except at the ground near the pole, the nearly constant, adiabatic lapse rate throughout most of the atmosphere, the strong night-time inversions due to cooling of the ground, and the highly unstable temperature discontinuities (up to 70°K) at the ground during the day. Recently, Gierasch (1971) has examined a general class of radiating, convecting models of the Mars atmosphere. He shows that very low temperatures, below the CO_2 saturation line, are thermodynamically possible, but that the atmosphere must function as an extremely efficient heat engine for these low temperatures to occur. That is, rising convective elements, driven by heating at the surface, would have to overshoot the height at which they are neutrally buoyant, with little dissipative energy loss. There is no evidence that such efficient convection takes place in laboratory or terrestrial situations.

STRUCTURE OF THE UPPER ATMOSPHERES

At sufficiently high atmospheric levels, local thermodynamic equilibrium breaks down and energy cannot be efficiently radiated away by molecular vibration-rotation bands in the thermal infrared. This breakdown occurs when the spontaneous emission probability per molecule exceeds the rate of excitation by collisions. On the Earth, Venus, and Mars, the LTE transition level is determined by relaxation of the lowest vibrational mode of CO_2, centered at 667 cm^{-1}. This occurs at a pressure of about 0.3×10^{-2} mb on all three planets. Since energy absorbed above the transition level can only be removed by conduction or mass transport downward, this level is also a temperature minimum, which, in the case of the Earth, is called the *mesopause*. In this section we shall consider the characteristics of the Mars and Venus atmospheres from the mesopause upward. Following usage for the Earth, we shall sometimes refer to two subdivisions of this upper atmosphere region: the *thermosphere*, in which the mean free path of molecules is less than a scale height, and the *exosphere*, in which the mean free path exceeds a scale height.

Observations.—The atmospheric refractive indices deduced from radio occultation of Mariners 4, 5, 6, and 7 yield information on upper-atmosphere structure as well as on the density profile near the ground. Refractive indices less than one arise from ionization, and electron densities can be inferred from the portions of the occultation data exhibiting such refractive indices. Figure 9 shows electron densities on the day sides of Venus and Mars in the height range 90 to 200 km deduced from Mariner 4, 5, and 6 S-band occultation data (Kliore et al 1967, 1969a, 1969b, Fjeldbo & Eshleman 1968, Fjeldbo, Kliore & Seidel 1970). Electron densities at higher levels on the day and night sides of Venus, inferred from the dual frequency radio occultation experiment of Mariner 5, are shown in Figure 10 (Mariner Stanford Group 1967). No ionization was detected on the night sides of either Venus or Mars by the S-band technique.

Information on the state of the upper atmosphere also comes from ultraviolet photometry on Mariner 5, and ultraviolet spectrometry on Mariners 6 and 7. The intensity of emissions can be directly related to number densities of ground states as well as of the emitting states, provided that the excitation mechanisms are known (Barth 1969). Mariner 5 detected emission in the 1050–1250 Å band, presumably due to Ly α (Barth, Pearce & Kelley et al 1967, Barth, Wallace & Pearce 1968). These data are shown in Figure 11. Ultraviolet photometers carried by Venera 4 also detected Ly α emission near Venus; intensities agree with the data shown in Figure 11 at distances exceeding 12,000 km on the day side, but showed lower emission rates closer to the planet. An attempt to detect the 1304–7 Å atomic oxygen emission by Venera 4 yielded negative results, from which an upper-limit atomic oxygen abundance of 2×10^8 atoms/cm^3 at 300 km height has been inferred (Kurt, Dostolow & Sheffer 1968). Ly α was also detected near Mars (Barth et al 1969), as were the spectral features of CO, O, CO$_2^+$ shown in Figure 4.

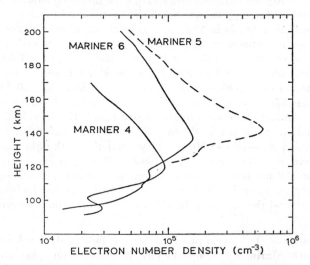

FIGURE 9. Electron number densities for Mariner 4 (replotted from Fjeldbo & Eshleman 1968), Mariner 5 (replotted from Kliore et al 1967), and Mariner 6 (replotted from Fjeldbo, Kliore & Seidel 1970). The solar zenith angle was 67° at Mars for Mariner 4, 33° at Venus for Mariner 5, and 57° at Mars for Mariner 6. The Mariner 7 electron density was nearly identical to that of Mariner 6.

An absolute density datum for the Venus upper atmosphere was obtained from the 1959 occultation of Regulus (deVaucouleurs & Menzel 1960, Hunten & McElroy 1968). Density 118 km above the surface was approximately 6×10^{13} molecules/cm^3. This density value agrees with the upward extrapolation of density inferred from the Mariner 5 radio occultation (Figure 12). No comparable neutral density datum is available for the Mars mesopause region.

Composition and thermal structure.—When theoretical models of the Mars and Venus thermospheres are compared with observation, only those models involving essentially pure CO_2 give good agreement. Compared with all combinations of H, C, O, N, their compounds of moderate molecular weight, and rare gases, pure CO_2 gives an extremum of thermospheric properties. Models involving pure CO_2 give the smallest electron densities, the lowest elevation for the electron density peak, and the smallest electron scale height above the peak. These properties of pure CO_2 thermospheres result from the high molecular weight and radiative efficiency of CO_2, which imply low temperatures and small scale heights.

Static thermal and ionization models of Venus illustrate this (Stewart 1968, Gross, McGovern & Rasool 1968, Stewart & Hogan 1969a, McElroy

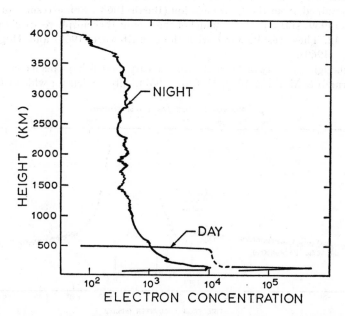

FIGURE 10. Day and night side electron densities for Venus from the Mariner 5 dual frequency occultation experiment (Mariner Stanford Group 1967). The dashed portion of the day curve indicates missing data due to the formation of caustics.

1967, 1969). These models show that the observed Venus electron density profile (Figure 9) can be explained by ionization of CO_2 and dissociative recombination $(CO_2^+ + e^- \rightarrow CO + O)$, with an upper thermosphere temperature near 700°K. They also show that as little as 2.5% dissociation $(O/CO_2 = 2.5\%)$ at 135 km, with diffusive equilibrium above 135 km, produces large increases over the observed electron densities at high elevation due to ionization of atomic oxygen. Similarly, only 10% N_2 by volume, mixed throughout the thermosphere, produces marked excesses of electrons relative to the observations. Using a complex model of diffusion, ionization charge exchange, and recombination processes, Shimizu (1969) has shown that electron density and height of the ionization peak increase dramatically as lighter gases are added to CO_2.

These results are consistent with the very low limit on atomic oxygen obtained from Venera 5 observations at 1304–7 Å (Kurt, Dostolow & Sheffer 1968). If we combine that negative result with the density datum obtained from the Regulus occultation, and assume an upper thermosphere temperature of 700°K, the O/CO_2 ratio at 300 km would be $<10^{-4}$. A rocket ultraviolet spectrum of Venus showing an atomic oxygen feature indicates that this limit may be too low, however (Moos, Fastie & Bottema 1969). The 700°K temperature inferred from the ionization profile agrees with that of atomic hydrogen in the exosphere at planetocentric distances exceeding 12,000 km, as determined from the Ly α emission (Barth 1968). Some results of McElroy's (1969) static thermal model of the Venus thermosphere are shown in Figure 12. These results agree with those of Stewart (1968) and Hogan & Stewart (1969).

Although a variety of Mars electron density models have been considered (see Brandt & McElroy 1968, McElroy 1967), only models for which CO_2^+ is

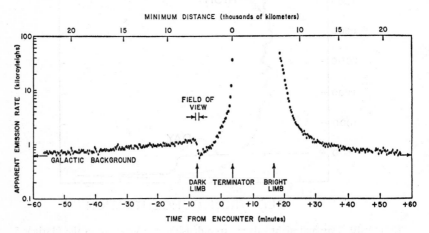

FIGURE 11. Lyman α emission observed at Venus by Mariner 5. From Barth et al (1967); copyright, the American Association for the Advancement of Science.

the predominant positive ion between 150 and 200 km appear to be acceptable when the most recent laboratory measurement of the rate of dissociative recombination of CO_2^+ is used (Weller & Biondi 1967). The ultraviolet spectra (Figure 4) confirm the predominance of CO_2^+. The O and CO emissions in the ultraviolet spectra have been shown to be consistent with an O/CO_2 ratio of 10^{-3} at 140 km, and 10^{-2} at 200 km (McConnell & McElroy 1970). The low, thin, and weak ionized layer on Mars is again an indication that concentrations of any lighter gases, such as N_2, are very low. As we have seen, the ultraviolet spectra provide even stronger evidence for a low N_2 abundance.

If we assume that the electron density profile is determined by ionization and dissociative recombination, the scale height above the peak indicates a Mars upper thermosphere temperature near 300°K in July 1965 (Stewart & Hogan 1969b, Hogan & Stewart 1969), and near 500°K in August 1969 (Fjeldbo, Kliore & Seidel 1970). The difference can be explained by the greater extreme ultraviolet (euv) solar flux in 1969.

According to McElroy (1969), the upper thermosphere temperature expected for Mars in July 1965 is about 450°K, significantly greater than the 300°K temperature inferred from the electron scale height. Cloutier, McElroy & Michel (1969) suggested that solar wind interaction with the Mars thermosphere could drive ionization downward on the day side, and might explain the low value of the electron scale height. They assume that the Venus thermosphere is shielded from this effect by atomic hydrogen in the

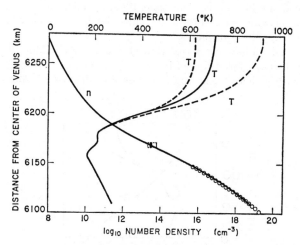

FIGURE 12. Temperature and density of the upper atmosphere of Venus according to McElroy's model. Solid curves are for Mariner 5 conditions; dashed curves are for solar maximum and minimum. Open circles are densities inferred from the S-band experiment on Mariner 5. The rectangle at 6170 km is density from the Regulus occultation data. From McElroy (1969); copyright, the American Geophysical Union.

exosphere. On the other hand, Hogan & Stewart (1969) argue that the photo-ionization heating efficiency in a CO_2 atmosphere may be smaller than the value assumed by McElroy, and that this may account for the 300°K temperature inferred from the electron scale height on Mars (Figure 13). They then account for the 700°K temperature inferred for Venus as an effect of that planet's slow rotation; relatively high temperatures develop on the dayside (Figure 12), in spite of the presumed low value of the heating efficiency, because the thermal response time of the thermosphere is short compared to the period of rotation. It remains to be seen whether a low value of the heating efficiency can be reconciled with laboratory and theoretical studies (Henry & McElroy 1968, McConnell & McElroy 1970).

Photochemistry of CO_2.—It is surprising that the upper atmospheres of Venus and Mars consist almost entirely of undissociated CO_2. Direct three-body recombination,

$$CO + O + M \rightarrow CO_2 + M \qquad \qquad 1.$$

has a rate coefficient less than 10^{-34} cm⁶/sec (Clyne & Thrush 1962). This is fast enough to account for recombination and predominance of CO_2 in the lower atmosphere, but not in the thermosphere. Turbulent mixing of dissociation products downwards, and of CO_2 upwards, could account for the undissociated state of the thermosphere, but only if the mixing rate were several orders of magnitude greater on Mars and Venus than it is on the Earth at comparable levels (Donahue 1968, Shimizu 1969). There is no obvious reason why both Mars and Venus should have such high mixing rates in their upper atmospheres.

Several authors have proposed that O and CO recombine in the upper atmosphere through formation of intermediate states or catalysis by minor constituents, but there are strong arguments against each proposed mech-

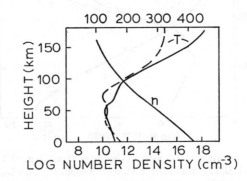

FIGURE 13. Temperature and density of the upper atmosphere of Mars computed from different atmospheric models. Solid lines are from McElroy (1969); the dashed line is from Hogan & Stewart (1969).

anism. McElroy (1967) and McElroy & Hunten (1970) suggested that O and CO would recombine in the following reactions:

$$O^1D + CO_2 \rightarrow CO_3^* \qquad \qquad 2.$$

$$CO_3 + CO \rightarrow 2CO_2 \qquad \qquad 3.$$

Excited atomic oxygen (O^1D) is produced in the dissociation of CO_2 by euv radiation. The excited CO_3 (CO_3^*) undergoes a radiative transition to a stable state. This state must last for at least 20 sec in order that the rate at which reaction (3) proceeds be sufficient to prevent accumulation of CO and O (Donahue 1968). However, there is now strong experimental evidence that the lifetime of CO_3 is negligibly small, and that the quantum efficiency for production of CO and O in the ground state is unity (DeMore 1970, Clark & Noxon 1970, Felder, Morrow & Young 1970, Slanger & Black 1970).

Donahue (1968) has suggested that recombination is catalyzed by hydrogen compounds produced in the dissociation of H_2O:

$$CO + OH \rightarrow CO_2 + H \qquad \qquad 4.$$

$$H + O_2 + M \rightarrow HO_2 + M \qquad \qquad 5.$$

$$HO_2 + CO \rightarrow CO_2 + OH \qquad \qquad 6.$$

With an ample supply of H and OH, this mechanism would be limited by the three-body reaction (5). Consequently, it would only be effective below the mesopause. Rapid vertical mixing of O, CO, and CO_2 would be required for these reactions to affect the state of CO_2 dissociation in the thermosphere.

None of these mechanisms appears to be satisfactory, especially in view of the fact that the real recombination mechanism works on both Venus and Mars, even though hydrogen compound concentrations and turbulent mixing rates are likely to be quite different on the two planets.

Hydrogen, deuterium, and helium on Venus.—The Mariner 5 Ly α observations at Venus (Figure 11) cannot be interpreted in terms of a single component in thermal escape. The dayside data are consistent with a number density distribution having a scale height of 800 km below 6000 km altitude, and a scale height of 1600 km above 6000 km. If the neutral exosphere densities are a consequence of thermal escape, two Ly α emitting components, whose ratio of scale heights is 2, are required to explain these data. Barth, Wallace & Pearce (1968) suggested that photodissociation of H_2 yielding $H(2P)$ followed by Ly α emission could explain the smaller scale height below 6000 km, but the high ratio of H_2 to H ($\sim 10^5$), which would be required to explain the observations, could not be maintained against photodissociation by any plausible transport mechanism (Donahue 1968, McElroy & Hunten 1969).

Alternatively, the small scale height below 6000 km could be due to deuterium (Donahue 1968). This hypothesis requires an H concentration of

10^4 cm^{-3} and a D concentration of 10^5 cm^{-3} at the base of the exosphere (Wallace 1969). Such a low H/D ratio need not require a low H/D ratio in the atmosphere as a whole. Because H escapes much more readily than D, the escape flux of H would exceed that of D by the factor 2×10^3, and if every H and D atom produced by primary photodissociation escapes, this would also be the ratio H/D in the atmosphere as a whole (Donahue 1968). The large change in the ratio H/D, from 2×10^3 in the lower atmosphere to 10^{-1} at the base of the exosphere, implies weaker vertical mixing in the Venus upper atmosphere than on the Earth (McElroy & Hunten 1969, Donahue 1969). This is at odds with the apparent requirement for intense mixing needed to maintain a CO_2 thermosphere. For the Earth, the ratio H/D is 6700. McElroy & Hunten suggest that H has been depleted relative to D on Venus as a result of fractionation by escape of large amounts of hydrogen. Since hydrogen is produced by photolysis of water vapor, a reliable determination of the H/D ratio would provide important information on the history of water on Venus.

The day and night ionization profiles determined by the Mariner Stanford Group (Figure 10) provide indirect evidence for abundant He^4. The daytime densities of 10^4 electrons cm^{-3} near 200 and 500 km can be easily explained by ionization of He^4, provided that the He^4 abundance at 200 km is near 5×10^7 cm^{-3}. This ionization cannot be easily accounted for by combinations of other likely constituents such as CO_2, H, D, or H_2 (Whitten 1970). Concentrations of He^4 of this magnitude have also been inferred by McElroy & Strobel (1969) as a plausible source for the observed night-side ionosphere. The latter could be maintained by transport of helium ions from the dayside by 100–200 m/s winds above 200 km. Knudsen & Anderson (1969) argue that the He^4 abundance of 5×10^7 cm^{-3} at the 200 km level on Venus implies that the rate of radiogenic production and outgassing of He^4 on Venus is at least as large as the terrestrial rate.

ATMOSPHERIC DYNAMICS

The large-scale dynamics of planetary atmospheres reflect a balance between the rate of generation of potential energy by radiation and the rate of destruction of mechanical energy by dissipation. Heat enters the system in the sunlit hemisphere and leaves the system from both hemispheres. An attempt to derive the scaling of velocity amplitudes and temperature differences from one planet to another was made by Golytsin (1970). He finds that the amplitude of the motion varies inversely as the mass of the atmosphere and directly as the solar heating. His theory does not account for the organization and direction of the motion, or for the occasional concentration of activity into intense local phenomena, such as dust devils on Mars, nor does it account for rapid rotation of part of the Venus atmosphere which we discuss below.

Venus lower-atmosphere dynamics.—Goody (1969) has summarized theories of the deep atmospheric circulation of Venus (Goody & Robinson 1966, Hess 1968, Stone 1968). The essential feature of these models is that there is no rotation of the atmosphere or the planet. The motion consists of a single overturning cell, symmetric about the subsolar and antisolar points, with rising motion at the subsolar point and sinking motion at the antisolar point. These models offer one explanation for the relatively small temperature contrast over the Venus clouds (Figure 8) and for the adiabatic temperature distribution in the deep atmosphere.

Goody's (1969) review contains only a brief discussion of the observations which suggest that near the tops of the Venus clouds, the atmosphere rotates with a 4 day period. The solid planet rotates with a 243 day period, and both rotations are retrograde, that is, opposite to the direction of orbital motion. The first evidence for the 4 day rotation came in photographs taken through an ultraviolet filter (wavelength 3500–4000 Å). At these wavelengths, amorphous patches of light and dark are detectable, and their motion can be measured by comparing photographs taken several hours apart (Smith 1967, Dollfus 1968, Boyer & Guérin 1969). Figure 14 shows one such sequence. Guinot & Feissel (1968) claim to have measured this motion spectroscopically. Ingersoll (1970b) has shown how the stratospheric oblateness associated with this motion could be detected with radio occultation data.

Schubert & Whitehead (1969) were the first to suggest that the 4 day circulation is the atmosphere's rectified response to the periodic thermal forcing of the Sun. This model was further elaborated by Schubert & Young (1970), Thompson (1970), and Malkus (1970). The only model which considers radiative heat transfer in a realistic way is Gierasch's (1970) model, which we will describe here. Gierasch assumes that heat is deposited as sun-

VENUS ON 21-22 MAY 1967

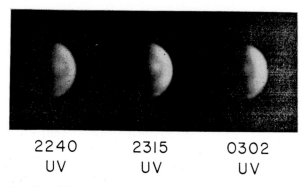

2240 2315 0302
UV UV UV

FIGURE 14. Ultraviolet photographs of Venus on 21–22 May 1967. Note the right to left (retrograde) drift of the dark circular marking during the 4.4 hr interval. Courtesy of the New Mexico State University Observatory.

light at some level within the cloud, where the atmosphere is presumed to rotate with the solid planet. Thus, the stratosphere above the cloud is warmed by radiation from below which varies periodically at the overhead frequency of the Sun. The radiative adjustment time of the stratosphere decreases as height increases, however, so the rotation of the stratosphere must increase with height in order that there be no phase shift in the thermal wave. If the phase of the wave were to vary with height, thermally driven motions would act to restore constant phase. Radiative adjustment times for CO_2 are known (Goody & Belton 1967), and with these Gierasch is able to compute the speed of the stratosphere as a function of height. He finds that the period is 4 days at the $P = 40$ mb level.

The Venus atmosphere may also play an important part in the dynamics of the planet's rotation. Gold & Soter (1969) showed that the effect of the thermal tide on Venus, if it were as large as the thermal tide on the Earth, might cancel the effect of the solid-body tide which tends to bring the planet into a prograde synchronous rotation. Ingersoll & Hinch (1971) have attempted to compute the atmospheric tide directly, and they find that the resulting torque is of the right order of magnitude to balance the solid-body torque. For their model to be valid, the lapse rate must be very close to the adiabatic lapse rate, but this is not an unwarranted assumption, according to recent occultation data (Figure 6).

Mars lower-atmosphere dynamics.—Similarities in rotation rates, axial tilts, and the small optical thicknesses for solar radiation for the Earth and Mars suggest that similar kinds of circulation are to be expected on the two planets (Leovy 1969). A major difference arises in the response of the surface temperature to seasonal changes. On the Earth, the oceans tend to maintain the average surface temperature in any latitude belt close to its annual average, and the atmosphere tends to distribute heat evenly with respect to latitude. On Mars, the thermal inertia of the ground and the heat capacity of the atmosphere are both small, and the surface temperature tends to be close to its local radiative equilibrium value at each latitude. Other differences include: the shorter radiative adjustment time for temperature perturbations (Goody & Belton 1967), absence of energetically significant latent heat effects due to water vapor on Mars, and the existence of very large-amplitude, large-scale Mars topography (Pettingill et al 1969, Belton & Hunten 1969b).

Atmospheric wind systems are driven by the uneven distribution, in time and space, of the solar heating. The dynamical behavior of large-scale wind systems depends on whether the thermal driving mechanism has a period longer or shorter than the planet's rotation period. On the Earth, the mean latitudinal difference in temperature drives the mean zonal (east-west) winds at midlatitudes, as well as the large cyclones and anticyclones which convect heat poleward. Such systems are expected on Mars, but because latitudinal temperature differences are larger than on Earth, larger velocities are expected. On the Earth, large-scale wind systems associated with diurnal temperature differences are relatively weak, because the diurnal thermal response of the atmosphere is small. On Mars, where diurnal temperature changes are larger, the associated diurnal wind systems are also expected to be larger.

Computer simulations of large-scale Mars wind systems have been carried out by Leovy & Mintz (1969). They used finite-difference analogues of the fluid mechanical equations to derive time-dependent temperatures and wind profiles on a coarse grid (15° longitude by 7° latitude) at two levels, at approximately 3 and 13 km heights. Some time-averaged temperature values obtained from these simulations are compared with temperatures inferred from the Mariner S-band occultation data, and with temperatures computed by Gierasch & Goody in Figure 7. The agreement is reasonably good, considering the many uncertainties in the theories and in the interpretation of the occultation data. Figure 15 shows the zonally averaged zonal and meridional winds computed by Leovy & Mintz. Their model also predicts large-scale cyclonic disturbances in all seasons except summer, and significant diurnal and semidiurnal wind oscillations. The diurnal heating, which is responsible for these "thermal tides," and the nature of the resulting oscillation have also been discussed by Gierasch & Goody (1968), Leovy (1969), and Lindzen (1970). The analogous problem for the Earth has been described in a recent monograph by Chapman & Lindzen (1970).

A recent systematic study of bright objects suspected of being clouds in the Lowell Observatory collection of Mars plates (5000 plates dating back to 1907) by Martin & Baum (1969) revealed only a few moving features, and predominantly low speeds (<5 m/s for most of these). Since high winds are expected, we infer that most of the features seen by Martin and Baum either were not clouds, or if clouds, were fixed to local surface features in the same way that wave clouds on the Earth remain stationary in flow over mountains.

The topography of Mars must exert a strong influence on the circulation. A partial theory of motions driven by large-scale topography, as observed over the Great Plains, has been given by Holton (1967). Gierasch & Sagan (1971) have pointed out that such topography can substantially intensify thermally driven winds on Mars.

Upper-atmosphere circulation and mixing.—Comparatively little work has been done on possible circulations of the upper atmospheres of Mars and Venus, even though, as we have seen, the validity of inferences drawn from static models depends partly on the degree to which circulation maintains local conditions in the thermosphere close to global average conditions. Dickinson (1969) has presented analytical and numerical solutions for flow in the thermosphere of a nonrotating planet. His solutions may be applicable to Venus. The results show that horizontal winds ~100–200 m/s are to be expected and are capable of maintaining a fair degree of temperature equilization between day and night sides (Dickinson 1970). These wind speeds are

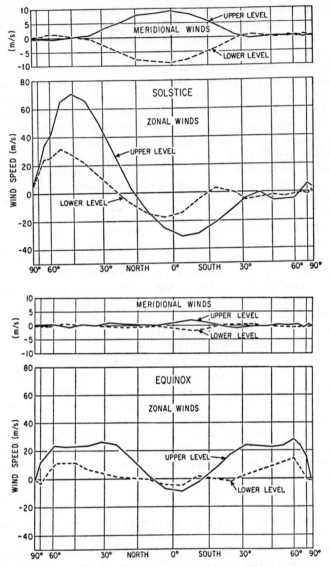

FIGURE 15. Zonally averaged zonal and meridional winds for Mars at two levels: approximately 3 and 13 km. Zonal winds are reckoned positive from the west, meridional winds positive from the south. Reproduced by permission of the American Meteorological Society from Leovy & Mintz (1969).

comparable to those required by McElroy & Strobel (1969) to account for Venus' nightside ionosphere.

We have seen that the intensity of vertical mixing by turbulent eddies is a key to the understanding of several aeronomic problems: Why does CO_2 remain undissociated in the upper atmospheres? What is the D/H ratio on Venus? What is the N_2 concentration on Mars? Lindzen (1970) speculates that the vertically propagating diurnal tide provides a major mechanism for generating turbulence and mixing in the Earth's upper atmosphere at latitudes between $\pm 30°$. He extends this reasoning to Mars to suggest that the region of turbulent mixing would extend 14 ± 1 scale heights above the surface (about 100 km). Lindzen also estimates that if the vertical energy transported upward by the terrestrial semidiurnal tide is dissipated in the thermosphere, it would maintain the thermosphere at 600°K without any other heating mechanism. Evidently possible dissipative heating of the Mars thermosphere by tides must be seriously considered. Other types of large-scale tropospheric disturbances may also propagate energy upward (Matsuno 1970), but our understanding of their role in upper-atmosphere phenomena is still rudimentary.

EVOLUTION OF THE ATMOSPHERES

Studying the atmospheres of the terrestrial planets provides information which is essential in our understanding of the Earth's atmosphere and its history. Clearly the composition of a planetary atmosphere is determined by several factors, including the mass and composition of the solid planet, the amount of incident sunlight, and conditions at the time of planetary formation. One of the most intriguing questions is whether the Earth is unique, or whether it is simply a member of the series of terrestrial planets, somehow filling the gap between Venus and Mars. We would like to know, for instance: to what extent do the outgassing histories of the terrestrial planets differ and to what extent have their atmospheres been modified after outgassing?

Reactions between atmosphere and crust.—We mentioned in the introduction that most of the Earth's CO_2 which has been released from the interior is now buried in sedimentary rocks (Table 1). The concentration of CO_2 in the oceans, and therefore the CO_2 partial pressure in the atmosphere, is set by reactions of the form

$$CaSiO_3 + CO_2 \rightleftharpoons CaCO_3 + SiO_2 \qquad 7.$$

Such reactions are at equilibrium in shallow seas where carbonate sediments are being deposited (Rubey 1951).

Mueller (1964) pointed out that reactions between CO_2 and common silicates might also set the CO_2 partial pressure on Venus. Lewis (1968) attempted to locate the Venus surface by computing the intersection of the observed pressure-temperature curve of the atmosphere and the equilibrium pressure-temperature curve of the relevant chemical reactions. Difficulties arise because there are several different reactions, involving different silicate compounds, and any of them might determine the equilibrium point. In any case, it is possible that there are extensive carbonate deposits on Venus, and that the amount of CO_2 released by outgassing is greater than the amount observed in the atmosphere today.

Lewis (1969, 1970) has also considered the chemistry of other volatile elements on Venus; his choice of elements is based mainly on terrestrial and solar abundances. He argues that because of the high surface temperatures, a large number of compounds may be present in the atmosphere in detectable quantities. Besides H_2O, CO, HCl, and HF, which have already been detected, he adds COS, H_2S, NH_3, HBr, Hg, $HgBr_2$, and many others. He has proposed several of these as likely constituents of the Venus clouds. As we have seen, there is some evidence for a multilayered cloud structure (Fjeldbo, Kliore & Eshleman 1971, Rasool 1970), but no further evidence that Lewis' model is correct.

The other important volatile on Venus is water. Mueller (1964) feels that the water vapor abundance is set by a balance between the rate of outgassing and the rate of photodissociation and escape. This is also the point of view taken by Walker, Turekian & Hunten (1970). Lewis (1968) claims that water may be in equilibrium with surface rocks, which raises the possibility that vast amounts of water are locked up chemically at the surface. Others (Fricker & Reynolds 1968, Anders 1968) have proposed that the amount of water outgassed from the interior is a small quantity, essentially equal to the present-day atmospheric abundance. The other possibility, that large amounts of water (equivalent to the Earth's oceans) have been lost on Venus, will be discussed later.

On Mars, both H_2O and CO_2 occur in solid and vapor form, so that it is impossible to estimate the amounts outgassed from atmospheric abundances alone. However, the upper limit to the amount of N_2 is about 0.3 mb (Dalgarno & McElroy 1970), which implies either that the amount of outgassing on Mars is 10^{-3} that of the Earth, or that nitrogen is bound chemically at the surface of Mars. The upper limit $N_2/CO_2 < 5\%$ in the Mars atmosphere is roughly equal to the ratio N_2/CO_2 for the Earth's volatiles (Table 1). However, if the amount of solid CO_2 in the Mars polar caps is much greater than the amount in the atmosphere, then the ratio of N_2 to CO_2 outgassed on Mars might be less than that outgassed on the Earth. The upper limit to the ratio of N_2 to CO_2 on Venus is also about 5%, and similar remarks apply. A positive detection of N_2 on either Mars or Venus would be an important aid to understanding the outgassing histories of the terrestrial planets.

Escape of atmospheric gases.—A classic description of escape by thermal acceleration of molecules at the top of an atmosphere was given by Spitzer (1952). The escape flux depends on the ratio of the planetary escape velocity to the mean thermal velocity of molecules in the exosphere. If the temperature of the neutral atmosphere and the density of each component are known at the base of the exosphere, the escape flux can be computed. Where direct measurements are not available, the escape flux may be computed from theoretical models of the upper atmosphere.

For the lightest gas, hydrogen, the probability of escape is so great that the actual escape flux usually depends on the rate of supply from levels below the exosphere. On the Earth, hydrogen is produced mainly by photodissociation of water vapor near the mesopause (Berkner & Marshall 1965, Brinkmann 1969). An important question is whether photodissociation of water and escape of hydrogen could have accounted for the initial accumulation of oxygen in the Earth's atmosphere. It is clear, from the oxidation state of gases in equilibrium with basaltic melts, that oxygen could not have been released directly into the atmosphere by outgassing (Holland 1964). Berkner & Marshall (1965) claim that the amount of oxygen in the Earth's atmosphere began to rise rapidly as a result of photosynthetic activity by green plants. But Brinkmann (1969) has shown that the rate of photodissociation of water and escape of hydrogen may have been sufficient to account for the early rise of oxygen in the absence of biological activity. The question has important implications for theories of the origin of life.

Walker, Turekian & Hunten (1970) deduce an upper limit to the escape of hydrogen from Venus, based on Mariner 5 Ly α observations (Figure 11). They assume that the escape flux is controlled by the rate of photodissociation of water in the Venus atmosphere; from the inferred upper limit to the escape flux they compute the maximum rate of supply of water from the interior. The rate of supply is 5 orders of magnitude less than the rate needed to supply an amount of water equivalent to the Earth's oceans during 4.5×10^9 years. Then, assuming that the outgassing histories of the Earth and Venus are the same, they conclude that the rate of supply was much greater in the past than it is at present on both planets.

A major difficulty with the assumption that the outgassing histories of the Earth and Venus are the same is that Venus probably has no more than 10^{-3} the amount of water that the Earth has. (Here we are assuming that large amounts of water are *not* locked up in crustal rocks on Venus.) Many authors (Hoyle 1955, Sagan 1960, Gold 1964, Shimazu & Urabe 1968, Ingersoll 1969, Rasool & de Bergh 1970) have discussed the possibility that this difference is due to the different escape histories of Venus and the Earth. The hypothesis to be tested is that water was once abundant on both planets and that the rate of photodissociation of water on Venus was then 3 or 4 orders of magnitude greater than it is on the Earth at present. If an ocean of water were added to Venus today, it is likely that water vapor would be a major atmospheric constituent. It can also be shown that if the present Earth were moved to the orbit of Venus, water vapor might soon become the major atmospheric constituent. Several authors (Sagan 1960, Gold 1964, Donahue 1968) argued that this would not significantly increase the rate of photodissociation of water, because water would still be a trace constituent at the high altitudes where photodissociation takes place. Later, Sagan

(1968) and Jastrow & Rasool (1969) argued that the fraction of water vapor in the stratosphere would have been large enough to permit rapid photodissociation, provided the stratospheric temperature were sufficiently large. However, Ingersoll (1969) showed that, regardless of the stratospheric temperature, water vapor would be a major constituent at all atmospheric levels if it were a major constituent at the surface. In short, it appears possible that the rate of photodissociation of water on Venus was once 3 or 4 orders of magnitude greater than it is on the Earth today.

If Venus has lost its oceans in this way, the amount of hydrogen which has escaped is large. Since deuterium escapes more slowly than ordinary hydrogen, the ratio D/H would be larger on Venus than it is on the Earth. As we have seen, there is some evidence, based on the Ly α emission near Venus, that this is the case (Donahue 1969, McElroy & Hunten 1969, Wallace 1969). The absence of free oxygen is then a mystery. If the primary gases from the interior of Venus were H_2O, CH_4, and CO, in relative abundances such that the elemental ratio O/C were less than 2, then only CO_2, CO, and/or C would remain. However, for the Earth's volatiles, the elemental ratio O/C is greater than 2. Another alternative is that ferrous iron in the crust of Venus has been oxidized to a depth of $\sim$10 km, sufficient to accommodate the liberated oxygen. No definitive test of these hypotheses has been proposed.

McElroy & Hunten (1970) have estimated the rate of photodissociation of water on Mars. They claim that at the present rate several meters of precipitable water might have been lost in 4.5×10^9 years. This is equivalent to a partial pressure at the surface of about 100 mb, or 20 times the mass of CO_2 observed in the atmosphere.

The escape of CO_2 and other gases from Mercury was analyzed by Rasool, Gross & McGovern (1966). Since their review, there has been a considerable increase in understanding of upper atmospheres in which CO_2 is a major constituent. Perhaps the subject should be reexamined. If the terrestrial planets are indeed closely related, the absence of an atmosphere on Mercury may prove to be the most puzzling observation of all.

ACKNOWLEDGMENT

We should like to thank Drs. Peter Goldreich, Duane O. Muhleman, and Carl Sagan for offering useful comments and suggestions. The participation of A. P. Ingersoll was supported in part under NASA grant NGL 05-002-003, and the participation of C. B. Leovy was supported in part under NASA grant 48-002-073.

LITERATURE CITED

Abhyankar, K. D. 1968. *Icarus* 9:507

Anders, E. 1968. *Accounts Chem. Res.* 1: 289

Arking, A., Potter, J. 1968. *J. Atm. Sci.* 25:617

Ash, M. E. et al 1968. *J. Atm. Sci.* 25:560

Avduevsky, V. S., Marov, M. Ya., Noykina, A. I., Polezhaev, V. I., Zavelevich, F. S. 1970. *J. Atm. S ci.* 27:569

Avduevsky, V. S., Marov, M. Ya., Rozhdestvensky, M. K. 1970. *J. Atm. Sci.* 27:561

Avduevsky, V. S., Marov, M. Ya., Rozhdestvensky, M. K., Borodin, N. F., Kerzhanovich, V. V. 1971. To be published

Barth, C. A. 1968. *J. Atm. Sci.* 25:564

Barth, C. A. 1969. *Appl. Opt.* 8:1259

Barth, C. A., Hord, C. W. 1971. To be published

Barth, C. A., Pearce, J. B., Kelley, K. K., Wallace, L., Fastie, G. 1967. *Science* 158:1675

Barth, C. A., Wallace, L., Pearce, J. B. 1968. *J. Geophys. Res.* 73:2541

Barth, C. A. et al 1969. *Science* 165:1004

Belton, M. J. S. 1968. *J. Atm. Sci.* 25:596

Belton, M. J. S., Hunten, D. M. 1966. *Ap. J.* 146:307

Belton, M. J. S., Hunten, D. M. 1968. *Ap. J.* 153:963

Belton, M. J. S., Hunten, D. M. 1969a. *Ap. J.* 156:797

Belton, M. J. S., Hunten, D. M. 1969b. *Science* 166:225

Belton, M. J. S., Hunten, D. M., Goody, R. M. 1968. *The Atmospheres of Venus and Mars*, ed. J. C. Brandt, M. B. McElroy, 69. New York: Gordon & Breach

Belton, M. J. S., Hunten, D. M., McElroy, M. B. 1967. *Ap. J.* 150:1111

Bergstralh, J. T., Gray, L. D., Smith, H. J. 1967. *Ap. J.* 149:L137

Berkner, L. V., Marshall, L. C. 1965. *J. Atm. Sci.* 22:225

Bottema, M., Plummer, W., Strong, J. 1965. *Ap. J.* 140:1640

Bottema, M., Plummer, W., Strong, J., Zander, R. 1965. *J. Geophys. Res.* 70: 4401

Boyer, C., Guérin, P. 1969. *Icarus* 11:338

Brancazio, P. J., Cameron, A. G. W., Eds. 1964. *The Origin and Evolution of Atmospheres and Oceans*. New York: Wiley

Brandt, J. C., McElroy, M. B., Eds. 1968.

The Atmospheres of Venus and Mars. New York: Gordon & Breach

Brinkmann, R. T. 1969. *J. Geophys. Res.* 74:5355

Brown, H. 1952, *The Atmospheres of the Earth and Planets*, ed. G. P. Kuiper, 258. Chicago: Univ. Chicago Press. 2nd ed.

Capen, C. F. 1970. *Icarus* 12:118

Carleton, N. P., Sharma, A., Goody, R. M., Liller, W. Roesler, F. L. 1969. *Ap. J.* 155:323

Chamberlain, J. W. 1965. *Ap. J.* 141:1184

Chapman, S., Lindzen, R. S. 1970. *Atmospheric Tides*. Dordrecht, Holland: Reidel

Clark, I. D., Noxon, J. F. 1970. *J. Geophys. Res.* 75:7307

Cloutier, P. A., McElroy, M. B., Michel, F. C. 1969. *J. Geophys. Res.* 74:6216

Clyne, M., Thrush, B. A. 1962. *Proc. Roy. Soc. London A* 269:404

Coffeen, D. L. 1969. *Astron. J.* 74:446

Connes, P., Connes, J., Benedict, W. S., Kaplan, L. D. 1967. *Ap. J.* 147:1230

Connes, P., Connes, J., Kaplan, L. D., Benedict, W. S. 1968. *Ap. J.* 152:731

Dalgarno, A., McElroy, M. B. 1970. *Science* 170:167

De More, W. B. 1970. *J. Geophys. Res.* 75:4898

de Vaucouleurs, G. 1968. *Icarus* 9:598

de Vaucouleurs, G., Menzel, D. H. 1960. *Nature* 188:28

de Wolf, D. A. 1970. *J. Geophys. Res.* 75:1202

Dickel, J. R. 1967. *Icarus* 6:417

Dickinson, R. 1969. *J. Atm. Sci.* 26:1199

Dickinson, R. 1970. *Eos* 51:766

Dollfus, A. 1961. *Planets and Satellites*, ed. G. P. Kuiper, B. M. Middlehurst, 534. Chicago: Univ. Chicago Press

Dollfus, A. 1963. *C. R. Acad. Sci.*, 256: 3250

Dollfus, A. 1968. *The Atmospheres of Venus and Mars*, ed. J. C. Brandt, M. E. McElroy, 133. New York: Gordon & Breach

Dollfus, A., Focas, J. H. 1969. *Astron. Ap.* 2:63

Donahue, T. M. 1968. *J. Atm. Sci.* 25:568

Donahue, T. M. 1969. *J. Geophys. Res.* 74:1128

Eshleman, V. R. 1970. *Radio Sci.* 5:325

Eshleman, V. R., et al 1968. *Science* 162: 661

Felder, W., Morrow, W., Young, R. A. 1970. *J. Geophys. Res.* 75:7311–15

Fjeldbo, G., Eshleman, V. R. 1968. *Planet. Space Sci.* 16:1035

Fjeldbo, G., Kliore, A., Eshleman, V. R. 1971. To be published, *Astron. J.*

Fjeldbo, G., Kliore, A., Seidel, B. 1970. *Radio Sci.* 5:381

Fricker, P. E., Reynolds, R. T. 1968. *Icarus* 9:221

Gale, W., Liwshitz, M., Sinclair, A. C. E. 1969. *Science* 164:1059

Gale, W., Sinclair, A. C. E. 1969. *Science* 165:1356

Gierasch, P. J. 1970. *Icarus* 13:25

Gierasch, P. J. 1971. To be published, *J. Atm. Sci.*

Gierasch, P. J., Goody, R. M. 1968. *Planet Space Sci.* 16:615

Gierasch, P. J., Goody, R. M. 1970. *J. Atm. Sci.* 27:224

Gierasch, P. J., Sagan, C. 1971. To be published, *Icarus*

Gillett, F. C., Low, F. J., Stein, W. A. 1968. *J. Atm. Sci.* 25:594

Giver, L. P., Inn, E. C. Y., Miller, J. H., Boese, R. W. 1968. *Ap. J.* 153:285

Gold, T. 1964. *The Origin and Evolution of Atmospheres and Oceans*, ed. P. J. Brancazio, A. G. W. Cameron, 249. New York: Wiley

Gold, T., Soter, S. 1969. *Icarus*: 11:356

Goldstein, R. M. et al 1970. *Radio Sci.* 5:475

Golytsin, G. H. 1970. *Icarus* 13:1

Goody, R. M. 1967. *Planet Space Sci.* 15:1817

Goody, R. M. 1969. *Ann. Rev. Astron. Ap.* 7:303

Goody, R. M., Belton, M. J. S. 1967. *Planet. Space Sci.* 15:247

Goody, R. M., Robinson, A. R. 1966. *Ap. J.* 146:339

Gross, S. H., McGovern, W. E., Rasool, S. I. 1968. *The Atmospheres of Venus and Mars*, ed. J. C. Brandt, M. B. McElroy, 103. New York: Gordon & Breach

Guinot, B., Feissel, M. 1968. *Publ. Obs. Haute-Provence* 9:No. 361

Hanel, R., Forman, M., Stambach, G., Meilleur, T. 1968. *J. Atm. Sci.* 25:586

Hansen, J. E., Arking, A. 1971. *Science* 171:669

Hansen, J. E., Cheyney, H. 1968. *J. Atm. Sci.* 25:629

Hansen, J. E., Matsushima, S. 1967. *Ap. J.* 150:1139

Henry, R. W., McElroy, M. B. 1968. *The Atmospheres of Venus and Mars*, ed. J. C. Brandt, M. B. McElroy, 251. New York: Gordon & Breach

Herr, K. C., Pimentel, G. C. 1969. *Science* 166:496

Herr, K. C., Pimentel, G. C. 1970. *Science* 167:47

Hess, S. L. 1968. *The Atmospheres of Venus and Mars*, ed. J. C. Brandt, M. B. McElroy, 109. New York: Gordon & Breach

Hogan, J. S., Stewart, R. W. 1969. *J. Atm. Sci.* 26:322

Holland, H. D. 1964. *The Orgin and Evolution of Atmospheres and Oceans*, ed. P. J. Brancazio, A. G. W. Cameron, 86. New York: Wiley

Holton, J. R. 1967. *Tellus* 19:199

Hoyle, F. 1955. *Frontiers in Astronomy*, 68. London: William Heinemann

Hunten, D. M., McElroy, M. B. 1968. *J. Geophys. Res.* 73:4446

Hunten, D. M., Goody, R. M. 1969. *Science* 165:1317

Ingersoll, A. P. 1969. *J. Atm. Sci.* 26:1191

Ingersoll, A. P. 1970a. *Science* 169:972

Ingersoll, A. P. 1970b. *Icarus* 13:34

Ingersoll, A. P. 1971. *Ap. J.* 163:121

Ingersoll, A. P., Hinch, E. J. 1971. To be published

Irvine, W. M. 1968. *J. Atm. Sci.* 25:610

Jastrow, R. 1968. *Science* 160:1403

Jastrow, R., Rasool, S. I. 1969. *The Venus Atmosphere*, 1. New York: Gordon & Breach

Kaplan, L. D., Connes, J., Connes, P. 1969. *Ap. J.* 157:L187

Kaplan, L. D., Munch, G., Spinrad, H. 1964. *Ap. J.* 139:1

Kieffer, H. 1970a. *J. Geophys. Res.* 75:501

Kieffer, H. 1970b. *J. Geophys. Res.* 75:510

Kliore, A., Cain, D. L. 1968. *J. Atm. Sci.* 25:549

Kliore, A., Levy, G. S., Cain, D. L., Fjeldbo, G., Rasool, S. 1967. *Science* 158:1683

Kliore, A., Cain, D., Fjeldbo, G., Rasool, S. I. 1969a. *Space Research* IX. Amsterdam: North-Holland

Kliore, A., Fjeldbo, G., Seidel, B. L., Rasool, S. I. 1969b. *Science* 166:1393

Knudsen, W. C., Anderson, A. D. 1969. *J. Geophys. Res.* 74:5629

Kurt, V. G., Dostolow, S. B., Sheffer, E. K. 1968. *J. Atm. Sci.* 25:668

Kuiper, G. P. 1969a. *Comm. Lunar Planet. Lab.* 100:209

Kuiper, G. P. 1969b. *Comm. Lunar Planet. Lab.* 101:229

Leighton, R. B., Murray, B. C. 1966. *Science* 153:136

Leighton, R. B. et al 1969. *Science* 166:49

Leovy, C. 1969. *Appl. Opt.* 8:1279

Leovy, C., Mintz, Y. 1969. *J. Atm. Sci.* 26:1167

Leovy, C. B., Smith, B. A., Young, A. T., Leighton, R. B. 1971. *J. Geophys. Res.* 76:297

Lewis, J. S. 1968. *Icarus* 8:434

Lewis, J. S. 1969. *Icarus* 11:367

Lewis, J. S. 1970. *Radio Sci.* 5:363

Lindzen, R. S. 1970. *J. Atm. Sci.* 27:537

Malkus, W. V. R. 1970. *J. Atm. Sci.* 27:529

Mariner Stanford Group 1967. *Science* 158:1678

Martin, W. L., Baum, W. A. 1969. *A Study of Cloud Motions on Mars.* Flagstaff, Ariz: Planet. Res. Cent., Lowell Obs.

Matsuno, T. 1970. *J. Atm. Sci.* 27:871

McConnell, J. C., McElroy, M. B. 1970. *J. Geophys. Res.* 75:7290

McCord, T. B. 1969. *Ap. J.* 156:79

McCord, T. B., Westphal, J. A. 1971. To be published, *Ap. J.*

McElroy, M. B. 1967. *Ap. J.* 150:1125

McElroy, M. B. 1969. *J. Geophys. Res.* 74:29

McElroy, M. B., Hunten, D. M. 1969. *J. Geophys. Res.* 74:1720

McElroy, M. B., Hunten, D. M. 1970. *J. Geophys. Res.* 75:1188

McElroy, M. B., Strobel, D. F. 1969. *J. Geophys. Res.* 74:1118

Melbourne, W. G., Muhleman, D. O., O'Handley, D. A. 1968. *Science* 160:987

Moos, H. W., Fastie, W. G., Bottema, M. 1969. *Ap. J.* 155:887

Moroz, V. I. 1965. *Sov. Astron.-AJ* 8:566

Moroz, V. I. 1968. *Sov. Astron.-AJ* 11:653

Morozhenko, A. V. 1970. *Sov. Astron.-AJ* 13:852

Morrison, D. 1970. To be published

Morrison, D., Sagan, C. 1967. *Ap. J.* 150:1105

Mueller, R. F. 1964. *Icarus* 3:285

Muhleman, D. O. 1969. *Astron. J.* 74:57

Murdock, T. L., Ney, E. P. 1970. *Science* 170:535

Murray, B. C., Wildey, R. L., Westphal, J. A. 1963. *J. Geophys. Res.* 68:4813

Neugebauer, G. et al 1969. *Science* 166:98

Ohring, G. 1969. *Icarus* 11:171

O'Leary, B. T. 1967. *Ap. J.* 149:L147

O'Leary, B. T., Rea, D. G. 1967. *Ap. J.* 148:249

Pettengill, G. H., Councilman, C. C., Rainville, L. P., Shapiro, I. I. 1969. *Astron. J.* 74:461

Plass, G. N., Stull, V. R. 1963. *J. Geophys. Res.* 68:1355

Plummer, W. T. 1970. *Icarus* 12:233

Pollack, J. B. 1969. *Icarus* 10:314

Pollack, J. B., Morrison, D. 1970. *Icarus* 12:376

Pollack, J. B., Sagan, C. 1968. *J. Geophys. Res.* 73:5943

Pollack, J. B., Sagan, C. 1969. *Space Sci. Rev.* 9:243

Pollack, J. B., Sagan, C. 1970. *Radio Sci.* 5:443

Pollack, J. B., Pitman, D., Khare, B. N., Sagan, C. 1970. *J. Geophys. Res.* 75:7480

Rasool, S. I. 1970. *Radio Sci.* 5:367

Rasool, S. I., de Bergh, C. 1970. *Nature* 226:1037

Rasool, S. I., Gross, S. H., McGovern, W. E. 1966. *Space Sci. Rev.* 5:565

Rasool, S. I., Hogan, J. S., Stewart, R. W., Russell, L. H. 1970. *J. Atm. Sci.* 27:841

Rea, D. G., O'Leary, B. T. 1968. *J. Geophys. Res.* 73:665

Ronov, A. B., Yaroshevsky, A. A. 1967. *Geochim. Int.* 4:1041

Rubey, W. W. 1951. *Bull. Geol. Soc. Am.* 62:1111

Sagan, C. 1960. *The Radiation Balance of Venus. Tech. Rep. 32–34.* JPL, Calif. Inst. Technol.

Sagan, C. 1966. *Ap. J.* 144:1218

Sagan, C. 1968. *International Dictionary of Geophysics,* ed. S. K. Runcorn, 2049. London: Pergamon

Sagan, C., Kellogg, W. W. 1963. *Ann. Rev. Astron. Ap.* 1:235

Sagan, C., Owen, T., Smith, H. 1971. *Proc. IAU Symp. 40 Planetary Atmospheres.* Dordrecht, Holland: Reidel

Samuelson, R. E. 1967. *Ap. J.* 147:782

Samuelson, R. E. 1968. *J. Atm. Sci.* 25:634

Schilling, G. F., Moore, R. C. 1967. *The Twilight Atmosphere of Venus. Memo. RM-5386-PR.* Santa Monica, Calif: RAND Corp.

Schorn, R. A., Barker, E. S., Gray, L. D., Moore, R. C. 1969. *Icarus* 10:98

Schorn, R. A., Farmer, C. B., Little, S. J. 1969. *Icarus* 11:283

Schubert, G., Whitehead, J. 1969. *Science* 163:71

Schubert, G., Young, R. E. 1970. *J. Atm. Sci.* 27:523

Shimazu, Y., Urabe, T. 1968. *Icarus* 9:498

Shimizu, M. 1969. *Icarus* 10:11

Sinton, W. M. 1963. *The Physics of Planets,* 300. Belgium: Univ. Liège

Slade, M. A., Shapiro, I. I. 1970. *J. Geophys. Res.* 75:3301

Slanger, T. G., Black, G. 1970. *Eos* 51:766

Slipher, E. C. 1962. *The Photographic Story of Mars,* Chap. 5. Flagstaff, Ariz: Northland Press

Smith, B. A. 1967. *Science* 158:114

Spinrad, H. 1966. *Ap. J.* 145:953

Spinrad, H., Field, G. B., Hodge, P. W. 1965. *Ap. J.* 141:1155

Spinrad, H., Shawl, S. J. 1966. *Ap. J.* 146:328

Spitzer, L. 1952. *The Atmospheres of the Earth and Planets,* ed. G. P. Kuiper, 211. Chicago: Univ. Chicago Press. 2nd ed.

Stewart, R. W. 1968. *J. Atm. Sci.* 25:578

Stewart, R. W., Hogan, J. S. 1969a. *J. Atm. Sci.* 26:330

Stewart, R. W., Hogan, J. S. 1969b. *Science* 165:386

Stone, P. H. 1968. *J. Atm. Sci.* 25:644
Thompson, R. 1970. *J. Atm. Sci.* 27:1107
Vinogradov, A. P., Surkov, U. A., Florensky, C. P. 1968. *J. Atm. Sci.* 25:535
Walker, J. C. G., Turekian, K. K., Hunten, D. M. 1970. *J. Geophys. Res.* 75: 3558
Wallace, L. 1969. *J. Geophys. Res.* 74:115
Weller, C. S., Biondi, M. A. 1967. *Phys. Rev. Lett.* 19:59
Westphal, J. A. 1966. *J. Geophys. Res.* 71:2693
Whitten, R. C. 1970. *J. Geophys. Res.* 75: 3707
Wood, A. T. Jr., Wattson, R. B., Pollack, J. B. 1968. *Science* 162:114
Young, L. D. G. 1969. *Icarus* 11:386
Young, L. D. G. 1970. To be published, *Icarus*
Young, L. D. G., Schorn, R. A., Smith, H. 1970. *Icarus* 13:74

EVOLUTIONARY PROCESSES IN CLOSE BINARY SYSTEMS

B. Paczyński

Institute of Astronomy, Polish Academy of Sciences
Warsaw, Al. Ujazdowskie 4, Poland

Introduction

The structure and evolution of the stars constitute one of the main problems of modern astrophysics. Binary stars are so common that it is not possible to develop a comprehensive theory of stellar structure and evolution without taking into account the duplicity of so many objects. According to Jaschek & Gomez (1970) 50% of the main-sequence stars are binaries. Van Albada & Blaauw (1967) estimated that 60% of the early-type stars are double with mass ratio larger than 0.2. Some types of objects are found in binary systems only: metallic-line stars (Abt 1961, 1965, Abt & Bidelman 1969), U Germinorum variables and novae (Kraft 1962, 1963, 1964), a large fraction of Wolf-Rayet stars (Underhill 1968). Hoyle (1964) and McCrea (1964) suggested that "blue stragglers" in open and globular clusters are the products of mass exchange in close binaries. This idea found some observational support in the case of open clusters (Cannon 1968, Deutch 1969, Strom & Strom 1970). Blaauw (1961) suggested that the early-type "runaway" stars had escaped from close binaries disrupted as a result of a supernova explosion of another component. Binaries are considered to be possible candidates for X-ray sources (Shklovsky 1967, Cameron & Mock 1967, Prendergast & Burbidge 1968). The general evolution of Population I objects is strongly affected by the commonness of binaries within this population. It should be emphasized that duplicity is not found among the Population II stars. It is also remarkable that none of the 50 known pulsars is in a binary.

The most fundamental physical parameter of a star, its mass, may be reliably measured for components of binary systems only. Eclipsing variables allow us to study in some detail the surface of their components. Distribution of mass in a stellar envelope affects the orbital motion and leads to apsidal motion. This phenomenon makes it possible to check the structure of stellar models. It is generally accepted now that mass transfer in a binary from one component to another is the most basic phenomenon in the evolution of such a system. Therefore matter with a chemical composition modified by nuclear reactions may be expected to be lifted up to the stellar surface. The two outstanding examples are the binaries HD 30353 and ν Sgr (Wallerstein 1968), deficient in hydrogen and overabundant in nitrogen. The matter that is visi-

183

ble now at the surface must have been processed through the CNO bi-cycle in the past.

The aim of this review is to present current ideas about the evolution of close binaries. Model computations with mass exchange are emphasized. The list of references is probably complete in this field and it is hoped that no important result has been neglected; it is certainly not complete in regard to such related topics as the stability of the components, changes of the orbital parameters, the physics of gaseous streams, and observational examples of various phenomena. An attempt has been made to find and list the recent surveys and the proceedings of recent conferences that dealt with those subjects.

HISTORICAL OUTLINE OF THE PROBLEM

A close binary system is a pair of stars that are close enough to each other for mass transfer between them to occur in the course of their evolution. Kuiper (1941) was the first to recognize the importance of the existence of Roche lobes for the evolution of close binaries. Parenago & Massevitch (1951) discovered that the subgiant components of Algol-type systems obey not the usual mass-luminosity relation, but rather a different mass-luminosity-radius relation. The luminosities and radii of those subgiants were significantly larger than those of the main-sequence stars of the same masses. Struve (1954) found that the luminosity excesses of subgiants were correlated with the mass ratios of the two components. The subgiants appeared as stars of advanced evolution, while their more massive companions were still on the main sequence. This appeared to be a paradox: it had been well established that the more massive stars should evolve faster. Crawford (1955) suggested the following solution of this so-called *Algol paradox:* the subgiants had originally been more massive and had lost a large fraction of their mass in the course of evolution. Kopal (1955b) divided close binaries into three groups: detached systems with both components smaller than their Roche lobes, semidetached systems with one star filling up its Roche lobe (this star is called a "contact" component), and contact binaries with the two stars filling their Roche lobes and touching each other. Many subgiants are the contact components of semidetached systems, but some are considerably smaller than their Roche lobes. These are called "undersize" subgiants (Kopal 1956). The early development of the theory was presented by Kopal (1959) in his comprehensive monograph.

The first model computations were done by Morton (1960). He found that a large fraction of the mass of the originally heavier component was transferred to the second star on a Kelvin-Helmholtz time scale. This process takes place as soon as the star fills up its Roche lobe. It is very difficult to observe because of the short time scale involved. In this phase of evolution the mass ratio of the two components is reversed. This result supported the idea due to Crawford. Two years later Smak (1962) suggested that subgiants with very large luminosity excesses were in the helium-burning phase.

Approximately 5 years ago it became clear that the numerical techniques developed to study the evolution of single stars might be used to compute evolution of binaries. Such studies began almost simultaneously in 1966 in Göttingen, in Warsaw, and in Ondřejov. The first results were presented at the IAU Colloquium "On the Evolution of Double Stars" in Uccle, in 1966 (Dommanget 1967). The development of the theory up to 1967 was thoroughly reviewed by Plavec (1968a), who also described the basic observational facts. Less thorough reviews and discussions were published by Plavec (1967a, 1968b), Piotrowski (1969), Weigert (1969), and Paczyński (1967a, e, 1970c). Techniques used to calculate the rate of mass loss from contact components were described in the last article. Details of numerical techniques were given by Kříž (1968), Harmanec (1970a), and Ziółkowski (1970a). Many contributions on binaries were presented at the Joint Discussion on Close Binaries during the thirteenth General Assembly in 1967 in Prague (Perek, 1968), at the Trieste Colloquium on "Mass Loss from Stars" in 1968 (Hack 1969), and at the IAU Colloquium No. 6 on "Mass Loss and Evolution in Close Binaries" held in 1969 in Elsinore (Gyldenkerne & West 1970).

Problems of angular momentum transfer in close binaries were reviewed recently by Huang (1966). Kruszewski (1966) wrote a comprehensive article on the changes of orbital elements, mass exchange, and mass loss in binaries. Similar topics were described by Hadjidemetriou (1967). Period changes were calculated for 333 eclipsing systems by Wood & Forbes (1963). Observational data were critically discussed by Plavec (1967b) and Ziółkowski (1971). The problem of mass determination of eclipsing variables was reviewed by Popper (1967). The catalogues of Koch et al (1963, 1970) and Batten (1967) are very useful as sources of information about the eclipsing and spectroscopic binaries. Novae and U Geminorum type stars as members of close binaries were reviewed by Kraft (1963) and Mumford (1967). Batten (1970) presented the available information about circumstellar matter in close binaries.

BASIC CONCEPTS AND ASSUMPTIONS

A double star is considered to be a close binary system if the evolution of the components is affected by the transfer of matter between them. Let us consider a pair of stars moving around each other in a circular orbit. Let the originally more massive star be always referred to as the primary, even if it becomes less massive than the secondary as a consequence of mass transfer. The masses of these stars are M_1 and M_2, respectively, and the separation between their centers is A. The gravitational potential of each component is nearly identical with that of a mass point. From Kepler's law we have

$$\left(\frac{2\pi}{P}\right)^2 A^3 = G(M_1 + M_2) \qquad\qquad 1.$$

where P is the orbital period. The orbital angular momentum J is given as

$$J^3 = \frac{G^2(M_1M_2)^3}{2\pi(M_1 + M_2)} P \qquad\qquad 2.$$

or

$$J^2 = \frac{G(M_1M_2)^2}{M_1 + M_2} A \qquad\qquad 3.$$

where G is Newton's constant.

Given a binary with a circular orbit, we may place the center of the co-ordinate system at the center of mass of the pair of stars and rotate this system with the binary. In such a frame of reference equipotential surfaces close to the center of either component are almost spherical. The larger the equipotential surface, the more distorted it is. Finally, for a certain value of the potential the surfaces surrounding the two components touch at a point between the two stars. This is called the inner Lagrangian point L_1, and the equipotential is called the critical Roche surface. It is composed of the two Roche lobes surrounding the two components and touching at the point L_1. For larger values of the potential the two stars are enveloped by a common, highly distorted equipotential surface.

It is convenient to measure the size of a Roche lobe by means of its average radius r which is defined so that $4/3\pi r^3$ is equal to the volume within the Roche lobe. This radius was tabulated by Kopal (1959, p. 136) as a function of the mass ratio M_1/M_2, and may be approximated to within 2% of its value with the simple formulae

$$\frac{r_1}{A} = 0.38 + 0.2 \log \frac{M_1}{M_2} \text{ for } 0.3 < \frac{M_1}{M_2} < 20$$

$$\frac{r_1}{A} = \frac{2}{3^{4/3}} \left(\frac{M_1}{M_1 + M_2}\right)^{1/3} \qquad\qquad 4.$$

$$= 0.46224 \left(\frac{M_1}{M_1 + M_2}\right)^{1/3} \quad \text{for } 0 < \frac{M_1}{M_2} < 0.8$$

The radius r_1 calculated by the first formula is larger than that obtained from the second relation for $M_1/M_2 > 0.523$. Taking the larger of the two values, we may have accurate results for the whole range of interesting mass ratios.

The most fundamental theoretical concept for the studies of the evolution of close binaries is the assumption that there exists a certain critical surface with an average radius R_{cr}. It is postulated that matter flows out from a star which has a radius R larger than R_{cr}. The precise value of R_{cr} is not so important for the theory as is the very existence of such a radius. All the evolutionary computations published so far were based on the assumption that the critical surface is identical with the Roche lobe. This assumption is well justified when the angular velocities of stellar rotation and orbital motion are identical, and the star has a well-defined surface. In such a case the rate of mass outflow depends very strongly on $\Delta R = R - R_{cr}$. Jędrzejec (1969) found that the rate of mass outflow dM_1/dt for a polytropic star is given as

$$\frac{dM_1}{dt} \sim \left(\frac{\Delta R_1}{R_1}\right)^{n+1.5} \qquad \qquad 5.$$

where n is the polytropic index. $\Delta R_1/R_1$ was smaller than 0.03 during the most rapid phases of mass exchange described in the literature. Unfortunately, the estimate of the constant of proportionality in the relation 5 was very rough.

All the model computations have been done under the assumption that ΔR_1 is equal to zero during mass exchange. It seems that the error caused in R_1 by this assumption should be less than 3% for a star with well-defined surface, and rotation synchronized with the orbital motion.

It is not so easy to define the critical radius if the stellar rotation is not synchronous with the orbital revolution. Plavec (1958) and Kruszewski (1963) discussed and calculated the shape and size of a critical surface for the case of nonsynchronous rotation. Their conclusion was that the critical radius was not changed very much, provided the departure from synchronism was not very large. According to Huang (1966) the observed rotation of contact components is only a little slower than synchronous. It is therefore reasonable to assume that the effects studied by Plavec and Kruszewski do not greatly affect the evolution.

Another difficulty arises when a star has a very extended atmosphere. It is not possible to define a stellar radius meaningfully under such circumstances. In particular the photospheric radius is not relevant for the problem of mass exchange in a binary with a supergiant component, as a nonnegligible fraction of the stellar mass may be found in the optically thin atmosphere of the red supergiant (Paczyński 1969, Uus 1970). Obviously in such a case mass exchange may occur while the supergiant's photosphere is much below the critical Roche surface. No theoretical models of this kind have been published.

Let us suppose that the initially more massive component of a binary system, the primary, fills up the Roche lobe in the course of its evolution, and mass transfer is started. Provided its velocity is small compared to the orbital velocities of the two stars, matter streaming through the vicinity of the inner Lagrangian point will not be able to escape from the binary directly. It will fall onto the secondary or it will form a gaseous disk rotating around the secondary. The flow velocities near the inner Lagrangian point were estimated by Jędrzejec (1969) and Kříž (1970), who found them to be small. Under such circumstances the total mass of a binary M remains constant, but the mass ratio M_1/M_2, and possibly the orbital angular momentum J, change. Almost all the model computations were based on the assumption that the total mass of the system and the orbital angular momentum are conserved during the mass exchange. If this assumption is accepted, the separation between the centers of the two components is given as

$$A = \frac{MJ^2}{G(M_1M_2)^2} = \frac{\text{const}}{M_1{}^2(M - M_1)^2} \qquad 6.$$

For a given binary the separation A is least when the mass ratio M_1/M_2 is unity. The radius of the primary's Roche lobe has a minimum for a M_1/M_2 = 0.8.

One may ask how justified is the assumption that the total mass and the orbital angular momentum are conserved. There is observational evidence for mass loss from binaries (Kruszewski 1966, Huang 1966, Batten 1970), but unfortunately no quantitative estimate, either observational or theoretical, is available. It is natural, therefore, to adopt the simplest possible assumption, and this has been done by conserving M and J during the evolution. One attempt has been made to estimate the effect of arbitrary mass and angular momentum loss (Paczyński & Ziółkowski 1967) on the evolution of a binary. There was no basic difference from the "conservative" case, but the quantitative results were, of course, somewhat changed. Nonconservation of the orbital angular momentum will be discussed again in the section on secondary components.

Let us summarize the main assumptions usually made in the studies on the evolution of close binaries.

1. A component of a close binary may be treated as a spherically symmetric star, even if it fills up the Roche lobe. This is rather well justified insofar as the mass-losing component is concerned. The effects of distortion due to tidal interactions and moderate rotation were studied by Dziembowski (1963), Kippenhahn & Thomas (1970), Jackson (1970), and Benson (1970), and were found to be small. The assumption is much worse for the component-accreting mass as the matter falling down from the inner Lagrangian point carries much angular momentum (Huang 1966, Kruszewski 1967). This problem will be discussed in the section devoted to secondary components.

2. There is a certain critical radius R_{cr}, such that mass exchange takes place when the stellar radius R exceeds R_{cr}. The critical radius is identified with the Roche lobe radius r, given by Equation 4. It should be emphasized that the relevant radius of the star is the one within which almost all the stellar mass is contained. In the case of main-sequence stars or giants this is practically the photospheric radius R_{ph}. The relevant radius of a red supergiant or a star with a strong prominence activity may be much larger than R_{ph}. The binaries such as ϵ and ζ Aurigae and VV Cephei (Batten 1970) are possible examples of the situation where mass transfer takes place and R_{ph} is smaller than the Roche lobe radius r.

3. The orbit of a binary is circular. This is a very important assumption as there is no equivalent of a Roche lobe in the case of an elliptic orbit, and it might be very difficult to calculate a critical radius. There is theoretical indi-

cation that mass exchange diminishes the initial ellipticity, provided it was not too large to begin with (Piotrowski 1965). No semidetached systems are known to have definitely noncircular orbits. The same is true for almost all binaries believed to be in post-mass-exchange evolution. However, many binaries with the two components on the main sequence have elliptic orbits.

4. The radius of a star is always smaller than, or equal to, the Roche lobe radius. In other words the mass outflow is assumed to keep the star within the critical surface. This seems to be well justified and reasonable, although it cannot be satisfied if the primary component has a deep convective envelope when mass loss commences (Paczyński 1965, Paczyński et al 1969). Another kind of impossible situation arises when the two Roche lobes are filled up by the two components which display a tendency towards expanding even more.

5. The star is assumed to be in a hydrostatic equilibrium. This assumption is violated in a small region close to the stellar surface and may be improper for primaries with deep convective envelopes, for secondaries while the rate of mass accretion is very high, and for contact systems with components that tend to expand beyond their Roche lobes.

6. The total mass of the binary M, and the angular orbital momentum J are conserved during evolution. This is a very controversial assumption and something should be done to improve the situation. Note that any mass loss from the system as a whole must be accompanied by a loss of angular momentum. However, angular momentum may be lost as a result of gravitational radiation without affecting the total mass of the binary (in the nonrelativistic limit). This process may be important for W Ursa Majoris type stars and other short-period systems (Paczyński 1967d).

EVOLUTION OF PRIMARY COMPONENTS

A few dozen papers have been published since 1966 on the model computations for close binaries with mass exchange. The assumptions listed in the previous section were adopted in almost all those investigations. The general results are presented in this section; many details in the original papers are omitted.

Let us consider the evolution of a single star first. Let it be a 5 $M_\odot$ object. The variation of the stellar radius as a function of time is shown in Figure 1 (following Paczyński 1970b). If this star was a component of a binary, mass exchange could occur provided the orbital period was not too long, i.e. the separation between the two stars was not too large. The Kepler law may be written as

$$\log P \text{ (days)} = 1.5 \log \left(\frac{A}{R_\odot} \right) - 0.5 \log \left(\frac{M_1 + M_2}{M_\odot} \right) - 0.936 \qquad 7.$$

for a binary with a circular orbit. Let us take $M_1 = 5\ M_\odot$ and $M_1/M_2 = 2$. The Roche lobe will be filled by the primary for $R_1/A = 0.44$ (cf Equation 4). The period of such a binary is given by the relation

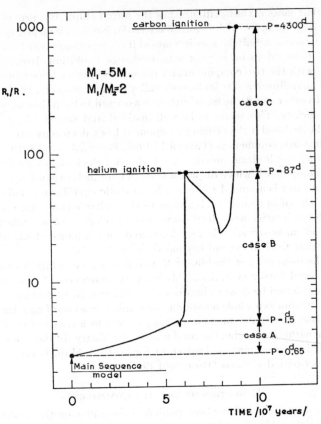

FIGURE 1. The time variation of the radius of a 5 $M_\odot$ star. The ranges of orbital periods corresponding to the evolution with mass exchange in cases A, B, and C are indicated. A mass ratio of $M_1/M_2 = 2$ is adopted.

$$\log P \text{ (days)} = 1.5 \log \frac{R_1}{R_\odot} - 0.84 \qquad\qquad 8.$$

There are three known evolutionary phases when the star expands and may fill up the Roche lobe: the first is associated with hydrogen burning in the core, while the second and third are caused by the rapid core contraction preceding helium ignition and carbon ignition. Let us follow Kippenhahn & Weigert (1967) and Lauterborn (1969) in using the terminology case A, case B, and case C, respectively, for the type of evolution associated with the on-set of mass exchange during the three stages of the primary's expansion (cf Figure 1). In our binary, case A may be obtained for the initial orbital periods in the range of 0.65–1.5 days, case B for periods of 1.5–87 days, and case C for periods of 87–4300 days, or even longer if the extension of the red supergiant's atmosphere is taken into account.

The range of periods corresponding to cases A and B for various stellar masses was discussed by Plavec (1968a). Let us estimate the longest period that corresponds to the case C. Let us assume that the primary evolves without any mass loss until it fills up the Roche lobe. The largest radius is attained by a star in the red supergiant phase. The theoretical radii of such models are very uncertain as they are sensitive to the mixing length in the convection theory, but it is reasonable to estimate the highest possible luminosities on the basis of model computations. To get the largest radii we may take the lowest effective temperatures from observations. An effective temperature of 2900°K, one-half the effective temperature of the Sun, may be assumed for simplicity.

Estimates of the highest luminosities may be taken from the publications of Stothers & Simon (1970) for the most massive, stable Population I star (100 $M_\odot$), from Barburo et al (1969) for a star of 30 $M_\odot$, from Kippenhahn (1969) for 25 $M_\odot$, and from Paczyński (1970a) for $M \leq 15 M_\odot$. The low-mass stars never ignite carbon, and their maximum luminosity is achieved in the double-shell-source-burning phase and is equal to $L_{max}/L_\odot = 5.9 \times 10^4$ ($M/M_\odot - 0.52$) for the mass range of 0.6–1.4 $M_\odot$. Stars in the mass range of 1.4–8 $M_\odot$ ignite carbon explosively, and when this happens maximum luminosity of $5.2 \times 10^4 L_\odot$ is reached, irrespective of the total mass of the star. Stellar luminosities at the onset of carbon burning are available for models of 10 and 15 $M_\odot$, and have to be extrapolated for 25, 30, and 100 $M_\odot$ models from the earlier evolution of those stars. No reliable models have been published for $M > 15 M_\odot$ at the onset of carbon burning. It is not clear whether those stars become red supergiants at all. Therefore the estimates of the longest periods of the most massive close binaries are very uncertain.

Perhaps the most convincing argument for mass transfer in binaries of very long period is the existence of VV Cephei with a period of 20 years. According to Peery (1966) one component of this system is an M-type star with a mass of 80 $M_\odot$ and a radius of 1600 $R_\odot$. Its extended atmosphere fills up the Roche lobe. The B-type secondary of 40 $M_\odot$ is surrounded by a shell of radius 90 $R_\odot$. That shell is probably formed by gas falling from the M-type component. Unfortunately, the masses of the two components are not reliable, but the length of the orbital period and the evidence of the mass transfer cannot be questioned.

The estimates of the highest luminosities and the largest radii of stars are given in Table 1 as function of stellar mass. With the assumption of unity mass ratio M_1/M_2 and a circular orbit it is possible to calculate the longest possible period of a close binary. This period, P_m, and the corresponding sum of the orbital velocities of the two components K_m are given in Table 1. The longest period P and the corresponding orbital velocities of the two stars K_1 and K_2 may be calculated for an arbitrary mass ratio as

TABLE 1. Extreme parameters for close binaries

$M_1/M\odot$	$\log L_{1,\,max}/L\odot$	$\log R_{1,\,max}/R\odot$	P_m (years)	K_m (km/sec)
0.6	3.67	2.44	5.7	18
0.8	4.22	2.71	13	15
1.0	4.45	2.82	16	15
1.2	4.60	2.90	20	15
1.4	4.72	2.96	22	15
8	4.72	2.96	9.6	35
10	4.42	2.81	5.0	47
15	4.83	3.02	8.5	46
25	5.3	3.25:	15:	45:
30	5.6	3.4:	22:	42:
100	6.4:	3.8:	49:	48:

$$P = P_m \times p$$
$$K_1 = K_m \times k_1 \qquad\qquad 9.$$
$$K_2 = K_m \times k_2$$

where p, k_1, and k_2 are given in Table 2 as a function of M_1/M_2.

Model computations for the case A of evolution were performed by Morton (1960), Paczyński (1966:8+5.3, 8+3.6, 1967b:8+5.3, 16+10.7, 1967c: 16+10.7), Paczyński & Ziółkowski (1967:16+10.7), Kippenhahn & Wiegert (1967: 9+5), Snezhko (1967), Plavec (1968b: 5+3, 1968c), Plavec et al (1968: 7+5, 9+8), Plavec et al (1969: 5+many), Plavec & Horn (1969: 5+many) Ziółkowski (1969a,b, 1970b: 4, 2,1+many), and Benson (1970: 5+2.5). The numbers following the year of publication and colon give the initial masses of the two components of a binary in units of solar mass. Horn

TABLE 2. Coefficients p, k_1, and k_2 as a function of the mass ratio

M_1/M_2	p	k_1	k_2
6.67	0.760	0.119	0.791
5.00	0.796	0.152	0.758
3.33	0.851	0.210	0.700
2.50	0.891	0.263	0.657
1.67	0.944	0.356	0.594
1.25	0.975	0.436	0.544
1.00	1.000	0.500	0.500
0.80	1.019	0.572	0.458
0.60	1.042	0.681	0.409

et al (1970) published a large number of static models of the primary component in the evolutionary stage following rapid mass exchange.

Case B of evolution was studied by Kippenhahn & Wiegert (1967: 9+3.1), Kippenhahn et al (1967: 2+1), Paczyński (1967e: 16+10.7), Giannone & Giannuzzi (1968, 1970: 2+1.5), Kippenhahn (1969: 25+15), Barburo et al (1969: 30+10), Ziółkowski (1969a,b, 1970a,b: 4+2.7, 3+2), Křiž (1969a,b: 5+4), Harmanec (1970a,b: 4+3.2, 4+1.6, 1970c: 7+5.6), Refsdal & Weigert (1969a,b: 1.8+0.7, 1.4+1.1), Giannone et al (1970: 1.4+1.1.) A large number of static models burning helium in a core and hydrogen in a shell were constructed by Giannone (1967) and Giannone et al (1968). These models may be used to approximate the primaries of massive systems that have completed case B of mass exchange. The primaries of low-mass systems after the rapid phase of mass transfer are described well by static models with a degenerate, isothermal helium core and a hydrogen-burning shell source (Refsdal & Weigert 1970).

Evolution with mass exchange in case C was computed by Lauterborn (1969, 1970: 5+2) and by Barburo et al (1969: 30+10).

In addition to cases A and B, evolution with mass transfer in case AB was introduced by Ziółkowski. This corresponds to the situation when the primary component fills up its Roche lobe while in the core hydrogen-burning phase, and evolution with mass outflow proceeds up to the hydrogen shell burning stage. Computations of this kind were performed by Ziółkowski (1969a,b, 1970a,b: 4, 2, 1+many) and by Horn (1970: 5+4, 1971: 5+4, 5+3).

The main results of all these computations may be summarized as follows. Mass transfer usually proceeds in two steps. The first one is very rapid and takes place on a Kelvin-Helmholtz (i.e. thermal) time scale given as

$$t_{\text{K-H}} = \frac{G M_1^2}{R_1 L_1} = 3.1 \times 10^7 \text{ years} \times \frac{M^{*2}}{R^* L^*} \qquad 10.$$

where $M^* = M_1/M_\odot$, $R^* = R_1/R_\odot$, and $L^* = L_1/L_\odot$ are the mass, radius, and luminosity of the primary at the onset of mass outflow. The time scales obtained with formula 10 may be compared with those published for 26 binaries in cases A and B. The results are shown in Figure 2. It is clear that the simple estimate agrees with the results of model computations to within a factor of 3 in most cases. The maximum rates of mass exchange during the evolution of 24 model binaries were found in the literature and compared with the formula

$$\dot{M}^* \equiv \frac{M^*}{t_{\text{K-H}}} = 3.2 \times 10^{-8} \ M_\odot/\text{year} \ \frac{R^* L^*}{M^*} \qquad 11.$$

The comparison is shown in Figure 3, and again the agreement is suprisingly good.

It is not commonly recognized how many binaries are observed in a process of rapid mass transfer. β Lyrae is the best-known example; V367 Cygni

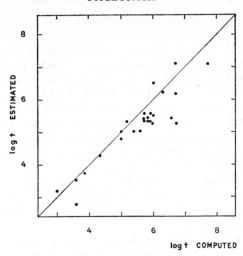

FIGURE 2. The relation between the time scale (in years) obtained from model computations for the rapid mass transfer in binaries, and the time scale estimated with formula 10.

is the next candidate (Plavec 1968a). The orbital period of SV Centauri has decreased by one thousandth of its value in the last 50 years. This implies a time scale of only 10^5 years for period changes. The diagram with the observed minus calculated $(O-C)$ times of eclipses versus the epoch E is very spectacular in the case of SV Centauri and is shown in Figure 4. Another star, W Serpentis, increases its period on a time scale of 10^5 years according to Kruszewski (1970; private communication). RX Cassiopeiae and SX Cassiopeiae are very likely in a phase of rapid evolution, too. The more massive component of V380 Cygni (14 $M_\odot$) is expanding rapidly and the mass transfer will begin soon, i.e. within the Kelvin-Helmholtz time scale (Semeniuk & Paczyński 1968). In general, all binaries with β Lyrae type light curves and periods longer than 10 days may be suspected of being in a rapid phase of evolution.

Case A evolution proceeds in a similar fashion in binaries of all masses. The primary is burning hydrogen in a core and it is in thermal equilibrium before filling up its Roche lobe. The mass ratio M_1/M_2 is more than reversed as a result of rapid mass exchange, while the star departs greatly from thermal equilibrium. Subsequent evolution proceeds on a slow, nuclear time scale and the binary is semidetached. The primary, now the less massive component, burns hydrogen in its core and fills up the Roche lobe as a subgiant. This phase of evolution is roughly as long as the main-sequence lifetime of the original primary. The subgiant is overluminous up to 3 mag. Overluminosity is defined as the magnitude difference between the star considered and the main-sequence object of the same mass. The mass ratios

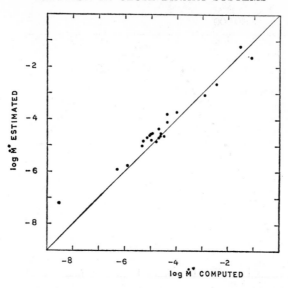

FIGURE 3. The relation between the highest rate of mass transfer (in solar masses per year) obtained from model computations of the rapid mass exchange in binaries, and the rate estimated with formula 11.

obtained in case A after the rapid mass exchange are never very large and the two stars do not differ in brightness by much. As a result of tidal distortion of the subgiant, a β Lyrae type light curve is likely to be observed in such a system. Many eclipsing variables of this kind are known, particularly among the early-type stars. It should be emphasized that almost all the evolutionary calculations carried for the case A led to the contact systems sooner or later. No attempts were made to study such systems.

Case B of evolution proceeds differently in massive binaries and in low-mass systems. The difference is due to the competition between helium ignition and the onset of electron degeneracy in the contracting helium core. The first, very rapid mass exchange develops in the same way in binaries of all masses. It takes place on the thermal time scale of the primary's radiative envelope, and may be estimated by means of formula 10. The mass may be exchanged on a dynamical time scale of $(G\,M_1/R_1{}^3)^{-1/2}$ if the envelope is convective when the Roche lobe is approached. The subsequent phase of a slower mass transfer proceeds on the thermal time scale of the contracting helium core in massive binaries, and on the nuclear time scale of the hydrogen-burning shell source in low-mass systems. If the mass of the original primary exceeded 3 $M_\odot$ or so (Kippenhahn & Weigert 1967), the mass exchange is terminated by the helium ignition. The core begins to expand, the envelope contracts, and the mass exchange is stopped. Stripped of almost all its hydrogen, the primary settles down close to the helium main sequence. It

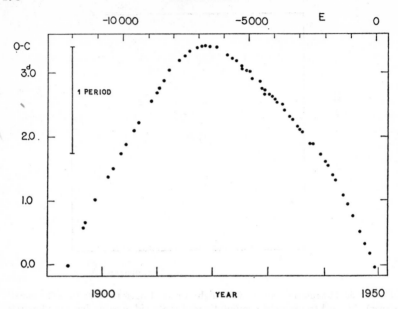

FIGURE 4. Observed minus calculated moments of eclipses versus time for the binary SV Centauri. The observed moments were taken from the compilation by Wood & Forbes (1963), while the formula Primary Minimum $= JD$ 24 33053.217 $+1.6604206 \times E$ was used to calculate those moments. This diagram implies a time scale of period changes of $P/\dot{P} = 10^6$ years, and indicates the possibility of rapid mass transfer in this binary.

may look like a Wolf-Rayet type star (Paczyński 1967a,e, 1968, Kippenhahn 1969). If the original mass of the primary was less than 3 $M_\odot$, the rapid contraction of the stellar core is stopped by the onset of electron degeneracy and the rapid mass transfer is terminated. A slow mass exchange follows with the primary burning hydrogen in a shell. The time scale involved is typically ~10% of the main-sequence lifetime of that star. Very high overluminosities are reached, up to 10 magnitudes. At the same time the mass ratio M_1/M_2 becomes as low as 1:5 or 1:10, and the secondary, being a main-sequence star, is much brighter than the "overluminous" primary. This kind of a binary will exhibit an Algol-type light curve if eclipses are observed. Theoretical masses, luminosities, and radii are in fairly reasonable agreement with the observed properties of the Algol-type systems. The most recent comparison has been published by Ziółkowski (1971). This phase of evolution is terminated when the hydrogen-rich envelope of the primary is exhausted. This envelope is burned at the bottom by the shell source, and flows out from the top towards the secondary component. When very little matter is left in the envelope the stellar radius decreases and a white dwarf is formed (Kippenhahn et al 1967, 1968, Weigert 1968, Kippenhahn & Weigert 1969).

The resulting systems may appear as single-line spectroscopic binaries with the secondaries, their only visible components, having very small amplitude of radial velocities.

Case C has been very little studied, and a large variety of situations is anticipated. It should be stressed again that little is known about the evolution of massive single stars ($M \geq 15\ M_\odot$) during the core helium-burning and more advanced evolutionary phases. This is so because of the difficulty entailed in dealing with semiconvection. The evolution of the most massive binaries in case C, and to some extent in case B, is strongly affected by this phenomenon. Perhaps further theoretical and observational studies of this kind of binaries may help to solve the problem of semiconvection. In systems of moderate mass, white dwarfs may be produced. Lauterborn (1970) suggested that ν Sagittarii and HD 30353 are two examples of systems that had undergone case C mass exchange. The primary components of these binaries are very luminous now, display an overabundance of helium and nitrogen, and are presumably evolving towards the white dwarf stage.

The evolution of close binaries in cases A and B is fairly well understood, at least in regard to the mass-losing components, the subgiants. The "Algol paradox" is solved: Crawford's original idea was correct. The phase of evolution with the more massive component filling its Roche lobe is short lived compared to the opposite situation. In the former the mass exchange proceeds on a thermal, or even on a dynamical time scale, while in the latter a nuclear time scale is involved. During the slow phase of mass exchange the luminosity of the subgiant increases while matter is being transferred to the secondary, and the mass ratio M_1/M_2 decreases. This explains the correlation between the luminosity excesses of the subgiant components and the observed mass ratios (Struve 1954). The mass-radius-luminosity relation found by Parenago & Massevitch (1951) for low-mass subgiants is also easy to interpret. Most likely the relevant semidetached systems are the products of case B mass exchange. The subgiants are burning hydrogen in a shell surrounding a degenerate helium core. Many models were constructed for such stars by Refsdal & Weigert (1970). It turns out that the luminosity L_1 of such a model depends on the core mass M_c, but not on M_1, the total mass of the subgiant. We have $L_1 = L_1(M_c)$. The stellar radius R_1 is a function of M_c and M_1. Eliminating M_c from these two relations, we obtain $f(M_1,\ L_1,\ R_1) = 0$. Quantitative agreement with observations is only fair (Ziółkowski 1971).

Two classes of low-mass binaries are similar to Algol systems but have not been explained until now: R Canis Majoris type stars and detached systems with undersize subgiants (Kopal 1956). The more massive components of R Canis Majoris type binaries look like the main-sequence stars on the H-R diagram, but their masses are too small for their spectral type or luminosity. The less massive components fill up their Roche lobes and have the appearance of normal subgiants. There is some doubt as to whether the anomalies of these systems have really been established, as the interpretation of observations is not too straightforward. Perhaps in some cases we may see the secondary component, now the more massive star, evolving off the main sequence. The binary RZ Cancri (Popper 1967, p. 102) is probably such an object but it is not classified as an R Canis Majoris star.

Undersize subgiants are less massive components that are significantly smaller than their Roche lobes. In many cases the interpretation of the observations may be questioned, but in some systems, such as Z Herculis and KO Aquilae, the existence of undersize subgiants is certain. Plavec (1958) suggested that in such binaries the size of the critical surface is reduced sharply by nonsynchronous rotation of the subgiant components. According to Smak (1962) these stars either are contracting towards the main sequence, or are in the post-semidetached phase of evolution. Pre-main-sequence contraction was also considered by Roxburgh (1966b), but Hall (1968) and Field (1969) argued that this explanation may be valid for a few binaries only. The most obvious difficulty is due to the insufficient concentration of those objects towards the galactic plane, and the relatively large number of such systems. Paczyński (1967a) and Refsdal & Weigert (1969b) suggested that undersize subgiants had already ignited helium. All available models however are too hot after helium ignition. Clearly, there is no generally accepted explanation of the nature of these stars. This problem has been discussed recently by Ziółkowski (1971). In the next section, the origin of undersize subgiants is considered in relation to angular momentum transfer in close binaries.

The Evolution of Secondary Components

Almost the entire effort of the theorists has hitherto gone into studying the structure of mass-losing components. Work on the mass-accreting components has been restricted to the case of spherically symmetric mass-fall onto a white dwarf. This problem will be considered in the section on cataclysmic variables. The first and, so far, only attempt to follow in detail the evolution of the secondary component from the main sequence and through the beginning of the mass-transfer stage has been the work of Benson (1970). He studied case A of evolution with the two stars having initial masses 5 and 2.5 $M_\odot$, and assumed the rotation of the two components to be always synchronized with the orbital motion. The Roche lobe was filled up by the primary when the central hydrogen content dropped from 0.75 to 0.36. The mass loss led to a decrease in the luminosity of the primary inasmuch as part of the radiative flux was used to heat up the expanding stellar envelope. The mass falling onto the secondary was heated up in the shockfront as well as during the subsequent contraction. The liberated heat prevented the matter from settling down quickly as part of the secondary component. This process increased the brightness and radius of this star to such an extent that it filled up the Roche lobe when 0.1 $M_\odot$ had been transferred from the primary. It is very probable that the majority of binaries following case A mass exchange form contact systems as the Benson model did.

Contact configuration may be avoided in a binary with a longer period, following case B mass exchange. Under such circumstances transfer of angular momentum is more likely to be involved. Some aspects of this problem have been reviewed by Huang (1966). The amount of angular momentum carried by matter flowing from the inner Lagrangian point towards the secondary was calculated by Kruszewski (1967) and tabulated as h_2 in his dimensionless units. It may be shown that

$$d \log J = h_2 \times d \log q \qquad 12.$$

where J is the orbital angular momentum, and q is the mass ratio M_1/M_2. (Note: Kruszewski used $q = M_2/M_1$.) The quantity h_2 is given in Table 3 as a function of q. In the course of mass exchange the mass ratio M_1/M_2 decreases, and as h_2 is positive, J must decrease. The angular momentum is stored in a gaseous disk spinning around the secondary.

Let us consider an extreme situation when all the matter and all the angular momentum transferred remain within the secondary and/or disk. The resulting changes of J with q are presented in Table 3. The variation of the separation between the two components is also given for the two extreme cases. First, A was calculated directly from Equation 6. Later, the variation of the orbital angular momentum was computed by taking Equation 12 into account, and the separation A was evaluated. In the second case, one half of the orbital angular momentum is transferred into the spin of the disk while the mass ratio is reversed from 5:1 to 1:5. The decrease in J by a factor of 2 reduces the separation A and the period P by factors of 4 and 8, respectively. A very rough estimate of the size of a rapidly spinning disk which contains two thirds of the total mass of the binary and one half of the total angular momentum indicates that it may be smaller than its Roche lobe, and that the formation of a contact system is avoided.

So little is known about angular momentum transfer that these considera-

TABLE 3. Variation of the orbital angular momentum (J) and size of the orbit (A) with the mass ratio

$q = M_1/M_2$	h_2	$\Delta \log q$	$\Delta \log J$	A/A_0 J variable	A/A_0 J constant
5.0	0.09			1.00	1.00
		−0.4	−0.05		
2.0	0.15			0.31	0.39
		−0.3	−0.05		
1.0	0.20			0.20	0.31
		−0.3	−0.07		
0.5	0.27			0.18	0.39
		−0.4	−0.13		
0.2	0.37			0.25	1.00

tions should not be taken seriously as far as numerical values are concerned. But there is indeed a large amount of angular momentum transferred with matter, and it may be stored, at least temporarily, within the disk. This may affect the orbital period and the evolution of a binary very strongly, since $P \sim J^3$. A decrease in J leads to a reduction of the distance between the two stars and may speed up the mass transfer. Suppose that during the phase of rapid mass exchange a considerable fraction of the total angular momentum is accumulated within the disk. This angular momentum may be transferred back into the orbital motion during the subsequent slow phase of evolution. As a result the separation between the two stars will increase, and the Roche lobes will expand. The primary may become detached from the critical surface and appear as an undersize subgiant.

There is some observational evidence in favor of the hypothesis outlined above. The existence of gaseous rings and/or disks is well established in semi-detached systems (Huang 1966, Batten 1970, Shakovskoy 1964, Ruciński 1966). In some systems the presence of opaque disks was postulated (Kopal 1955a, Huang 1963, 1965, Hall 1969, Smak 1971). In the case of β Lyrae, Woolf (1965) suggested that half of the mass of the secondary is contained within the disk supported by the rapid rotation. That was supposed to explain the low luminosity of that particular star. If the Woolf model is accepted, a large fraction of the total angular momentum of the binary is stored within the disk. β Lyrae is well known to be in a phase of rapid transfer of matter. The secondary (more massive) components of "normal" Algol-type systems appear as ordinary main-sequence stars, and do not show excessively large rotational velocities (Huang 1966). It would seem that the angular momentum collected by those stars during the rapid mass exchange is lost during the subsequent slow evolution. This may be due to the tidal interaction between the two components, and probably takes place on the thermal time scale (Zahn 1966a,b,c, Dziembowski 1967).

The matter flowing onto the secondary component of a close binary might have been affected by nuclear reactions within the primary. It has been suggested that Am and Ap stars show anomalous abundances because they have been exposed to supernova explosions of nearby companions (van den Heuvel 1968). Stothers & Simon (1969) advanced the hypothesis that mass exchange in a massive binary leads to the inversion of the mean molecular weight in the envelope of the secondary, with instability ensuing and the star appearing as a β Cephei type variable. These two suggestions are very controversial for a variety of reasons. But the abundance anomalies are more likely to be observed at the surface of the primary, which exposes the layers that have been lifted up from the deep stellar interior to the surface. The secondary is likely to be mixed throughout as a result of rapid rotation caused by the infalling matter. However, even without rotation, mixing will take place if there is an inversion of the mean molecular weight μ. A layer in a spherical star with a positive $d\mu/dr$ is secularly (i.e. thermally) unstable to nonradial, nonadiabatic perturbations (Kippenhahn et al 1970, Dziembowski 1970: private communication).

CONTACT SYSTEMS

In one class of binaries observed, the two components are believed to be in contact with each other. These are called W Ursa Majoris eclipsing variables. The mass ratio of the two components is close to $2:1$, on average. It is well known (Kuiper 1941) that the equilibrium configuration of the two contact components with radiative envelopes exists only for unit mass ratio. Lucy (1968) noticed that W UMa stars should have convective envelopes, as their spectral types are F, G, or K. He was able to find equilibrium configurations for model binaries with components of unequal masses. In addition, the luminosity ratio of the two stars was reasonably close to what is observed $(L_1:L_2 \approx M_1:M_2)$, inasmuch as a large fraction of energy produced in the core of the more massive star was transferred to the secondary component through the common convective envelope. The main difficulty of the Lucy theory lay in the very narrow mass range of his model binaries for which he could find the equilibrium solution, and the extreme Population I composition that was required. This was true because if his model was to be an equilibrium model with nonunity mass ratio, the primary had to burn hydrogen in the carbon-nitrogen cycle, while the proton-proton reaction had to be the main energy source in the secondary. Similar models have been discussed recently by Hazlehurst (1970) and Moss & Whelan (1970) who concluded that it is easier to get agreement with observations if the primary component is assumed to be somewhat evolved. Whelan (1970) considered the problem of contact systems in the pre-main-sequence contraction phase.

Altogether not very much is known about contact systems, in particular about those with very thick common envelopes and those with envelopes in radiative equilibrium. Such configurations are frequently encountered at the end of model computations following case A, and sometimes case B, of mass exchange. The formation of a contact binary in model computations is due to the decrease of separation between the two stars, when the large initial mass ratio changes to unity in a process of mass transfer. In some cases the secondary evolves off the main sequence and expands while the primary is still filling its Roche lobe (RZ Cnc). It is likely that the formation of a contact system under such circumstances does not necessarily prevent further expansion of the common envelope. So far no model computations are available for such evolutionary configurations, and nobody has suggested how to treat them.

CATACLYSMIC VARIABLES

Very little is known about the origin and nature of cataclysmic variables: novae and U Germinorum type stars. It is generally accepted now that most, if not all, of these objects are binaries with orbital periods of a few hours. In many systems a late-type spectrum is visible together with very wide, sometimes double, emission lines. The G, K, or M type spectrum is thought to belong to a star filling its Roche lobe and losing mass through the vicinity of the inner Lagrangian point. This matter falls toward the second star and forms a gaseous disk rotating around it. The prominent emission lines are due to that disk (Kraft 1962, 1963, 1964). Kraft (1965) suggested that W Ursa Majoris stars are the ancestors of cataclysmic variables.

Explosions of novae are traditionally associated with the "hot" component (Schatzman 1965), i.e. the one surrounded with a gaseous disk, and presumably accreting hydrogen-rich matter. Recent studies of mass-accreting white dwarfs were published by Giannone & Weigert (1967), Saslaw (1968), Rose (1968), Secco (1968), and Starrfield (1969, 1970), but nobody has so far succeeded in getting dynamical effects and mass ejection. All the models considered so far were spherically symmetric, while the matter is probably accreted from the rapidly spinning disk. There is no observational evidence to show which component is responsible for a nova explosion.

U Geminorum type binaries are very similar to novae at minimum light but the outbursts are entirely different. In particular, there is no spectroscopic evidence of mass loss. Also, not a single binary of this type is known to have white dwarf lines in its spectrum, or, as a matter of fact, any lines coming from a so-called "hot" component. The prominent emission lines come from a gaseous disk as in novae. Krzemiński (1965) demonstrated that the eclipses visible at minimum light of the binary U Geminorum itself disappear at the time of outbursts. He believed that a white dwarf was eclipsed at minimum light, and concluded that the "cool" star brightens and is responsible for the outbursts. As a result, theorists (Paczyński 1965, Paczyński et al 1969, Bath 1969, Osaki 1970) tried to associate U Gem type activity with the instabilities of mass outflow from the cool component. To their enbarrassment, Warner & Nather (1970) and Smak (1971) reinterpreted Krzemiński's observations and found that the selfconsistent model places the outbursts in the "hot" star, or more precisely in the central region of the spinning disk. Perhaps one of the most mysterious features in the U Gem type spectra, very shallow absorption lines visible at the beginning of outbursts, may be so wide because they originate in the rapidly rotating object. One of the most important problems to be understood theoretically about these peculiar stars is that of the structure and properties of the rapidly spinning gaseous disks.

The Origin and Fate of Close Binaries

The origin of close binaries is not known. From time to time fission of a rapidly spinning protostar is suggested as a mechanism by which close binaries come into existence (Stoeckly 1965, Roxburgh 1966a, and many others), but no models of this process have been actually calculated, and the hypothesis is still controversial. Another idea has been around for some years but has probably never been published. Suppose that a single, rapidly rotating protostar contracts towards the main sequence. Matter is left in the equatorial plane as the star shrinks. A considerable fraction of the stellar mass may be left in this way in the form of a rotating gaseous disk. It is not clear whether the disk will be dispersed into outer space or not, but the configuration is likely to be unstable. If the mass available is large enough, one may speculate that a second star could be formed. The formation of a planetary system under such circumstances has frequently been suggested, but very little, if any attention has been paid to the possibility of a close binary being formed in this way.

Disruption of a massive binary by a supernova explosion of the more massive component has been suggested by Blaauw (1961) to be the birth of an early-type runaway star. Such stars are observed to have velocities ~100 km/sec, supposedly a relic of their former orbital motion. The problem of runaway stars was recently considered by Gott et al (1970) and Trimble & Ress (1970) in connection with pulsars 0527 and 0531 near the Crab Nebula. Let us discuss the possible final phases in the evolution of a close binary and the possible origin of runaway stars.

It is a rule that case A evolution leads to a contact system. As nothing is known about the fate of such binaries, we have to exclude case A from our considerations. If the orbital period of a binary is shorter than that indicated in Table 1, mass transfer between the two components should take place. If the process of mass transfer leaves the primary with $M_1 < 1.4 M_\odot$, this star will evolve finally to a white dwarf. In due course the secondary, now the more massive star, will expand and rapid mass exchange will proceed in the reverse direction. Nothing is known about this kind of repeated mass exchange. It is not unlikely that a contact system will be formed after all.

The situation is entirely different if a more massive binary is considered, and after the first mass transfer the primary is left as a helium star of $M > 1.4$ $M_\odot$. The evolutionary lifetime of this star is equal to that of the primary's post-main-sequence evolution in the case of no mass loss. The primary is, therefore, likely to complete its nuclear evolution before the hydrogen is exhausted in the core of the secondary, even though the secondary is now the more massive of the two stars. This means that the primary will be the first to explode as a supernova. Symmetric explosion of the less massive component cannot disrupt a binary with a circular orbit. The situation will not be basically changed if the secondary explodes first, provided it has become a red supergiant and has transformed much mass back to the primary before the explosion. The only easy way for the more massive component to explode and to disrupt a binary is to do this in a system with a period longer than P_m as given in Table 1. In such case the orbital velocity of the secondary K_2 is given by Equation 9 and Table 2, and is under 40 km/sec.

To produce runaway stars with significantly larger velocities, some of our assumptions have to be modified. Perhaps not all the massive stars become red supergiants before supernova explosion. The orbital periods for binaries without mass exchange may then be reduced and the orbital velocities may be increased considerably. If the supernova explosion was not spherically symmetric, the explosion of the less massive star could disrupt the binary. Finally, if the explosion was spherically symmetric with respect to the center of the star, it would not be symmetric with respect to the center of gravity of the binary system. The binary as a whole may consequently be given a considerable momentum as a result of the disruption of the less massive component. Let us consider a relatively short-period binary, i.e. a close binary system with high orbital velocities. A spherically symmetric explosion of the less massive star will produce a rapidly moving binary with an eccentric orbit. If the bound remnant of the disrupted component has a small mass, the duplicity of the system may be difficult to detect. It may look like a single runaway star, but in fact have a hidden less massive companion, perhaps a neutron star or a collapsed object.

If one accepts the hypothesis that not all the binaries are disrupted by supernova explosions and that some supernovae produce collapsed objects (*black holes*), one may look for such objects among single-line spectroscopic binaries (Zel'dovich & Guseynov 1965). No such objects were found in a search by Trimble & Thorne (1969).

None of the 50 known pulsars appears to be a member of a binary. This may indicate either that binaries are disrupted at the time of pulsar formation, or that pulsars are not produced in close binaries. According to Gunn & Ostriker (1970), stars in the mass range of 4–10 $M_\odot$ become pulsars at death. Wheeler (1970) suggested that pulsars may be produced as a result of explosive carbon ignition in degenerate cores that are formed in stars with initial masses in the range of 3.5–8 $M_\odot$ (Paczyński 1970a). It is rather well established now that the primary components of massive binaries become helium stars as a result of mass exchange. Helium stars ignite carbon non-explosively (Paczyński 1971). If Wheeler's hypothesis was correct, pulsars could not be born in close binaries.

Gas Dynamics in Close Binaries

The evolution of the mass-losing component of a close binary has been sucessfully studied without considering the dynamical effects of gas outflow from the star. However, it proved impossible to follow in detail the evolution of more massive components with deep convective envelopes under the standard assumptions presented in this review (Paczyński et al 1969). This difficulty has not been explicitly mentioned in papers by Refsdal & Weigert (1969b) and Giannone et al (1970), but it has been acknowledged by Lauterborn (1970, p. 152). It seems that mass outflow takes place on a dynamical time scale (Paczyński 1965) and hydrodynamic treatment of this process is necessary. The first approaches to this problem, very limited in scope, have been attempted by Jędrzejec (1969), Osaki (1969), and Křiž (1970).

The evolution of the mass-accreting component is more explicitly affected by the gas flows. There have been some studies approximating gaseous rings with particle orbits, e.g. by Huang (1966), Kruszewski (1967), Piotrowski & Ziołkowski (1970). And perhaps one single attempt has been made to investigate the flow of matter through the disk with the gas pressure taken into account (Prendergast & Burbidge 1968). Turbulence was assumed to be the cause of exchange of angular momentum between the nearby rings within the gaseous disk. As a result the inner parts of the disk were forced to spiral onto the star, while the outer regions were driven away. In this picture the turbulence was far more important than the tidal interaction, and the characteristic time scale was very short.

The problem of matter falling onto the secondary and the formation of a shockfront was considered by Starrfield (1969) and by Benson (1970), who also discussed the properties of the gaseous stream itself. Prendergast (1960) and Biermann (1971) discussed the general properties of gas flows in binary systems. The number of papers published on this topic is fairly large but I was unable to find positive and reliable results. It is probably fair to say that this field has remained entirely unexplored.

It might be convenient to divide the gas dynamical problems in close binaries into a number of specific topics: slow, nonsynchronous rotation and tidal interactions within the main body of the two stars; outflow of mass from the contact component; flow of gas towards the secondary; collision of this stream with the star or with the disk; rapidly rotating disk of small or large optical thickness, and of small and large mass; turbulence within the disk; merging of the disk with the secondary component at the inner boundary and disruption at the outer boundary; optically thin corona and mass loss from the system; and contact binaries with two stellar cores embedded in a common, perhaps very thick and expanding, envelope. These problems are difficult to tackle, but further development of the theory of close binaries will not be possible unless at least some of them are solved. The transfer of angular momentum and the loss of matter and angular momentum from binary systems also need to be studied.

ACKNOWLEDGMENT

It is a great pleasure to express my sincere thanks to all of my colleagues who helped me so much with their comments in the task of preparing this review, and to all of those astronomers who were so kind as to supply me with preprints of their contributions. Most of all, I am grateful to Dr. Andrzej Kruszewski and Dr. Janusz Ziółkowski for their patience and many day-to-day discussions that helped me clarify many problems.

LITERATURE CITED

Abt, H. A. 1961. *Ap. J. Suppl.* 6:37
Abt, H. A. 1965. *Ap. J. Suppl.* 11:429
Abt, H. A., Bidelman, W. P. 1969. *Ap. J.* 158:1091
van Albada, T. S., Blaauw, A. 1967. *Comm. Obs. Roy. Belg., Uccle* B17:44
Barburo, G., Giannone, P., Giannuzzi, M. A., Summa, C. 1969. See Hack, M., p. 217
Bath, G. T, 1969. *Ap. J.* 158:571
Batten, A. H. 1967. *Publ. Dom. Ap. Obs. Victoria* 13:119
Batten, A. H. 1970. *PASP* 82:574
Benson, R. S. 1970. PhD thesis. Univ. California, Berkeley
Biermann, P. 1971. *Astron. Ap.* 10:205
Blaauw, A. 1961. *Bull. Astron. Inst. Neth.* 15:265
Cameron, A. G. W., Mock, M. 1967. *Nature* 215:464
Cannon, R. D. 1968. *Observatory* 88:206
Crawford, J. A. 1955. *Ap. J.* 121:71
Deutch, A. J. 1969. See Hack, M., p. 260
Dommanget, J., Ed. 1967. *Comm. Obs. Roy. Belg., Uccle* B17
Dziembowski, W. 1963. *Acta Astron.* 13:157
Dziembowski, W. 1967. *Comm. Obs. Roy. Belg., Uccle* B17:105
Field, J. V. 1969. *MNRAS* 144:419
Giannone, P. 1967. *Z. Ap.* 65:226
Giannone, P., Giannuzzi, M. A. 1968. *Oss. Astron. Roma. Contrib. Sci.* Ser. III, No. 68
Giannone, P., Giannuzzi, M. A. 1970. *Astron. Ap.* 6:309
Giannone, P., Kohl, K., Weigert, A. 1968. *Z. Ap.* 68:107
Giannone, P., Refsdal, S., Weigert, A. 1970. *Astron. Ap.* 4:428
Giannone, P., Weigert, A. 1967. *Z. Ap.* 67:41
Gott, J. R., Gunn, J. E., Ostriker, J. P. 1970. *Ap. J. Lett.* 160:L91
Gunn, J. E., Ostriker, J. P. 1970. *Ap. J.* 160:979
Gyldenkerne, K., West, R. M., Eds. 1970. *Proc. IAU Colloq. No. 6.* Copenhagen Univ. Publ. Fund
Hack, M., Ed. 1969. *Mass Loss from Stars.* Dordrecht-Holland: Reidel
Hadjidemetriou, J. 1967. *Advan. Astron. Ap.* 5:131
Hall, D. S. 1968. *PASP* 80:477
Hall, D. S. 1969. *Bull. AAS* 1:345
Harmanec, P. 1970a. *Bull. Astron. Inst. Czech.* 21:113

Harmanec, P. 1970b. *Ap. Space Sci.* 6:497
Harmanec, P. 1970c. *Bull. Astron. Inst. Czech.* 21:316
Hazlehurst, J. 1970. *MNRAS* 149:129
van den Heuvel, E. P. J. 1968. *Bull. Astron. Inst. Neth.* 19:326
Horn, J. 1970. *Ap. Space Sci.* 6:492
Horn, J. 1971. *Bull. Astron. Inst. Czech.* 22:37
Horn, J., Křiž, S., Plavec, M. 1970. *Bull. Astron. Inst. Czech.* 21:45
Hoyle, F. 1964. *Roy. Obs. Bull.* 82:90
Huang, S.-S. 1963. *Ap. J.* 138:342
Huang, S.-S. 1965. *Ap. J.* 141:976
Huang, S. -S. 1966. *Ann. Rev. Astron. Ap.* 4:35
Jackson, S. 1970. *Ap. J.* 160:685
Jaschek, C., Gómez, A. E. 1970. *PASP* 69:546
Jędrzejec, E. 1969. MS thesis. Warsaw Univ.
Kippenhahn, R. 1969. *Astron. Ap.* 3:83
Kippenhahn, R., Kohl, K., Weigert, A. 1967. *Z. Ap.* 66:58
Kippenhahn, R., Meyer-Hofmeister, E., Thomas, H. C. 1970. *Astron. Ap.* 5:155
Kippenhahn, R., Thomas, H. C. 1970. *Stellar Rotation,* ed. A. Slettebak, 20. Dordrecht-Holland:Reidel
Kippenhahn, R., Thomas, H. C., Weigert, A. 1968. *Z. Ap.* 69:265
Kippenhahn, R., Weigert, A. 1967. *Z. Ap.* 65:251
Kippenhahn, R., Weigert, A. 1969. *Low Luminosity Stars,* ed. S. S. Kumar, 373. New York: Gordon & Breach
Koch, R. H., Sobieski, S., Wood, F. B. 1963. *Publ. Univ. Pa. Astron. Ser.* 9
Koch, R. H., Plavec, M., Wood, F. B. 1970. *Publ. Univ. Pa. Astron. Ser.* 10
Kopal, Z. 1955a. *6th Liège Colloquium. Mem. 8° Soc. Roy. Sci. Liège, 4th Ser.* 15:241
Kopal, Z. 1955b. *Ann. Ap.* 18:379
Kopal, Z. 1956. *Ann. Ap.* 19:298
Kopal, Z. 1959. *Close Binary Systems.* New York: Wiley
Kraft, R. P. 1962. *Ap. J.* 135:408
Kraft, R. P. 1963. *Advan. Astron. Ap.* 2:43
Kraft, R. P. 1964. *Ap. J.* 139:457
Kraft, R. P. 1965. *Ap. J.* 142:1588
Křiž, S. 1968. *Bull. Astron. Inst. Czech.* 19:248
Křiž, S. 1969a. See Hack, M., p. 257
Křiž, S. 1969b. *Bull. Astron. Inst. Czech.* 20:127

Kříž, S. 1970. *Bull. Astron. Inst. Czech.* 21:211

Kruszewski, A. 1963. *Acta Astron.* 13:106

Kruszewski, A. 1966. *Advan. Astron. Ap.* 4:233

Kruszewski, A. 1967. *Acta Astron.* 17:297

Krzemiński, W. 1965. *Ap. J.* 142:1051

Kuiper, G. P. 1941. *Ap. J.* 93:133

Lauterborn, D. 1969. See Hack, M., p. 262

Lauterborn, D. 1970. *Astron. Ap.* 7:150

Lucy, L. B. 1968. *Ap. J.* 151:1123

McCrea, W. H. 1964. *MNRAS* 128:147

Morton, D. C. 1960. *Ap. J.* 132:146

Moss, D. L., Whelan, J. A. J. 1970. *MNRAS* 149:147

Mumford, G. S. 1967. *PASP* 79:283

Osaki, Y. 1969. 1970. *Ap. J.* 162:621

Paczyński, B. 1965. *Acta Astron.* 15:89

Paczyński, B. 1966. *Acta Astron.* 16:231

Paczyński, B. 1967a. *Comm. Obs. Roy. Belg., Uccle* B17:111

Paczyński, B. 1967b. *Acta Astron.* 17:1

Paczyński, B. 1967c. *Acta Astron.* 17:193

Paczyński, B. 1967d. *Acta Astron.* 17:287

Paczyński, B. 1967e. *Acta Astron.* 17:355

Paczyński, B. 1968. See Perek, L., p. 409

Paczyński, B. 1969. *Acta Astron.* 19:1

Paczyński, B. 1970a. *Acta Astron.* 20:47

Paczyński, B. 1970b. *Acta Astron.* 20:195

Paczyński, B. 1970c. *Proc. IAU Colloq. No. 6*, ed. K. Gyldenkerne, R. M. West, 139. Copenhagen Univ. Publ. Fund

Paczyński, B. 1971. *Acta Astron.* 21:1

Paczyński, B., Ziółkowski, J. 1967. *Acta Astron.* 17:7

Paczyński, B., Ziółkowski, J., Żytkow, A. 1969. See Hack, M., p. 237

Parenago, P. P., Massevitch, A. G. 1951. *T. Astron. Inst. Sternberg* 20:81

Peery, B. F. 1966. *Ap. J.* 144:672

Perek, L., Ed. 1968. *Highlights of Astronomy.* Dordrecht-Holland: Reidel

Piotrowski, S. 1965. *Bull. Acad. Pol. Sci. Ser. Math. Astron. Phys.* 12:419

Piotrowski, S. 1969. *Zvezdy, Tumannosti, Galaktiki* 109. Erevan, Armianskoi SSR: Izdat. Akad. Nauk

Piotrowski, S., Ziołkowski, K. 1970. *Ap. Space Sci.* 8:66

Plavec, M. 1958. *8th Liège Colloquium. Mem. 8° Soc. Roy. Sci. Liège, 4th Ser.* 20:411

Plavec, M. 1967a. *Bull. Astron. Inst. Czech.* 18:253

Plavec, M. 1967b. *Bull. Astron. Inst. Czech.* 18:334

Plavec, M. 1968a. *Advan. Astron. Ap.* 6:201

Plavec, M. 1968b. *Ap. Space Sci.* 1:239

Plavec, M. 1968c. See Perek, L., p. 396

Plavec, M., Horn, J. 1969. See Hack, M., p. 242

Plavec, M., Kříž, S., Harmanec, P., Horn, J. 1968. *Bull. Astron. Inst. Czech.* 19:24

Plavec, M., Kříž, S., Horn, J. 1969. *Bull. Astron. Inst. Czech.* 20:41

Popper, D. 1967. *Ann. Rev. Astron. Ap.* 5:85

Prendergast, K. H. 1960. *Ap. J.* 132:162

Prendergast, K. H., Burbidge, G. R. 1968. *Ap. J. Lett.* 151:L83

Refsdal, S., Weigert, A. 1969a. See Hack, M., p. 253

Refsdal, S., Weigert, A. 1969b. *Astron. Ap.* 1:167

Refsdal, S., Weigert, A. 1970. *Astron. Ap.* 6:426

Rose, W. K. 1968. *Ap. J.* 152:245

Roxburgh, I. 1966a. *Ap. J.* 143:111

Roxburgh, I. 1966b. *Astron. J.* 71:133

Ruciński, S. M. 1966. *Acta Astron.* 16:127

Saslaw, W. C. 1968. *MNRAS* 138:337

Schatzman, E. 1965. *Stars and Stellar Systems*, ed. L. H. Aller, D. B. McLaughlin 8:327. Chicago: Univ. Chicago Press

Secco, L. 1968. *Publ. Oss. Astron. Padova* No. 145

Semeniuk, I., Paczyński, B. 1968. *Acta Astron.* 18:33

Shakovsky, N. M. 1964. *Astron. Zh.* 41:1042

Shklovskii, I. S. 1967. *Ap. J. Lett.* 148:L1

Smak, J. 1962. *Acta Astron.* 12:28

Smak, J. 1971. *Acta Astron.* 21:15

Snezhko, L. I. 1967. *Perem. Zv.* 16:253

Starrfield, S. G. 1969. PhD thesis. Univ. California, Los Angeles

Starrfield, S. G. 1970. *Ap. J.* 161:361

Stoeckly, R. 1965. *Ap. J.* 142:208

Stothers, R., Simon, N. R. 1969. *Ap. J.* 157:673

Stothers, R., Simon, N. R. 1970. *Ap. J.* 160:1019

Strom, K. M., Strom, S. E. 1970. *Ap. J.* 162:523

Struve, O. 1954. *5th Liège Colloquium. Mèm. 8° Soc. Roy. Sci. Liège, 4th Ser.* 14:236

Trimble, V. L., Thorne, K. S. 1969. *Ap. J.* 156:1013

Trimble, V., Rees, M. J. 1970. Preprint

Underhill, A. B. 1968. *Ann. Rev. Astron. Ap.* 6:39

Uus, U. 1970. *Sci. Inf.* 17:1. Astron. Counc. Acad. Sci. USSR

Wallerstein, G. 1968. *Science* 162:625

Warner, B., Nather, R. E. 1970. Preprint

Weigert, A. 1968. See Perek, L., p. 414

Weigert, A. 1969. *Astron. Ges. Mitt.* 25:19

Wheeler, J. C. 1970. *Nature* 226:1043
Whelan, J. A. J. 1970. *MNRAS* 149:167
Wood, D. B., Forbes, J. E. 1963. *Astron. J.* 68:257
Woolf, N. J. 1965. *Ap. J.* 141:155
Zahn, J.-P. 1966a. *Ann. Ap.* 29:313
Zahn, J.-P. 1966b. *Ann. Ap.* 29:489
Zahn, J.-P. 1966c. *Ann. Ap.* 29:565

Zel'dovich, Ya. B., Guseynov, O. H. 1965. *Ap. J.* 144:841
Ziółkowski, J. 1969a. See Hack, M., p. 231
Ziółkowski, J. 1969b. *Ap. Space Sci.* 3:14
Ziółkowski, J. 1970a. *Acta Astron.* 20:59
Ziółkowski, J. 1970b. *Acta Astron.* 20:213
Ziółkowski, J. 1971. *Acta Astron.* 21: In press

ULTRAVIOLET STUDIES OF THE SOLAR ATMOSPHERE

Robert W. Noyes

Smithsonian Astrophysical Observatory and Harvard College Observatory, Cambridge, Massachusetts

INTRODUCTION

Extreme ultraviolet (EUV) solar observations from space have made great progress in recent years, largely from the development of new instrumentation with improved spatial or spectral resolution. These observational advances have allowed EUV solar astronomy to progress from a young and largely qualitative discipline to a mature and quantitative one. Different investigations now appear to be converging on a common description of the average physical conditions in the solar atmosphere. At the same time, high-resolution observations have begun to reveal the scarcely surprising existence of inhomogeneous structure, both in the active and in the quiet atmosphere. Solar EUV research in the next few years will undoubtedly focus on the physical nature of the inhomogeneous solar atmosphere, in order to determine for individual features the run of temperature, density, magnetic field, thermal conduction, mechanical flux, and other dynamic behavior. This in turn may be basic for understanding the physical nature of the atmosphere as a whole.

Although we will discuss below the evidence for inhomogeneous structure and some of its implications, the main emphasis of this review will be on the mean structure of the solar atmosphere, in both quiet and active regions. Of course, the mean solar atmosphere, like all idealizations, is in some sense fictitious. Nevertheless, it serves as a very useful zero-order framework for investigation of the problem of overall energy balance in the solar atmosphere, and, we hope, will be a valid starting point for the more sophisticated studies that surely will follow.

In this review, we rather arbitrarily define the EUV portion of the solar spectrum to extend from about 300 Å—the approximate limit of usefulness of normal-incidence optics—to about 3000 Å, where ground-based ultraviolet observations become possible. For a recent review of X-ray solar astronomy at wavelengths shorter than several hundred angstroms, see Neupert (1969).

Figure 1 shows schematically the run of temperature and density with height in the quiet and active solar atmosphere, based on some of the work that will be discussed in this review. The figure illustrates the importance of EUV observations, for emission in the region 300 to 3000 Å completely spans the temperature range from 4×10^3 to 3×10^{6}°K, and therefore it is in many

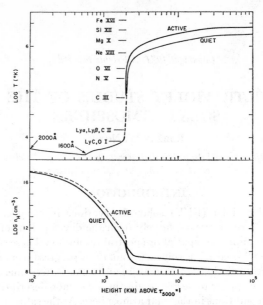

FIGURE 1. Schematic run of temperature and total hydrogen density with height in quiet and active regions. The structure of active regions below $10^5°$K is not well established. The location of the steep transition zone for both regions is arbitrarily put near 2000 km in the figure. Also shown are temperatures of emission for various EUV sources discussed in this review.

cases possible to probe the entire three-dimensional atmosphere by use of the same instrument.

We begin by reviewing the major observational advances in the 4 years since the last review of the subject on these pages (Goldberg 1967). In subsequent sections, we shall discuss the interpretation of these observations in terms of the physical structure of the quiet Sun and of active regions.

EUV SOLAR OBSERVATIONS

OBSERVATIONS EMPHASIZING SPATIAL RESOLUTION

The Harvard College Observatory spectroheliometer aboard OSO 4 (Goldberg et al 1968) has obtained two-dimensional images of the Sun in individual EUV emission lines in the range 280 to 1400 Å. During October and November 1967, the OSO 4 instrument recorded roughly 4000 spectroheliograms of the entire disk in about 50 individual emission lines or continuum wavelengths, with a spatial resolution of about 1 arc min (Figure 2). The instrument was rather versatile and could be commanded to obtain spectroheliograms in arbitrary sets of related lines in successive orbits. This versatility made the resulting data extremely valuable, in spite of the only mod-

OSO-IV SPECTROHELIOGRAMS-Nov 26,1967

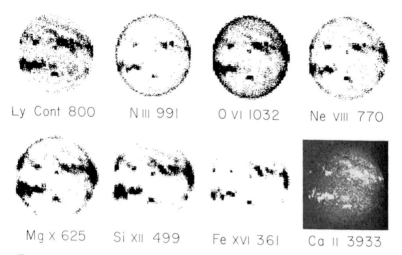

Ly Cont 800   N III 991   O VI 1032   Ne VIII 770

Mg X 625   Si XII 499   Fe XVI 361   Ca II 3933

FIGURE 2. Photoelectric spectroheliograms obtained by the Harvard spectrometer aboard OSO 4. Lines are in order of increasing temperature of formation (see Figure 1). Note the increase of the active/quiet intensity ratio with increasing ionization, the pronounced limb brightening in the transition zone (N III, O VI, Ne VIII), and the weakened coronal emission at the poles. See Reeves & Parkinson (1970) for a complete atlas of spectroheliograms. Courtesy of *Solar Physics*.

erate spatial resolution obtained. The averaged data from each orbit have been published in atlas form (Reeves & Parkinson 1970).

Recently, a similar spectrometer with somewhat better spatial resolution (35 arc sec) was orbited aboard the OSO 6 spacecraft. The spacecraft had the capability of obtaining spectroheliograms not only of the full disk but also of an area 7 arc min² centered on some feature of interest. The smaller area permitted rapid data acquisition, such that a spectroheliogram was recorded every 30 sec. The Harvard instrument operated successfully from August 13, 1969, through March 7, 1970, and obtained a vast body of data, both spectroheliograms and spectra (see below).

Tousey and his collaborators at the Naval Research Laboratory (NRL) have continued to increase the spatial resolution of their EUV slitless spectrograph. For a review of observations with this instrument to 1967, see Tousey (1967). In more recent work (Tousey 1971a, b, Purcell & Tousey 1969), the NRL group has achieved spatial resolution in the EUV of 3 to 5 arc sec (see Figure 9, p. 234). These observations reveal the extension of the chromospheric emission network into the transition zone; they also show important variations with height of the fine structure of active regions. We shall discuss these points in more detail later.

Blamont and his colleagues have carried out two different programs aimed at studying the spatial properties of the very high photosphere and low chromosphere. One program (Blamont & Carpentier 1967, 1968) took advantage of an opacity minimum in the Earth's atmosphere near 2000 Å to photograph the solar granulation at that wavelength from a balloon at 100,000 ft. Angular resolution of about 1 arc sec was achieved through a 125 Å bandpass, and the authors were able to measure the size and temperature excess of both the solar granulation ($\Delta T \sim 60°K$) and somewhat larger structures, which they call microfaculae, having a temperature excess of about 300°K. The other program (Bonnet & Blamont 1968, Bonnet 1968) used both balloon and rocket platforms to obtain broadband filtergrams at 2885, 2665, 2235, 2190, and 1980 Å. These data show great differences in both fine structure and limb darkening, especially on either side of the Al I opacity edge at 2080 Å (see Figure 5 below).

Fredga (1969), using a rocket-borne birefringent Šolc filter to isolate a bandpass of 2.1 Å centered on the Mg^+ K line at 2795 Å, recorded solar images with an angular resolution of about 5 arc sec. Even with the relatively wide bandpass filter, the contrast of active regions in Mg^+ K equals that in the core of Ca^+ K, suggesting that the true contrast in the core of Mg^+ K considerably exceeds that in Ca^+ K.

Sloan (1968) has flown in a rocket an ionization chamber sensitive mainly to H Ly α, behind a 1.6 arc sec pinhole at the focal plane of a small telescope. The image was allowed to wander over a small area of the quiet Sun in an uncontrolled fashion, but postflight reconstruction of the pointing permitted mapping the area with a resolution that in some places is as good as 2.5 arc sec. Bright features, 6 to 20 arc sec in size, are found with intensity typically 20% brighter than the background.

EUV observations were made during both the November 12, 1966, and the March 7, 1970, eclipses in an attempt to locate absolute heights of emission with precision. Blamont & Malique (1969) analyzed a 1966 eclipse measurement of the brightness of H Ly α at the extreme limb; they found that the Ly α limb lies about 2500 km above the visible limb and that the intensity shows 30% limb brightening that peaks 4600 km inside the limb.

Recently, Brueckner et al (1970) and Speer et al (1970) carried out separate observations of the EUV flash spectrum during the March 7, 1970, total solar eclipse. The former was only partially successful owing to a parachute failure that necessitated retrieving the payload from the ocean floor, but usable spectra were obtained. The second experiment (Speer et al 1970) obtained more than 50 spectra from 850 to 2150 Å, throughout a 3 min period including totality, with angular resolution of the order of a few seconds of arc. The moon progressed at an average rate of about 1 arc sec over the limb between exposures. Figure 3 shows part of two frames from the sequence, the first taken a few seconds after second contact, and the second near midtotality. The emission remaining at midtotality consists of (a) pure coronal emission lines such as Fe XII 1243 and Fe XIII 1350, which are

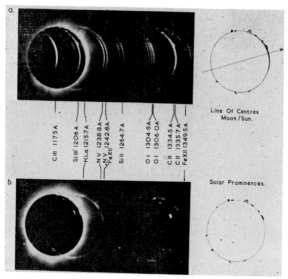

FIGURE 3. Frames from the March 7, 1970, eclipse flash spectrum (Speer et al 1970), (*a*) a few seconds after second contact, and (*b*) near midtotality. The bright Ly α image is due to resonance scattering of chromospheric Ly α by neutral hydrogen in the corona (see text).

identifiable as coronal by the emission forming a complete ring; (*b*) chromospheric lines in prominences that extend into the corona; and (*c*) a bright extended ring of emission at the wavelength of Ly α. Speer et al (1970) conclude this is due to resonance scattering of chromospheric Ly α by the neutral hydrogen in the corona; although only about 10^{-7} of the coronal hydrogen is neutral, this is enough to produce the observed intensity. A more detailed study of this and other aspects of the data has been carried out by Gabriel et al (1971).

OBSERVATIONS EMPHASIZING SPECTRAL RESOLUTION

The Harvard spectrometer aboard OSO 4 (Goldberg et al 1968) was able for the first time to obtain photoelectric spectra of a restricted area of the solar disk, namely, a 1 arc min² at the quiet-disk center. The resulting data are uncontaminated by effects of active regions, variations of limb brightening at different wavelengths, or variations from equator to poles. A number of scans have been averaged to get a mean spectrum of the quiet Sun (Dupree & Reeves 1971).

A similar spectrometer aboard OSO 6 recorded spectra at many different points on the solar disk, including active and quiet regions, both on the disk and above the limb, as well as prominences, filaments, sunspots, flares, etc. Figure 4 (Dupree et al 1970) compares parts of two typical spectra of active

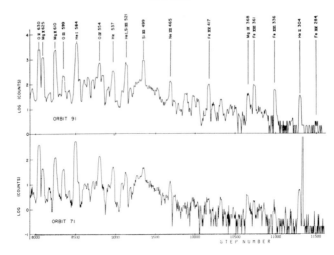

FIGURE 4. Typical photoelectric scans of active (*top*) and quiet (*bottom*) regions, from the Harvard OSO 6 spectrometer (Dupree et al 1970). Only part of each scan is shown. Note the increased active-region enhancement of coronal lines relative to the transition zone and chromospheric lines.

and quiet regions obtained by this instrument. The active region is about a factor of 10 brighter in lines formed below a temperature of about $10^{5°}$K, but coronal lines such as Mg X, Si XII, Fe XV, and Fe XVI are enhanced by factors of 30 to 50.

Burton, Ridgeley, and their co-workers have continued their program, already reviewed by Goldberg (1967), in which they have obtained photographic spectra from a slit 4 arc sec wide placed parallel to the limb about 10 arc sec above it. These data have a spectral resolution of about 0.3 Å in the range 750 to 3000 Å. Burton & Ridgeley (1970) have published a list of wavelengths and intensities of about 600 emission lines, most of which are identified. The observations can be interpreted in terms of the geometry and optical depth of the fine structures in the chromosphere and transition zone, as we discuss later.

Bonnet & Blamont (Bonnet 1968, Bonnet & Blamont 1968) on two rocket flights obtained stigmatic spectra in the spectral range 1800 to 3000 Å with a spectral resolution of 0.4 Å. These data (Figure 5) show large variations in the spectrum from center to limb, especially on either side of the Al I edge at 2080 Å. However, the resolution is not quite high enough to distinguish a true continuum level.

Boland et al (1971) have obtained an echelle spectrum of the center of the disk in the wavelength region 2000 to 2200 Å, with very high spectral resolution of 0.03 Å. This new set of data reveals many individual lines that were unresolved in earlier work. The authors tabulate 660 lines within the

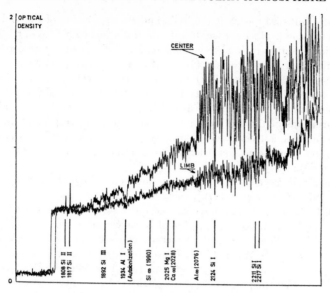

FIGURE 5. Microdensitometer tracings of the stigmatic spectra of Bonnet & Blamont (1968) in the region 1800 to 2300 Å at disk center and near the limb. Diagram courtesy of *Solar Physics*.

region and find that the average interval between absorption lines is only 0.2 Å. The continuum level on either side of the Al I discontinuity at 2080 Å can be identified, and the size of the intensity drop is found to be a factor of 5. The location of the drop 9 Å redward of the Al I ionization limit permits a determination of the pressure, and indirectly the temperature, in the high photosphere, as we shall mention later.

Very recently, Brueckner (1971) flew a rocket prototype of an NRL spectrometer scheduled to be operated in the Apollo telescope mount as part of the NASA Skylab mission. A series of spectra was obtained at four different limb positions, defined by $\cos \theta = 0.7$, 0.3, 0.2, and 0.1, where θ is the angle between the line of sight and the radial direction. The spectral resolution was about 75 mÅ and the wavelength range, 1170 to 1800 Å. These data promise to make a major contribution to our knowledge of chromospheric structure.

LINE-PROFILE STUDIES

Lemaire (1969) used a balloon-borne spectrograph to obtain high-resolution stigmatic spectra in the region of the 2800 Å Mg+ H and K lines. The spectral resolution was 45 mÅ, and the spatial resolution 10 arc sec. Spectra were obtained both at the disk center and at the limb; they show pronounced differences: an asymmetry in the K_2 and H_2 emission peaks is present at disk center but disappears at the limb, and the separation of the violet and the red emission peaks increases toward the limb, just as it does in the Ca^+ H and K lines. Active regions show enhanced emission in the line cores.

Bradley and his co-workers (Bates et al 1970) have used an ultraviolet Fabry-Perot interferometer in conjunction with an echelle spectrograph to obtain observations with a spectral resolution of 30 mÅ and a spatial resolution of 6 arc sec to obtain similar results. Both they and Lemaire also see large variations of the line profile from point to point, especially in active regions. In some active regions, the violet and the red wings of the profile merge into a single emission peak, as is sometimes found in the Ca^+ K line (Smith 1960).

Bruner and his colleagues (Bruner and Rense 1969, Berger, Bruner & Stevens 1970, Bruner et al 1970), using an echelle spectrograph with a resolution of from 20 to 35 mÅ, have carried out a series of observations aimed at exploring the detailed line profiles of H Ly α and nearby chromospheric resonance lines. The Ly α results generally confirm the earlier work of Purcell & Tousey (1960, 1961). The O I resonance triplet at 1305 Å is resolved for the first time and shows a strong reversal; however, the interpretation is complicated by the presence of a terrestrial O I absorption core in the line center, analogous to that seen in the H Ly α line and arising in the geocorona. Bruner et al (1970) assumed two extreme models of the atomic oxygen distribution in the Earth's atmosphere, differing by a factor of 10 in the oxygen density. Correction of the observation for the higher-density oxygen atmosphere leads to a nonreversed solar profile; correction for the low-density case produces a profile with a flat top or only a very slight reversal. In no case is the theoretical prediction of Athay & Canfield (1970) of a strong self-reversal confirmed. Jones & Rense (1970) interpret this result in terms of nonthermal velocities in the chromosphere with an amplitude of about 7 km/sec.

The C II resonance lines at 1335 Å were found by Berger et al (1970) to be essentially Gaussian, with no self-reversal. Their half width is measured as 0.22 Å, considerably greater than that expected for thermal broadening at the 20,000°K temperature of formation of C II. To explain the observed broadening requires about 30 km/sec turbulent velocities, comparable to the spicular velocities known to exist in the upper chromosphere.

ABSOLUTE-INTENSITY MEASUREMENTS

A number of problems in solar physics depend sensitively on absolute-intensity measurements in the EUV. Of these, the problem of the temperature minimum in the low chromosphere is one of the most important. Widing, Purcell & Sandlin (1970), Parkinson & Reeves (1969), and Carver et al (1969) have all carried out rocket observations to measure the solar flux in the region near 1600 Å, which arises from the level of the temperature minimum. The resulting data and their interpretation are discussed below.

The rocket observations of Hinteregger and his colleagues have been the basic source of absolute-flux data for other workers in the field for a number of years. These data have now been improved by the publication of absolute fluxes observed from the OSO 3 satellite (Hall & Hinteregger 1970), for which no correction for residual atmospheric absorption is necessary.

THE QUIET SOLAR ATMOSPHERE
THE LOW CHROMOSPHERE

Solar radiation in the range 1000 to 3000 Å originates in the high photosphere and low chromosphere, where the temperature reaches a minimum value and begins to increase outward (Figure 1). In principle, analysis of the emission in this wavelength region can lead to a description of physical conditions near the temperature minimum. (An alternative spectral range where emission originates from the temperature minimum region is the far infrared, but owing to the nature of the Planck function, the ultraviolet is much more sensitive to the temperature structure.)

However, to interpret the ultraviolet emission, we need a knowledge of the line and continuous opacity sources in this region of the spectrum. In spite of recent progress in laboratory and theoretical opacity determinations, discrepancies of as much as an order of magnitude still exist between these opacities and those inferred from the observed emission. We discuss the opacity problem below, after which we comment on the temperature structure implied by the observations.

Opacity sources.—Figure 5 (Bonnet & Blamont, 1968) shows densitometer tracings of the spectral region 1800 to 2850 Å at disk center and at minimum inside the limb. It is immediately apparent that the spectrum is exceedingly rich in absorption lines, to the extent that between 2100 and 2800 Å it is difficult or impossible to establish a true continuum level between the lines. As we discussed earlier, the high-resolution spectrum of Boland et al (1971) reveals an absorption line on the average every 0.2 Å, and thus a resolution of about 0.05 Å or better is required to separate lines from continuum. Even then, line-free regions probably do not show a true continuum level, owing to overlapping absorption from the far wings of neighboring lines.

The influence of continuous opacity is apparent in the spectrum of Figure 5, especially in creating the remarkable discontinuity at 2080 Å, which is due to the bound-free absorption edge of Al I. The wavelength corresponding to the ionization limit of Al I is 2071 Å, but Boland et al (1971) have shown that the difference between the observed edge and the ionization limit is due to overlapping of stark-broadened lines from high-lying levels. Their calculations based on the Bilderberg Continuum Atmosphere (Gingerich & de Jager 1968) predicted that the wavelength of complete level merging would be shifted 18 Å, twice the observed shift of 9 Å; however, a slight downward adjustment of the temperature in the region $0.001 < \tau_{5000} < 0.2$ would bring the observed and calculated edges into agreement.

Numerous other absorption edges are predicted to occur shortward of 3000 Å, owing to ground levels or low-lying excited levels of abundant elements such as Mg, Ca, Si, and Fe. Two of these absorption edges divide the spectrum into three qualitatively distinct regions (Gingerich et al 1971):

(*a*) $\lambda > 1683$ Å (the ^1D excited edge of Si I): Emission in this region originates below the temperature minimum, where the temperature is still decreasing outward. Hence, spectral lines appear strongly in absorption, continuum edges show lower intensities on the short-wavelength (higher opacity) sides, and the intensities in lines and continua show limb darkening.

(*b*) $1683 > \lambda - 1525$ Å (the ^3P ground-state edge of Si I): Emission in this region originates near the level of the temperature minimum, where the temperature gradient becomes zero. Therefore, spectral lines in this region (if they are in LTE) should be shallow, continuum emission jumps should be small, and the emission should show little or no limb darkening.

(*c*) $\lambda < 1525$ Å: Emission in this region originates from the chromosphere above the temperature minimum, where the temperature is increasing with height. Hence, lines should be in emission, continuum edges should be more intense on the short-wavelength side, and the emission should show limb brightening.

Figure 6 (Gingerich et al 1971) compares observed ultraviolet brightness temperatures with the theoretical brightness temperature from the Harvard-Smithsonian Reference Atmosphere (HSRA) model (Gingerich et al 1971).

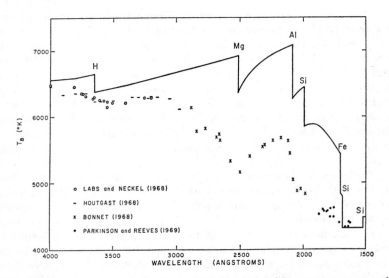

FIGURE 6. Comparison of observed and calculated brightness temperatures in the range 1500 to 4000 Å (Gingerich et al 1971). The discrepancy apparently reflects the absence of a major source of opacity in the calculations rather than a large error in the structure of the model atmosphere (see text). Courtesy of *Solar Physics*.

The model in the range 1700 to 3000 Å is determined primarily by observations of continuum emission in the near infrared and can scarcely be in error by more than a few hundred degrees. It is apparent, however, that in the ultraviolet there is an enormous discrepancy of about 1000°K in brightness temperature, or equivalently a factor of about 7 in the intensity at 2500 Å. In addition, the remarkable decrease toward the limb in the size of the Al I jump at 2080 Å (see Figure 5) is not reproduced in the HSRA model, which actually predicts a larger jump toward the limb.

Both these discrepancies could be due to the neglect of a major opacity source, which might be either a continuous source or possibly numerous unresolved absorption lines. The HSRA model contains as opacity sources H, H^-, Rayleigh scattering, absorption edges of abundant elements, absorption by quasimolecular hydrogen (Doyle 1968), and bound-free absorption near 1700 Å by several low-lying levels of Fe, with a revised Fe/H abundance ratio of 3×10^{-5} (Garz et al 1969). Cuny (1969) proposed that resonance broadening of the Ly α wing might cause widespread absorption at wavelengths as long as 3600 Å. However, Sando, Doyle & Dalgarno (1969) have shown that the conventional formula used by Cuny fails far from line center because of the failure of the quasistatic approximation at small internuclear distances; they find from a semiclassical calculation that the contribution to the opacity drops rapidly to zero beyond 1623 Å.

Direct empirical information on opacity in the region 1800 to 3000 Å has been obtained from analysis of the filtergrams described earlier, obtained at 1980, 2150, 2235, 2665, and 2885 Å. Bonnet & Blamont (1968; see also Bonnet 1968) have determined the depth variation of the source function $S_\lambda(\tau_\lambda)$ at each wavelength from an analytic Laplace inversion of the center-to-limb intensity variation and, assuming LTE, have found the run of temperature $T(\tau_\lambda)$. Comparison with $T(\tau_{5000})$ from visible observations yields $\kappa_\lambda/\kappa_{5000}$. They find that the resulting empirical opacity κ_λ is always larger than the theoretical one and also that the predicted abrupt drop of opacity longward of the Mg I edge at 2510 Å (see Figure 6) is totally absent, being replaced by a gradual decrease in opacity between 2500 and 3000 Å.

Another outstanding discrepancy between theory and observation occurs at 1683 Å, where a drop of a factor of 10 is predicted (Figure 6) but not observed. Bonnet (1970) suggests that line blanketing would decrease the size of the jump, because lines redward of the discontinuity tend to be in absorption, while lines to the violet are absent or in emission.

An attempt has been made (Bonnet 1968) to assess the effect of line blanketing quantitatively by synthesizing the spectrum, introducing more than 2400 individual lines with known gf values. In a recent computation, Sacotte (Bonnet, 1970) has found that agreement is reached if two additional opacity sources are included:

(a) One that increases linearly with decreasing wavelength and that could be due to the wings of numerous weak and unresolved lines (which increase in density toward the ultraviolet).

(*b*) A broad and symmetrical feature located at 2500 Å, with a full width at half-maximum of about 400 Å.

The problem of the missing opacity in the 1680 to 3000 Å range is of considerable importance, in view of the huge discrepancies that still exist. It is to be hoped that new high-resolution observations such as those described earlier (Boland et al 1971, Brueckner 1971) will clear up the question whether the extra opacity is supplied entirely by a multitude of absorption lines or whether an additional source of continuous opacity is necessary.

The temperature minimum.—As mentioned above, emission between 1525 and 1683 Å should originate in the temperature minimum region. The behavior of the Planck function is such that an uncertainty of $\pm 100°K$ at $T = 4500$ and $\lambda = 1600$ Å is equivalent to an uncertainty of about 50% in the absolute intensity, somewhat greater than the uncertainty quoted in the data of Widing et al (1970). Hence, if LTE is a valid assumption, one should be able to determine the value of the temperature minimum to within better than 100°K from the observed brightness temperature at 1600 Å. Determination of the value of the minimum temperature to this precision is important, for owing to the high density in the low chromosphere, the mechanical energy necessary to maintain a temperature excess of 100°K over the radiative equilibrium value is comparable to that needed to heat the entire corona to $10^{6°}K$ (cf Athay 1970).

At the time of the last review of the subject on these pages (Goldberg 1967), the temperature minimum was thought to have a value of about 4600°K and to extend over a rather broad region, as described by the Bilderberg Continuum Atmosphere (Gingerich & de Jager 1968). This model was based largely on the rocket data of Tousey and his colleagues (Tousey 1963), which indicated a brightness temperature of 4600°K at 1600 Å. More recent observations by the NRL group (Widing et al 1970) lend support to this value by yielding temperatures between 1570 and 1682 Å that lie in the range $4700 \pm 100°K$.

However, in another recent rocket flight, Parkinson & Reeves (1969) measured absolute intensities in the range 1400 to 1875 Å that were a full factor of 3 below NRL results and corresponded to a brightness temperature of 4400°K or lower. Carver et al (1969), using broadband filters in the range between 1400 and 1600 Å, obtained an intermediate minimum temperature of about 4500°K. At the time of this writing, there has been no resolution of these discrepancies, which serve at the very least to illustrate the difficulty of making absolute-flux measurements in the rocket ultraviolet.

Independent evidence from airborne infrared observations has recently become available (Eddy, Léna & MacQueen 1969) that indicates a brightness temperature at 300 μ of about 4300°K. In addition, analysis of Ca^+ H and K line profiles suggests the need for a minimum temperature near 4300°K (Linsky & Avrett 1970). This led Gingerich et al (1971) to adopt the lower value in their model. However, observations of the ultraviolet continuum intensity remain the most sensitive indicator of the precise value of the minimum temperature, and a definite conclusion awaits resolution of the observational discrepancy mentioned above.

It is doubtful that the difference between the radiative-equilibrium temperature and the true temperature at the minimum can be meaningfully determined to better than about $\pm 100°K$, even if the observations are improved, for the uncertainty in the theoretical radiative-equilibrium temperature is $\pm 150°K$, according to Athay's (1970) line-blanketed model calculations. It is interesting, but perhaps coincidental, that he obtains a radiative-equilibrium boundary temperature of $4330 \pm 150°K$, indicating that if the lower observationally determined temperature minimum is correct, there may be little or no nonradiative energy deposited below the temperature minimum.

The High Chromosphere

Shortward of 1525 Å, all the solar emission comes from the chromosphere and corona. Continuum edges now appear in emission, the most spectacular being the H I Lyman edge at 912 Å, the He I edge at 504 Å, and the C I edge at 1100 Å. Figure 7 (Noyes & Kalkofen 1970) illustrates the spectrum of the disk center as observed by the Harvard OSO 4 spectrometer. The brightness temperature at the head of the continuum is about 6450°K, but the emission is clearly not Planckian at that temperature; rather, the emission corresponds more closely to that of a blackbody with temperature about 8300°K, decreased by a factor of about 200.

The Harvard experiment on OSO 4 obtained in addition a set of limb-darkening data for six different wavelengths within the Lyman continuum. These data show a nearly flat center-to-limb variation at all wavelengths. However, there is a systematic trend with wavelength: there exists slight limb darkening at 912 Å, which gradually decreases with decreasing wavelength, becoming flat at about 750 Å and showing slight limb brightening below 750 Å. This implies that the color temperature of the continuum increases from center to limb, which is not surprising, for there is close agreement between the observed color temperature and the electron temperature at $\tau = \cos \theta$, and the continuum is formed where the electron temperature is increasing outward.

The above observations can be explained (Noyes & Kalkofen 1970) on the basis of a non-LTE model in which the source function $S_\nu = B_\nu(T)/b_1$ is decreased below the Planck function by an amount $1/b_1$, where b_1 is the departure coefficient of the ground level of hydrogen; $T(\tau_{Ly\,c}=1)$ is approximately the observed color temperature of the continuum; and $b_1(\tau=1)$ is approximately the amount by which the emission from a blackbody at the color temperature exceeds the observed emission. The model derived from the observed spectrum and the limb-darkening variations is characterized by a smooth increase in temperature with height above the temperature minimum, until the temperature reaches 8300°K at $\tau_{Ly\,c}=1$, at a height

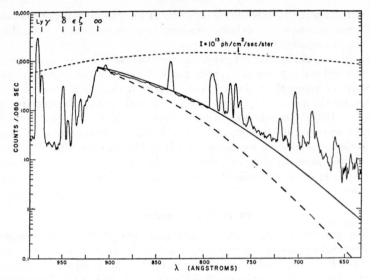

FIGURE 7. Disk-center spectrum of the Lyman continuum region obtained by the Harvard OSO 4 spectrometer (Noyes & Kalkofen 1970). *Dashed line* corresponds to a Planck function of temperature 6450°K. *Solid line* corresponds to a Planck function of temperature 8300°K, but decreased in intensity by a factor of 200. Courtesy of *Solar Physics*.

about 1800 km above $\tau_{5000} = 1$ (1500 km above the limb). A few tens of kilometers above this level, the temperature rises abruptly to a level sufficient to cause complete ionization of the overlying hydrogen. The departure coefficient b_1 is about 200 at $\tau_{Ly\ c} = 1$ and rises rapidly with height, so that the source function $S_\nu = B_\nu(T)/b_1$ is nearly constant.

The abrupt rise of temperature shortly above $\tau_{Ly\ c} = 1$ is required by the observation that the color temperature, i.e., the electron temperature at $\tau_{Ly\ c} = 1$, is as low as 8300°K. At this relatively cool temperature, with the pressure great enough to support the overlying corona (cf Athay 1969), optical-depth unity for the head of the Lyman continuum is reached in a few tens of kilometers, even for the 50% ionization at that level found in the above model. If the temperature did not rise sharply enough to ionize most of the hydrogen immediately above the 8300°K level, optical-depth unity would occur at a higher temperature.

The detailed shape of both the center-to-limb variation and its wavelength dependence (i.e., the center-to-limb variation of the color temperature) cannot be explained by a homogeneous one-component model unless one assumes an unrealistically low gas pressure. This implies that the inhomogeneous structure of the chromosphere must be considered in the construction of detailed models. We shall return below to a discussion of the inhomogeneous nature of the chromosphere and corona.

It is interesting to compare the above empirical model with that proposed by Thomas & Athay (1961), based entirely on visual continuum data. Their observations consisted mainly of the intensities of the chromospheric flash spectrum at eclipse, measured at 4700 and 3646 Å. When analyzed by use of a non-LTE calculation of the hydrogen ionization equilibrium, these data yielded the height variation of T_e and n_e, assuming spherical symmetry and hydrostatic equilibrium. Thomas & Athay (1961; see also Thomas 1960) derived a model with a gradual outward increase of temperature to a value between 8000 and 9000°K about 1300 km above the limb, at which there occurred an abrupt jump to temperatures in excess of 20,000°K.

The close agreement of two models derived from entirely different sets of data by different techniques is very encouraging.

THE TRANSITION ZONE

The transition region between the 10^{4}°K chromosphere and the 10^{6}°K corona is perhaps the most intriguing region of the observable solar atmosphere. It has become clear in recent years that this transition region is almost incredibly thin—the temperature rise from 10^5 to 2×10^5°K, for example, appears to take place over a distance of only about 20 km. (See, e.g., Athay 1971.) This sharp transition region immediately overlies the upper chromosphere, where large-scale velocities are observed up to several times the chromospheric sound velocity, where vertical displacement of chromospheric material reaches 10,000 km in spicules, and where the influence of magnetic fields is suggested by the channeled flow of spicules. (For a review of spicules and other dynamic phenomena in the upper chromosphere, see Beckers 1968.)

The steepness of the transition zone was already evident by the time of Goldberg's (1967) review, and we only summarize the main evidence here, referring the reader to that discussion and to Athay's recent review (1971) for details. The emission measure $\int n_e^2 dh$ of spectral lines emitted near 10^5°K was found from analysis of emission from the entire disk of the Sun (e.g., Pottasch 1964, Jordan 1965, Athay 1966, Dupree & Goldberg 1967) to be rather small, implying a restriction of the volume at temperatures near 10^5°K because of the large temperature gradient in this region. Athay (1966) showed that the emission measures were consistent with a temperature structure obeying $T^{5/2} (dT/dh) = \text{const}$, as would arise from a constant downward conductive flux of $F_c \gtrsim 3 \times 10^5$ ergs/cm²/sec. Such a large flux would dominate other energy fluxes in the transition zone, making it plausible that the conductive flux be constant throughout the region, since the divergence of the total energy flux is zero in a steady state. The conductive flux is also comparable to the flux radiated from the upper chromosphere, where the conducted energy is deposited, and it is a natural inference that the chromospheric temperature excess might be maintained by this downward conductive flux. However, the temperature gradient is so steep in the low transition zone that all the energy would be deposited in a very narrow region a few kilometers thick at the base of the transition zone, whereas the chromospheric radiation comes from a more extended region (Kuperus & Athay 1967, Athay 1971). As an alternative, Kuperus & Athay suggested that the conductive flux supplies the energy not for thermal radiation from the chromosphere but for kinetic energy lost to the chromosphere in the form of spicule ejection. For a more complete discussion of the energetics of the transition zone, see Athay (1971).

Since 1967, new EUV observations have been acquired with a spatial resolution permitting a fresh attack on the empirical description of the transition zone. Withbroe (1970a) has analyzed center-to-limb observations of the lithiumlike lines N V, O VI, Ne VIII, Mg X, and Si XII, obtained from OSO 4 data with 1 arc min resolution. Comparison of intensities above the limb with those at disk center enabled Withbroe to evaluate the conductive flux, on the assumption that it was constant throughout the transition zone. The method relies on the fact that for two of the lithiumlike ions, namely, O VI and Ne VIII, the emission arises partly from the transition zone and partly from the corona. The intensity well above the limb, above the height of the transition zone, arises only from the corona, while that observed at disk center comes from both the transition zone and the corona. The ratio of the intensity above the limb to that at disk center therefore decreases as the thickness of the transition zone increases, i.e., as the conductive flux decreases.

Withbroe (1970a) developed a simple model in which the transition-zone structure was defined by the following parameters:

(a) The pressure at the base at the transition zone given by

$$P_0 = (n_e T)_0 = 6 \times 10^{14} \text{ cm}^{-3}{}^\circ\text{K}$$

(b) The scaled conductive flux $1/C = T^{5/2} \, dT/dh$, given by

$$C = 10^{-12}, \quad \text{or} \quad F_c = 6 \times 10^5 \text{ ergs/cm}^2/\text{sec}$$

(c) The temperature T_c of the isothermal corona, which overlies the transition zone, determined to be $T_c = 2 \times 10^{6}{}^\circ\text{K}$.

The value of the conductive flux determined by Withbroe agrees well with others based on radio data (see Pottasch 1964, Athay 1971) and EUV observations of the integrated disk emission (Dupree & Goldberg 1967). The last determination, although it uses similar EUV data, depends upon an assumed value for the electron pressure and upon the abundance of the observed elements relative to each other and to hydrogen. Withbroe's treatment is independent of these assumptions, but must assume that the corona is spherically symmetric and in hydrostatic equilibrium.

Location of the abrupt rise in temperature.—There is considerable evidence from the visual spectrum that the abrupt rise of the temperature to coronal values must set in within a few thousand kilometers of the solar limb. This conclusion has been drawn from observations of forbidden lines and continua in eclipse and of hydrogen and helium emission both during and outside eclipse. Recent analyses of EUV data now add support to the evidence for a sharp temperature rise at low heights.

EUV observations during eclipse can in principle tie down the height of emission to rather good accuracy, because of the narrow height range over which transition-zone lines can emit. In practice, the analysis is limited by the irregular nature of the transition zone. If, as seems entirely reasonable, the transition zone parallels the surface of the Hα-emitting chromospheric elements, then a variation of some 10,000 km in the height of the transition zone can be expected. Nevertheless, slitless EUV spectra at eclipse can provide information on the relative volume of transition-zone material at each height, and in particular in the lowest height to which it extends. Such data were obtained for the first time during the March 7, 1970, eclipse (Speer et al 1970, Brueckner et al 1970), and one can look forward to the results of the analyses of these data presently.

Withbroe (1970b) has used the OSO 4 data in an interesting way to determine the height of the transition zone. Although these data have rather low resolution (1 arc min or 40,000 km), he was able to determine the mean location of the transition zone to within about 1500 km, by using the known geometrical properties of opaque spicules for height discrimination. He found that certain emission lines such as N IV 765, O III 702, O V 630, and O IV 555 show less limb brightening than would be expected, given the fact that they are optically thin. The anomalous lines all have wavelengths shorter than 912 Å, the wavelength of the hydrogen Lyman limit. One would expect absorption by spicules in the Lyman continuum to decrease the limb brightening of these lines, provided that they are formed below the tops of the spicules. The size of the effect depends, of course, on how far the emitting layer lies beneath the tops of the spicules. The OSO 4 observations show that lines such as O III, O IV, and O V experience considerable obscuration, while Ne VIII, which is formed somewhat higher, shows much less obscuration. This led Withbroe to place the steep rise in the low transition zone near 2000 km above the limb. If it were lower, Ne VIII would show more obscuration; if it were higher, O V would show less. The accuracy of the observation leads to an uncertainty of 1000 to 2000 km in this height.

Another approach to determining the height of the transition zone was made by Burton et al (1971), who analyzed a 1969 Skylark observation in which spectra were obtained both on the disk and just above the limb. The edge of the (straight) slit closest to the limb lies below the height of the transition zone at its closest point; the other edge lies some 6000 km higher, in the corona. It is assumed that lines are formed entirely in a narrow height range Δh at a height h, where h varies from line to line and is governed by the temperature structure. In the ratio of limb to disk intensities, the quantities Δh, pressure, and elemental abundance cancel out, leaving the ratio dependent only on the height h, through the geometry of the intersection of the spectrograph slit with the volume of the emitting shell at height h above the limb. Burton et al find that optically thin lines formed between $1.2 \times 10^{5}°K$ and $2.2 \times 10^{5}°K$ lie at the same height to an uncertainty ± 700 km, as expected for the sharp transition region. Unfortunately, the absolute height scale is not known. Optically thick lines have heights averaging about 7500 km higher. This sizable difference might be due in part to errors in the optical depth assigned to the lines, but the effect may well be real and due to the spicular structure of the chromosphere. The optically thick, higher-lying lines also are generally neutral or singly ionized and might be emitted by the relatively cool spicules at a height above the mean location of the hotter transition-zone ions.

Such an interpretation requires that the spicules themselves not emit significantly in transition-zone lines. In other words, the rise to coronal temperatures at the surface of spicules must be very rapid. This situation is also implied by Withbroe's work discussed above, for if spicules are to decrease the limb brightening by obscuration of the interspicular transition zone, they must not replace the obscured emission by emission from their own transition zone.

In summary, several lines of evidence point to a very steep temperature gradient for temperatures at or above $10^{5}°K$, located between the spicules about 2000 km above the limb. The contribution of the transition zone sheathing the spicules to emission at $10^{5}°K$ appears to be less (per unit area) than that of the interspicular regions.

The temperature gradient below $10^{5}°K$ is less well studied. Lines formed at lower temperatures are increasingly optically thick, and the consequent corrections for self-absorption and radiative-transfer effects make the observed intensity a less reliable indicator of the thickness of the emitting region.

As we mentioned earlier, the Lyman continuum data suggest that a sharp rise from chromospheric temperatures of about 8300°K sets in about 1500 km above the limb, assuming hydrostatic equilibrium. Turbulent pressure could raise this height by several hundred kilometers. The temperature must rise at least to the level of complete hydrogen ionization, but this need only be 20,000°K or so; there is no requirement from the Lyman continuum that the rise extend directly to the transition zone and corona. Indeed, Thomas & Athay (1961) suggested that there might be a "plateau," or a region of lesser temperature gradient, at temperatures between 15,000 and 60,000°K, due to the radiative efficiency of He I in this temperature range. This would be reflected in the optical thickness of lines formed in that region, such as the Lyman lines, C II, or C III. Analysis of existing detailed line profiles of Ly α, Ly β, and C II 1335 (Purcell & Tousey 1960, 1961, Bruner & Rense 1969, Berger et al 1970) should give information on their optical thickness. In addition, studies of center-to-limb variations of intensities in these lines can be made from several of the sources of data described at the beginning of this paper, in order to derive the optical thickness of the lines. For example, by such an analysis, Withbroe (1970b) determined the optical thickness of C III 977 to be about unity.

THE QUIET CORONA

EUV observations of coronal resonance lines provide a new and valuable tool for determining physical conditions in the corona, for their intensity depends in a straightforward and separable way on density and temperature. In the quiet corona, the ionization equilibrium of many ions is essentially density independent, and the population of the ion is concentrated almost entirely in the ground level. Examples of ions for which these assumptions are true include those of the lithium isoelectronic sequence. For these ions, the emission per unit volume can be expressed as

$$dE/dV = \text{const } f \cdot (n_{\text{El}}/n_{\text{H}}) n_e^2 F(T) \qquad 1.$$

where f is the oscillator strength, $(n_{\text{El}}/n_{\text{H}})$ is the abundance of the element relative to hydrogen, n_e is the electron density, and $F(T)$ contains the temperature-dependent terms:

$$F(T) = g(T) \exp(-E_{12}/kT) T^{-1/2}[n_i(T)/n_{\text{El}}]$$

where $g(T)$ is the gaunt factor, E the energy of the transition, and n_i/n_{El} the fraction of the element in the given abundance state (see Athay 1966, Withbroe 1970a). Withbroe (1971, 1970a) has separated the density and temperature effects in Equation 1 by comparing the intensities of two coronal lines. If both lines are optically thin and if the temperature along the line of sight is constant, then the ratio of the two intensities depends only on the temperature:

$$\frac{I_2}{I_1} = \frac{f_2}{f_1} \frac{n_{\text{El}_2}}{n_{\text{El}_1}} \frac{F_2(T)}{F_1(T)} \qquad 2.$$

Using intensity data from Harvard OSO 4 spectroheliograms, Withbroe found the temperature 2 arc min above the limb from several pairs of lithium-like ions. The results are given in the lower curve of Figure 8, which shows temperatures near $1.8 \times 10^6 \,^\circ\text{K}$ from all the line ratios.

The fact that the ratios $I(\text{Si XII})/I(\text{Mg X})$, $I(\text{Si XII})/I(\text{Ne VIII})$, and $I(\text{Si XII})/I(\text{O VI})$ all yield about the same temperature supports the notion that the corona 2 arc min above the limb is essentially isothermal along the line of sight, or at least along those parts of the line of sight that contribute significantly to the emission.

It is interesting that the corona 2 arc min above the limb also appears to be nearly isothermal at different positions around the limb. This is true although the observed emission intensity varies by a factor of 30 from point to point above the limb, as is shown by the solid line on the top of Figure 8. There is a slight tendency for the brightest (most active) positions to be somewhat hotter than their surroundings, by perhaps 1 to $2 \times 10^5 \,^\circ\text{K}$.

If the corona is indeed nearly isothermal, then the principal cause of intensity variation above the limb must be electron-density variations. Withbroe (1971) has used the temperature determined as we have described

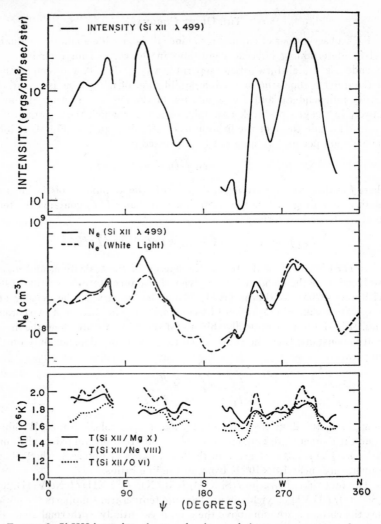

FIGURE 8. Si XII intensity, electron density, and electron temperature 2 arc min above the solar limb, derived from Harvard OSO 4 spectroheliograms and High-Altitude Observatory K-coronameter data (Withbroe 1971).

above and integrated the emission (Equation 1) along the line of sight, assuming hydrostatic equilibrium in a locally stratified atmosphere (i.e.,

$$n_e(r, \psi) = n_e(r_0, \psi) \exp \left(- \int_{r_0}^{r} \frac{dr}{H} \right)$$

where H is the scale height) for all points in the line of sight at position angle ψ. Comparison with the observed intensity then yields the density at (r_0, ψ), where r_0 is located 2 arc min above the limb. The resulting profile is plotted in the middle curve of Figure 8.

Also shown on the middle curve are the densities inferred by Withbroe from K-coronameter observations obtained by the High Altitude Observatory of the University of Colorado, by scaling the quiet coronal model of van de Hulst (1950). The intensity of the K corona is proportional to the integral of the electron density along the line of sight, whereas EUV emission is proportional to the integral of the square of the electron density along the line of sight. Therefore, it is somewhat surprising that the electron densities inferred by the two methods are in such good agreement. The agreement suggests that all locations along the line of sight that contribute substantially to the white-light or EUV intensity have approximately the same density.

We note that the density 2 arc min above the limb varies by a factor of about 5 between active regions near the equator and the quiet poles. The mean value of $10^8/cm^3$ can be used to infer the pressure in the transition zone, under the assumption of hydrostatic equilibrium. Withbroe found by this means a pressure $(n_e T)_0 = 7 \times 10^{14}$ in the equatorial transition zone, in good agreement with the results described earlier.

INHOMOGENEOUS STRUCTURE OF THE QUIET ATMOSPHERE

Inhomogeneities in the solar atmosphere certainly exist on a scale finer than most present EUV observations can resolve. Data from a few rather high-resolution observations can help establish whether neglect of inhomogeneities seriously affects the interpretation of lower-resolution observations.

The effects of photospheric granulation are noticeable in Blamont & Carpentier's (1967, 1968) observations near 2000 Å, corresponding to $\tau_{5000} \sim 0.1$. The 60°K amplitude of the fluctuations is less than half that found at $\tau_{5000} = 1$ (see Leighton 1963) and presumably decreases with increasing height. On the other hand, the microfaculae seen by Blamont & Carpentier with a temperature excess of about 300°K may represent the deep-lying part of the chromospheric emission network, which becomes more important with increasing height. The emission network is very bright in chromospheric lines ranging from the Ca^+ H and K lines through He^+ 304, and it extends into the transition region at least as high as the O V 630 line, formed at a temperature of about 250,000°K in the transition zone (Purcell & Tousey 1969, Tousey 1971b).

There is some evidence that emission in the corona maps out the local magnetic-field strength (Krieger, Viana & Van Speybroeck, 1971), just as it does in the chromosphere. It is currently thought that the magnetic network spreads out in the corona to become a nearly uniform field (see, e.g., Leighton 1963, Parker 1963, Kopp & Kuperus 1968). It is important to resolve the coronal network to see whether the emission does become more diffuse with increasing altitude.

Kopp & Kuperus (1968) developed a model of the inhomogeneous transition zone based on the funneling of conductive flux by magnetic-field lines converging into the network elements. In their picture, the temperature gradient must steepen over the network to force the conducted energy through the smaller cross section of the magnetic-flux tubes at low altitudes. From a very schematic model of the structure of the magnetic funnel, they predict that nearly all the emission in the transition zone would come from the non-network areas rather than from the network. This is in direct conflict with the O V observations of Purcell & Tousey (1969, see also Tousey 1971b), but the disagreement may be due more to an oversimplification of the model than to its basic incorrectness. Thus, the ratio of the density inside and outside the network was set equal to the square root of the ratio of the conductive flux incident from the corona on the network and non-network elements; however, this might well underestimate the density in the network. Greater densities could cause greater emission in the network in spite of the increased temperature gradient.

In any case, it is clear that interpretation of low-resolution observations of the transition zone must ultimately allow for variations in height of the physical and geometrical properties of the network.

THE STRUCTURE OF ACTIVE REGIONS

Active regions are visible from the ground as areas of enhanced magnetic field, slightly increased continuum brightness when seen as white-light faculae at the limb, and strongly increased brightness when observed in the cores of collision-dominated chromospheric absorption lines such as Ca^+ H and K. The contrast in intensity between active and quiet areas increases as observations are made in lines formed at higher temperatures in the atmosphere, until at a temperature of several million degrees, the emission comes essentially entirely from active regions.

Several studies of the structure of active regions have been carried out using the EUV data from the Harvard OSO 4 and OSO 6 experiments. Although these data suffer from rather low spatial resolution, they permit accurate photometric measurements in a wide variety of emission lines.

Noyes, Withbroe, & Kirshner (1970) have analyzed OSO 4 spectroheliograms (cf Figure 2) to determine the mean contrast of active regions for different temperatures in the transition zone and corona. The authors took Withbroe's (1970a) three-parameter model of the quiet atmosphere, described earlier, and calculated the contrast I_A/I_Q expected for different values of these parameters in active regions. The best fit to the data occurs for the following values:

$$C = T^{-5/2}/(dT/dh) = 2 \times 10^{-13}; \qquad C_A/C_Q = 1/5$$

$$P_0 = (n_eT)_0 = 3 \times 10^{15}/cm^3{}^\circ K; \qquad P_0(A)/P_0(Q) = 5$$

$$T_{cor} = 2.5 \times 10^{6}{}^\circ K; \qquad T_{cor}(A) - T_{cor}(Q) \sim 5 \times 10^{5}{}^\circ K$$

The value of C implies that the transition zone is five times steeper in active regions. This in turn implies a fivefold increase in downward conductive flux, provided that the magnetic field in active regions is primarily vertical and thus does not inhibit the conductive energy flow. Since thermal conduction is the most important mechanism for energy loss from the quiet corona (Kopp 1968), the required energy input would be increased at least fivefold.

The mean transition-zone pressure deduced for active regions is also about five times that of the quiet Sun, implying that the electron density in the low corona is increased by that ratio. The coronal temperature T_{cor} was determined to be about 0.5×10^{6}°K higher in active regions; however, this number is rather uncertain owing to the low sensitivity of the Harvard OSO 4 instrument at wavelengths below 500 Å, where the high-temperature lines Fe XV and Fe XVI are emitted.

Very recently, Reimers (1971) analyzed the Harvard OSO 4 data in terms of a model in which the mean structure of active regions is determined by a single parameter, the conductive flux F_c, which is assumed to be constant throughout the transition zone although it varies from point to point on the disk. Reimers effectively postulates an explicit dependence of the two other parameters P_0 and T_{cor} on the conductive flux. He assumes that the pressure P_0 at the base of the transition zone is everywhere proportional to F_c; pressures at higher levels are then determined by hydrostatic equilibrium. The temperature T_{cor} of the isothermal corona assumed to overlie the transition zone is equal to the temperature at the top of the transition zone, which in Reimer's model is that region bounded below by the steep temperature rise and above by the isothermal corona, and within which the conductive flux is constant. Reimers assumes that the thickness of the transition zone is everywhere constant. The value of the conductive flux defines the temperature structure in the transition zone, and hence the temperature of the isothermal corona, which overlies it. In spite of the extreme restrictions implied by these conditions, Reimers is able to fit the Harvard data rather well by varying the only remaining parameter, F_c. He finds F_c to increase by a factor up to 20 or even 40 in active regions, but with a typical increase of 5, in agreement with the results of the three-parameter calculation described above.

It is by no means clear that the rather surprising success of simple one-parameter or even three-parameter models is of physical significance. The constant thickness of the transition zone, for example, requires a positive correlation between T_{cor} and F_c, as is in fact observed. However, the data are not precise enough to warrant a stronger statement than that they are not inconsistent with a model in which the thickness of the transition zone is everywhere constant. What is clear is that in general the temperature gradient is several times greater in active regions than is the quiet Sun, and that the density is several times higher.

The treatments above, both for the active and for the quiet Sun, rely ultimately on observed line intensities that depend upon both $n_e(h)$ and $T_e(h)$. These parameters can be separated, for example, by assuming a value for the transition-zone pressure (e.g. Athay 1966), or by measuring line ratios that yield T_e in the corona, and extending the resulting densities downward by assuming hydrostatic equilibrium (Withbroe 1970a). A method of directly measuring n_e without restrictive assumptions would be very useful. One such method has been independently developed by Munro, Dupree & Withbroe (1971) and by Jordan (1971). It involves measurement of line ratios of triplet and singlet lines within Be-like ions such as C III, N IV, or O V. These ratios are independent of the absolute abundance of the element, and only weakly dependent on temperature. However, they depend strongly on the electron density at the temperature where the ion exists, for electron collisions are important in determining the population of the triplet levels relative to the singlet levels. Munro and his co-workers have carried out statistical-equilibrium calculations for the six lowest levels of the above ions as well as Mg IX and Si XI; they find that the populations and hence the triplet/singlet ratios in C III and O V are quite sensitive to electron density at typical densities occurring in the region of line formation. (The relevant lines in the N IV ion are badly blended.) In both cases, quiet-region densities inferred from the line ratios are in substantial agreement with those predicted in the model calculations described earlier. The triplet/singlet ratio increases in active regions, which implies a density increase of 5 to 10 in active regions. Jordan (1971) arrives at essentially the same result, using OSO 4 spectroheliograms.

A different method of employing line ratios to infer active-region densities has been explored by Hearn, Noyes & Withbroe (1969), using the ratio of the first two members of He I resonance series, $\lambda 584$ and $\lambda 537$. The method is based on calculations by Hearn (1969) of the statistical equilibrium of a 41-level model helium atom. The ratio $I(537)/I(584)$ was found from OSO 4 spectroheliograms to decrease regularly with increasing activity, i.e., increasing intensity in the He I 584 line. The detailed variation of the ratio with $\lambda 584$ intensity led Hearn et al to conclude that the effect was due primarily to a density increase in active regions rather than to an increase of the physical thickness at constant density (which would also decrease the ratio because of a difference in the optical depth dependence of the intensity in the two lines). For the quiet Sun, the density and temperature in the region of He I emission were determined to be $4.5 \times 10^{10}/\text{cm}^3$ and $3 \times 10^{4\circ}\text{K}$, respectively, giving a pressure about twice that found in the models described above. In active regions, the pressure increased by a factor 2.5, to a value comparable to that found above for active regions. The uncertainty in these determinations is about a factor of 3.

A detailed knowledge of the density and temperature structure of active regions is basic to the problem of energy balance in active regions, for the conductive flux that results may be directly related to the mechanical energy that must be deposited in the inner corona. The analyses just described, however, cannot provide the detailed information needed, because they are based on data with low spatial resolution. Active regions are seen from ground-based observations to have a fine structure on the scale of at most a few seconds of arc. Recent EUV and X-ray observations confirm that this fine structure extends into the corona as well. Figure 9 (Tousey 1971a) shows an active region as seen in the chromosphere, transition zone, and corona, with a resolution better than 5 arc sec. It is interesting to note the change with height of the active region as revealed by these high-resolution data:

(a) The chromospheric emission in He I and He II looks substantially the same as in the core of the Hα line, with perhaps an increase in the width of the emitting elements in the higher-lying He II 304 line ($T\sim80,000°$K).

(b) The Ne VII 465 line ($T\sim600,000°$K), formed entirely in the transition zone, shows less contrast in the active region, presumably because of the increased temperature gradient in the transition zone, as discussed above. The emission that is present is concentrated in bright points, of size 5 to 15 arc sec, where either the density is increased or the temperature gradient decreased. These bright points do not generally correspond to the brightest points in the chromospheric plage.

(c) The coronal active region as seen in Si XII, Fe XV, and Fe XVI exhibits much less fine structure than is seen in the chromosphere and transition zone. Krieger et al (1971) concluded from broadband X-ray spectroheliograms that emission from highly ionized species maps the coronal magnetic field, just as chromospheric lines like Ca^+ K map the chromospheric field; if this is the case, the coronal fields evidently have character entirely different from that of the chromospheric fields.

Observations such as those shown in Figure 9 clearly indicate the limitations of analyses based on low-resolution observations. More importantly, they serve as a strong incentive for new EUV observations of active regions combining the best possible spatial resolution with spectral resolution sufficient to isolate individual emission lines, wide wavelength coverage, and good photometric accuracy. The prospects seem good for obtaining real understanding of the energy balance in active regions, once such data are in hand.

SUMMARY

Recent EUV observations with improved spatial and spectral resolution lead to the following description of the solar atmosphere: The quiet Sun has a minimum temperature in the range 4200 to 4600°K at $\tau_{5000}\sim10^{-4}$. Above that level, a gradual increase occurs until a temperature of 8000 to 9000°K is reached some 1500 to 2000 km above the limb. At this level, an abrupt rise occurs, but it is not yet clear whether the temperature rises more or less directly to values in excess of 10^5°K or whether there may be a region between 15,000 and 60,000°K with a smaller temperature gradient. The gradient above 10^5°K is extremely steep, but decreases with increasing temperature, consistent with constant conductive flux, $F_c = \text{const} \times T^{5/2} dT/dh \sim 6 \times 10^5$

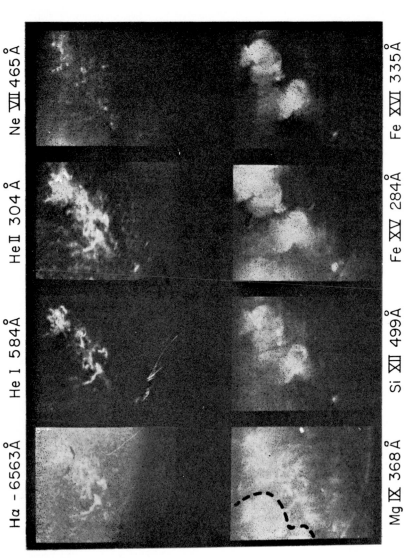

FIGURE 9. Naval Research Laboratory EUV spectroheliograms of an active region (Tousey 1971a) showing the variation in structure between chromospheric (Hα, He I, He II), transition-zone (Ne VII, Mg IX), and coronal (Si XII, Fe XV, Fe XVI) lines.

ergs/cm²/sec. This chromosphere-corona transition zone begins a few thousand kilometers above the limb, below the tops of the chromospheric spicules. The transition zone around spicules appears to be thinner or less dense than the nonspicular transition zone.

About 10^4 km above the limb, the transition zone flattens out into a rather isothermal inner corona with temperature in the range 1.5 to $2 \times 10^{6\circ}$K. The density 2 arc min (8×10^4 km) above the limb is about 1.5×10^8/cm³ and the distribution seems locally rather homogeneous. However, the mean density varies by a factor of about 5 from equator to poles. These figures refer to solar maximum and will vary during the solar cycle.

The quiet atmosphere exhibits large-scale inhomogeneity in the chromospheric emission network, which maps out the concentration of magnetic-field lines in the chromosphere. This network extends upward at least to a temperature of $2.5 \times 10^{5\circ}$K and probably reflects a higher density in network regions. It is not yet clear whether the general network extends into the corona.

The transition zone in active regions has a gradient even steeper than in quiet regions, and presumably carries a downward conductive flux several times greater than that in the quiet Sun. The density is also considerably enhanced, and the temperature in the overlying corona is increased. Active regions are complicated by much fine structure, which varies with height and probably reflects the extension of the magnetic field into the corona.

The author wishes to acknowledge helpful discussions with A. K. Dupree, L. Goldberg, and G. L. Withbroe.

LITERATURE CITED

Athay, R. G. 1966. *Ap. J.* 145:784
Athay, R. G. 1969. *Solar Phys.* 9:51
Athay, R. G. 1970. *Ap. J.* 161:713
Athay, R. G. 1971. *NATO Advanced Study Institute on the Physics of the Solar Corona,* Vouliagmeni, Greece. In press
Athay, R. G., Canfield, R. C. 1970. *Proc. IAU Colloq. No. 2,* Commi. 36, *NBS Spec. Publ. 332,* p. 65
Bates, B. et al 1970. *IAU Symp. No. 34,* 274. Dordrecht-Holland: Reidel
Beckers, J. M. 1968. *Solar Phys.* 3:367
Berger, R. A., Bruner, E. C. Jr., Stevens, R. J. 1970. *Solar Phys.* 12:370
Blamont, J. E., Carpentier, G. 1967. *Solar Phys.* 1:180
Blamont, J. E., Carpentier, G. 1968. *Ann. Ap.* 31:333
Blamont, J. E., Malique, C. 1969. *Astron. Ap.* 3:135
Boland, B. C., Jones, B. B., Wilson, R., Engstrom, S. F. T., Noci, G. 1971. *Phil. Trans. Roy. Soc.* In press
Bonnet, R. M. 1968. *Ann. Ap.* 31:597

Bonnet, R. M. 1970. *Lab. Phys. Stellaire Planet. CNRS Rep. A-42*
Bonnet, R. M., Blamont, J. E. 1968. *Solar Phys.* 3:64
Brueckner, G. E. 1971. Paper, Solar Phys. Div. AAS. *BAAS* 3:259
Brueckner, G. E., Bartoe, J. F., Nicolas, K. R., Tousey, R. 1970. *Nature* 226:1132
Bruner, E. C. Jr., Jones, R. A., Rense, W. A., Thomas, G. E. 1970. *Ap. J.* 162:281
Bruner, E. C. Jr., Rense, W. A. 1969. *Ap. J.* 157:417
Burton, W. M., Jordan, C., Ridgeley, A., Wilson, R. 1971. *Phil. Trans. Roy. Soc.* In press
Burton, W. M., Ridgeley, A. 1970. *Solar Phys.* 14:3
Carver, J. H., Edwards, P. J., Gough, P. L., Gregory, A. G., Rope, B., Johnson, S. G. 1969. *J. Atmos. Terr. Phys.* 31:563
Cuny, Y. 1969. *Theory and Observations of Normal Stellar Atmospheres,* ed O. Gingerich, 73. Cambridge, Mass: MIT Press

Doyle, R. O. 1968. *Ap. J.* 153:987

Dupree, A. K., Goldberg, L. 1967. *Solar Phys.* 1:229

Dupree, A. K. et al 1970. *BAAS* 2:191

Dupree, A. K., Reeves, E. M. 1971. *Ap. J.* 165:599

Eddy, J. A., Léna, P. S., MacQueen, R. M. 1969. *Solar Phys.* 70:330

Fredga, K. 1969. *Solar Phys.* 9:358

Gabriel, A. H. et al 1971. To be submitted to *Ap. J.*

Garz, T., Holweger, H., Kock, M., Richter, J. 1969. *Astron. Ap.* 2:446

Gingerich, O., de Jager, C. 1968. *Solar Phys.* 3:5

Gingerich, O., Noyes, R. W., Kalkofen, W., Cuny, Y. 1971. *Solar Phys.* In press

Goldberg, L. 1967. *Ann. Rev. Astron. Ap.* 5:279

Goldberg, L., Noyes, R. W., Parkinson, W. H., Reeves, E. M., Withbroe, G. L. 1968. *Science* 162:95

Hall, L. A., Hinteregger, H. E. 1970. *J. Geophys. Res.* 75:6959

Hearn, A. G. 1969. *MNRAS* 142:53

Hearn, A. G., Noyes, R. W., Withbroe, G. L. 1969. *MNRAS* 144:351

Jones, R. A., Rense, W. A. 1970. *Solar Phys.* 15:317

Jordan, C. 1965. *Thesis.* Univ. London

Jordan, C. 1971. *Highlights of Astronomy,* XIV IAU Gen. Assembly. Dordrecht-Holland: Reidel. In press

Kopp, R. A. 1968. *Thesis.* Harvard Univ. Cambridge, Mass.

Kopp, R. A., Kuperus, M. 1968. *Solar Phys.* 4:212

Krieger, A. S., Viana, G. S., Van Speybroeck, L. P. 1971. *IAU Symp. No. 43.* In press

Kuperus, M., Athay, R. G. 1967. *Solar Phys.* 1:361

Leighton, R. B. 1963. *Ann. Rev. Astron. Ap.* 1:19

Lemaire, P. 1969. *Ap. Lett.* 3:43

Linsky, J., Avrett, E. H. 1970. *PASP* 82:169

Munro, R. H., Dupree, A. K., Withbroe, G. L. 1971. *Tech. Rep. 23,* Harvard Coll. Obs. *Solar Phys.* In press

Neupert, W. M. 1969. *Ann. Rev. Astron. Ap.* 7:121

Noyes, R. W., Kalkofen, W. 1970. *Solar Phys.* 15:120

Noyes, R. W., Withbroe, G. L., Kirshner, R. P. 1970. *Solar Phys.* 11:388

Parker, E. N. 1963. *Ap. J.* 138:552

Parkinson, W. H., Reeves, E. M. 1969. *Solar Phys.* 10:342

Pottasch, S. R. 1964. *Space Sci. Rev.* 3:816

Purcell, J. D., Tousey, R. 1960. *J. Geophys. Res.* 65:370

Purcell, J. D., Tousey, R. 1961. *Mem. Soc. Roy. Sci. Liège* 274:283

Purcell, J. D., Tousey, R. 1969. *BAAS* 1:290

Reeves, E. M., Parkinson, W. H. 1970. *Ap. J. Suppl.* 21:1

Reimers, D. 1971. Preprint, submitted to *Astron. Ap.*

Sando, K., Doyle, R. O., Dalgarno, A. 1969. *Ap. J.* 157:L143

Sloan, W. A. 1968. *Solar Phys.* 5:329

Smith, E. van P. 1960. *Ap. J.* 132:202

Speer, R. J. et al 1970. *Nature* 226:249

Thomas, R. N. 1960. *Mem. Soc. Roy. Sci. Liège* 4:305

Thomas, R. N., Athay, R. G. 1961. *Physics of the Solar Chromosphere.* New York: Interscience

Tousey, R. 1963. *Space Sci. Rev.* 2:3

Tousey, R. 1967. *Ap. J.* 149:239

Tousey, R. 1971a. *IAU Symp. No. 41.* In press

Tousey, R. 1971b. *Phil. Trans. Roy. Soc.* In press

van de Hulst, H. C. 1950. *Bull. Astron. Inst. Neth.* 11:135

Widing, K. G., Purcell, J. D., Sandlin, G. D. 1970. *Solar Phys.* 12:52

Withbroe, G. L. 1970a. *Solar Phys.* 11:42

Withbroe, G. L. 1970b. *Solar Phys.* 11:208

Withbroe, G. L. 1971. *Solar Phys.* In press

THE FORMATION OF SPECTRAL LINES

D. G. Hummer[1]

Joint Institute for Laboratory Astrophysics, University of Colorado, Boulder, Colorado

G. Rybicki

Smithsonian Astrophysical Observatory, Cambridge, Massachusetts

1. INTRODUCTION

This review is addressed to the nonspecialist reader and is intended to provide an introduction to the physics and phenomenology of the radiation field in spectral lines. Results that give the maximum insight are stressed and selected, while technical details are suppressed. Because of restrictions on space and because of the appearance or promise of several other relevant books and reviews, we have omitted all discussion of direct astrophysical application. We especially recommend the recent books of Jefferies (1968) and Mihalas (1970) for discussions of line formation in stellar atmospheres, the review of the solar H and K lines by Linsky & Avrett (1970), the forth-coming review by Mihalas (1972) on theoretical stellar spectra, and the review of line formation in planetary nebulae by Hummer (1968a). We exclude all discussion of computational methods, for which see the reviews of Hummer & Rybicki (1967), Avrett (1971), and Auer (1971). The important work of the Leningrad school could not be given the prominence it deserves, but this omission will be partially redressed by the translation, now under way, of Ivanov's (1971) remarkable book. Discussion of the relevant aspects of collision-broadening theory and of atomic-collision theory has had to be omitted, but for the latter the reader can consult the excellent review by Bely & Van Regemorter (1970). Finally, we make no attempt to discuss or even to refer to all work on line formation. Whenever possible, we refer the reader elsewhere for comprehensive literature discussions.

Our attention is confined to problems involving the transfer of radiation in which optical-depth effects are, or can be, significant and in which it cannot be assumed a priori that atomic levels are populated according to the laws of Saha and Boltzmann. These problems are, of course, included in the category that has (unfortunately, in our opinion) become known as non-LTE or NLTE. Because the meaning of this expression continues to cause confusion, it is worth clarifying once again. Thermodynamic equilibrium

---

[1] Staff member, National Bureau of Standards and Department of Physics, University of Colorado.

(TE) implies equilibrium among all three components of the system: radiation, electrons, and the atoms and ions of the gas. Thus the distribution laws of Planck, Saha-Boltzmann, and Maxwell hold for radiation, atoms, and electrons, respectively. In local thermodynamic equilibrium (LTE), the atoms and electrons are in equilibrium with themselves and with each other; that is, the Saha-Boltzmann distribution laws are valid, while the radiation field is determined by the radiative-transfer equation and the boundary conditions on the system. Finally, in NLTE, only the electrons are assumed to be in equilibrium with one another, with velocities distributed according to Maxwell's law corresponding to a well-defined temperature that may vary from point to point. What might be called the restricted NLTE problem is to find the atomic populations and the radiation field at each point, given the run of kinetic temperature (or temperatures, if not all subsystems are in equilibrium), of electron density, of the density of each chemical species, and of the macroscopic velocity field. The general NLTE problem includes as well a self-consistent determination of the atmospheric parameters such as the temperature distribution.

It is important to understand that the NLTE approach does not deny the existence of LTE, any more than LTE denies TE, but only asserts that LTE cannot, in general, be assumed a priori. Under certain conditions, parts of a given system can be accurately described by the assumptions of LTE, and it is a proper part of NLTE theory to identify the circumstances in which this simplification occurs.

The NLTE program is a very demanding one. In the first place, it is essential that all physical processes of conceivable relevance be identified and included, at both microscopic and macroscopic levels. Thus, in addition to the radiative-transition probabilities and collision-broadening parameters required in LTE calculations, a large variety of collision cross sections, recombination coefficients, and the like must be known. This situation is acerbated by the need for atomic data for many transitions in each atom of interest. Although at present data of such completeness do not exist, recent developments in atomic physics offer promise that much of the necessary data will be forthcoming within the next few years. Among the macroscopic effects playing a role are those arising from velocity fields, including turbulent ones, and again we are brought into contact with aspects of physical theory that are not well developed.

Because of the number and diversity of the physical processes that generally must be considered simultaneously and because of the special problems relating to the solution of the transfer problems that arise in NLTE theory, very substantial technical problems have had to be solved before the NLTE methodology could usefully be applied to specific problems. Since so much time and effort have been expended in this phase of the work with little of clear and direct astronomical consequence resulting, at least in the context of stellar astronomy, a wide gap has appeared between astronomers on the one hand and astrophysical specialists in NLTE on the other. Extreme statements to the effect that "NLTE theory has made no contribution to astronomy," or that, conversely, "all interpretations made on the basis of LTE are wrong," illustrate the extent of that gap, which we hope this article will help to bridge. Part of the difficulty is that the expression NLTE is associated in the minds of many only with applications in stellar atmospheres and solar physics, when in fact theoretical work in planetary nebulae and the interstellar medium, based on the same approach, has been widely accepted for many years. Subject to the availability of atomic data, much more of direct relevance to astronomy will shortly be forthcoming, of the same nature as the fine work of Auer & Mihalas (1971) on the spectra of O stars. The more detailed nature of the new theory, on the other hand, will generally require for its application high-quality observational material, including high-dispersion profiles.

Following a discussion of the basic physics required for the theory, various aspects of that very useful protype, the *two-level atom*, are considered as an illustration of current theoretical ideas. The final major section extends the discussion to systems in which more than two levels are taken into account.

2. BASIC PROCESSES

Two general classes of physical processes are of interest here—those that transform kinetic energy, primarily that of the electron gas, into radiative or excitation energy and vice versa, and those that change the frequencies at which photons are absorbed and emitted by the particles composing the gas or plasma.

2.1 SOURCES AND SINK OF RADIANT ENERGY

The simplest mechanism for introducing energy into a spectral line is collisional excitation of the upper level of the transition, which subsequently decays with the emission of a photon. In astrophysical settings the excitation energy is usually supplied from the kinetic energy of the exciting particle, although direct transfer of excitation from an impinging atom or ion may be important in some cases where the relevant cross sections are large.

Radiative, dielectronic, and three-body recombination also lead to the formation of excited states at the expense of electron kinetic energy; the first two also make a direct contribution to the radiation field.

The conversion of radiative to kinetic energy occurs through the collisional deexcitation or ionization of an atom left in an excited state following the absorption of a photon. Photoionization is a direct process for converting radiative to kinetic energy. On the other hand, extinction of a photon by dust can be an effective loss mechanism for visual and ultraviolet photons, although the energy reappears in the infrared.

The balance of all these sources and sinks, including loss through the boundaries of the gas, provides the basis for our theory.

For electron and proton collisions, the rate constants for excitation and deexcitation are connected in a simple way through the well-known detailed balance relations. If the particles colliding with the atom in question have internal structure, quite large rate constants may be appropriate, Moreover, as the internal states of these particles may well not be in accord with the Saha-Boltzmann laws, the collisional *rates* for excitation and deexcitation cannot be determined a priori. Even the familiar detailed balance relations for the rate constants have to be replaced by much more complex ones.

Laboratory experience indicates that reactions of the following types may occasionally be important in the deexcitation of an excited atom A^*:

$$A^* + B \rightarrow A + B^* + (h\nu)$$

$$A^* + B \rightarrow AB^+ + e$$

$$A^* + B \rightarrow A + B^+ + e$$

where $(h\nu)$ indicates that the reaction may depart slightly from exact resonance. As examples are known of reactions of these types whose rate constants are competitive with those for electron excitations, in each application such possibilities should be kept in mind, especially if the degree of ionization is low.

2.2 RADIATIVE PROCESSES

The second category of important processes consists of those affecting the absorption, emission, and scattering of radiation. Although collision broadening of the absorption process is of substantial interest this topic cannot be discussed here. The situation regarding photon emission has received much less attention. (We remark that in nearly all work on collision broadening, the frequency dependence of the emission following "white" or nonresonant excitation of the upper level is discussed, with the absorption profile following from assumptions of thermal equilibrium.)

An aspect of this problem that is currently of central importance concerns the slight shift in frequency that a photon undergoing an absorption-emission or scattering process may experience. This effect is referred to as *frequency noncoherence* and is important because it allows a photon created in the central region of a line, where the opacity is large, to shift into the relatively transparent line wing, enabling them to move more easily to some remote part of the atmosphere or escape. The relation of the frequency and direction of the photon before scattering (ν',n') to the same quantities after scattering (ν, n) is expressed by the *redistribution function* $R(\nu'n'; \nu n)$. Obviously, this function must be known before a complete treatment of the transfer problem can be made.

Frequency noncoherence arises from several mechanisms. Under many circumstances the most important is the Doppler effect, since the projection of the atom's velocity on the initial direction of the photon is generally quite different from that on its final direction. Typical shifts for this process are of the order of a thermal Doppler width; much smaller shifts to longer wavelengths arise from atomic recoil, which has been studied in detail by Field (1959) (recoil effects for electrons are naturally more important—an interesting treatment has been given by Weymann 1970). The theory associated with Doppler noncoherence was worked out by Levich (1940) and Henyey (1941). This work is generalized and summarized by Hummer (1962), who gives a thorough discussion of the relevant literature.

Natural and collisional broadening of the atomic levels leads to noncoherence within the rest frame of the atom. For natural broadening, noncoherence can occur only if the lower level is broadened, so that the uncertainty in the energy of the lower level is reflected in the difference between the initial and the final frequency. Although this process was discussed long ago by Weisskopf (1931, 1933), Spitzer (1936), and Woolley (1938), the basic theory is not yet satisfactory. Further discussion of this point is deferred to the end of this section. Recently, House (1970a, b) has developed redistribution functions for linearly polarized radiation including the effects of a magnetic field.

Collisional broadening will be accompanied by noncoherence if the emitting atom suffers perturbations on a time scale shorter than the mean lifetime of the upper level. Although a number of investigations (Spitzer 1944, Zanstra 1941, 1946, Holstein 1950, Edmonds 1955, Towne 1954, Huber 1969) have been addressed to aspects of this problem, we do not yet have anything like a complete theory for this phenomenon. Unfortunately, space is not available to discuss the present position; Hummer (1965a, b) gives a discussion of work to that date.

Oxenius (1965) has suggested that frequency redistribution can occur if the velocity of the emitting atom is changed by collision during its lifetime. Following this suggestion, Rees & Reichel (1968) have developed redistribution functions incorporating the assumptions that the final velocity is completely uncorrelated with its initial value and that the internal state of the atom is unaffected by the velocity-changing collisions. This treatment is unsatisfactory in several respects: 1. the necessary momentum can be obtained only in heavy-particle collisions for which the rates under most astrophysical conditions are small; 2. it is hard to imagine that collisions sufficiently strong to transfer the required momentum could leave unchanged the atomic state; and 3. instantaneous changes of the atomic velocity introduce phase changes into the wavetrain similar to those arising in impact broadening, an effect entirely neglected by Rees & Reichel.

The importance of collisions in this theory justifies commenting on the four different roles they play: 1. *Source of radiant energy*. For lines in the visual and ultraviolet, these collisions arise primarily from electrons in the high-velocity tail of the Maxwellian distribution; transitions at longer wavelengths will be excited by electrons of low and moderate velocities and by high-velocity protons. Lines emitted in the far-infrared and microwave region will be excited by neutral particles, protons, and slow electrons. 2. *Sink of radiant energy*. 3. *Collisional broadening and redistribution*. Depending on circumstances, these collisions may involve neutral particles, ions, or slow electrons. In general, these particles will differ from those responsible for excitation and deexcitation of excited levels. 4. *Momentum-changing collisions*. Clearly, electrons cannot be effective; which particles, if any, do play a role here must be investigated.

2.3 PHYSICAL UNCERTAINTIES

It would be wrong to give the impression that the only remaining obstacles are technical. We are aware of a number of unsolved problems and would like to discuss them briefly here, without assessing their astrophysical implications. One difficulty made obvious by the succession of papers each rebutting the preceding is that of the transfer equation for dispersive media. The most recent work on this problem is that of Enome (1969).

A particular vexing problem is that of the redistribution function for transitions between two naturally broadening levels, mentioned above. By treating the broadened levels as made up of a continuum of physically meaningful substates, Woolley (1938) derived an expression relating the frequencies of the initial and final photons; the derivation is repeated by Woolley & Stibbs (1953). On the other hand, Weisskopf (1931, 1933) has derived a different expression regarding the act of scattering as a single quantum process; Heitler (1954) has obtained the same result from a different treatment. Woolley's result is symmetric in the two frequencies so that detailed balance can be satisfied and integration over the initial frequency gives the correct emission for white-light excitation, namely a Lorentz profile with a half-width equal to the sum of the half-widths of the two levels; on the other hand, Weisskopf's expression is not symmetric in the frequencies and integration over initial frequencies gives an emission profile with a half-width equal to that of the upper level alone. Recently, Dr. John Castor pointed out to us that Weisskopf's and Heitler's treatments are based on the unrealistic assumption that the final state does not interact with any other states and so has a perfectly sharp energy.

If resonance fluorescence is correctly treated as a single quantum process, it will depend on the particular "substate" of the naturally broadened level, so that any scattering depends on the history of the system; i.e., the scattering process becomes non-Markovian. In this situation, the meaning and use of statistical equilibrium equations are open to question. While it may be that the Markovian nature of the process can be restored by an appropriate choice of representations, this problem remains a nagging one.

It is also unsatisfactory that no convincing derivation of the transfer equation from first principles has been given, especially as no experimental verification of its validity has appeared. While the transfer equation has a strong intuitive appeal and seems likely to be adequate for low-intensity, phase-noncoherent radiation with reasonably uniform frequency and spatial variation, we have little reason to accept its validity otherwise. Recent attempts to provide a foundation for this theory have been presented, for example, by Gelinas & Ott (1970), who give references to similar work.

3. THE TWO-LEVEL ATOM

The model problem in which all but the upper and lower levels associated with the transition of interest are ignored has been a valuable prototype in the development of our understanding of line formation. Its simplicity allows one to examine directly the way in which various physical processes affect the radiation field and to elucidate the transfer process itself. The concepts and many of the techniques can be applied to the study of the more complicated problems in which three or more levels are considered simultaneously as well as to problems involving the scattering and absorption of continuum radiation. Finally, the two-level model is directly applicable, in certain circumstances, to resonance lines and to subordinate lines whose lower level is either metastable (e.g., He I $\lambda 10830$ Å) or in detailed balance with the ground level, as will be discussed in the final section.

The quantitative basis for the theory is provided by the equations of statistical equilibrium and of radiative transfer, which relate the specific intensity $I(\nu,\mathbf{n})$ of radiation of frequency ν flowing in direction $\mathbf{n}$ and the atomic densities of the lower and upper levels, N_1 and N_2, respectively. The interaction of the radiation field and matter is expressed in terms of the Einstein coefficients A_{21}, B_{12}, and B_{21}, referring to spontaneous emission, absorption, and stimulated emission, respectively, and the collisional excitation and deexcitation rate constants C_{12} and C_{21}, where

$$C_{12} \equiv N_e \langle v Q_{12} \rangle = (g_2/g_1) \exp\left(- h\nu_0/kT_e\right) C_{21} \qquad 1.$$

g_1 and g_2 being statistical weight factors and $h\nu_0$ the energy of the transition. Equating the rates of upward and downward transitions, we have the condition of statistical equilibrium

$$N_1(B_{12}\bar{J} + C_{12}) = N_2(A_{21} + C_{21} + B_{21}\bar{J}) \qquad 2.$$

where

$$\bar{J} \equiv \int d\nu \phi(\nu) J(\nu) = \int\int d\nu d\Omega \phi(\nu) I(\nu\mathbf{n})/4\pi \qquad 3.$$

Here $\phi(\nu)$ is the normalized absorption coefficient, i.e.,

$$k(\nu) = k\phi(\nu), \qquad \int \phi(\nu) d\nu = 1 \qquad 4.$$

The transfer equation

$$(d/ds)I(\nu\mathbf{n}) = - k(\nu)I(\nu\mathbf{n}) + \mathcal{E}(\nu\mathbf{n}) \qquad 5.$$

relates the change in the intensity of radiation of frequency ν along an increment of pathlengths ds in direction $\mathbf{n}$ to the phenomenological coefficients of absorption and emission, $k(\nu)$ and $\mathcal{E}(\nu\ \mathbf{n})$, respectively. For unpolarized radiation in the absence of an external electromagnetic field and with no macroscopic velocity fields, $k(v)$ does not depend on direction, whereas $\mathcal{E}(\nu\,\mathbf{n})$ may do so. If we neglect polarization and treat stimulated emission as negative absorption, the frequency-independent factor in $k(\nu)$ for a line with central frequency ν_0 is conventionally written as

$$k = (h\nu_0/4\pi)(N_1 B_{12} - N_2 B_{21}) \qquad\qquad 6.$$

Although this treatment of stimulated emission is in principle incorrect, as shown by Oxenius (1965), Hummer (1965a, b) gives reasons why no serious error results for most cases of interest in astrophysical applications. The treatment of macroscopic velocity fields has been reviewed by Rybicki (1970).

The expression for the emissivity $\mathcal{E}(\nu\,\mathbf{n})$ is rather more complex. The contribution to $\mathcal{E}(\nu\,\mathbf{n})$ from scattering is conveniently expressed in terms of the redistribution function introduced in the preceding section. $R(\nu'\mathbf{n}',\nu\mathbf{n})\ d\nu'$ $d\Omega'\ d\nu\ d\Omega$ gives the probability that a photon with frequency in $\nu',\nu'+d\nu'$ and direction within $d\Omega'$ about $\mathbf{n}'$ is absorbed from a radiation field of unit strength and reemitted in the same line with frequency in the range $(\nu,\nu+d\nu)$ and direction within $d\Omega$ about $\mathbf{n}$, under the assumption that no process destroying the excited atomic state intervenes. The first set of arguments of R will always refer to the incident photon. Since in this picture each photon absorbed must ultimately be emitted, we have

$$\iint d\nu' d\Omega' R(\nu\mathbf{n},\ \nu'\mathbf{n}') = \phi(\nu)/4\pi \qquad\qquad 7.$$

and from the normalization of $\phi(\nu)$,

$$\iiiint d\nu' d\Omega' d\nu d\Omega R(\nu\mathbf{n},\ \nu'\mathbf{n}') = 1 \qquad\qquad 8.$$

In what follows, we shall make the conventional assumption that collisional deexcitation destroys excited atoms without modifying the frequency dependence of the emission by the remaining atoms. That this cannot be strictly correct is apparent from the observation that if the collision deexcitation rate is comparable to that for spontaneous transitions, the lifetime of the upper level would be reduced significantly. This effect should, of course, lead to a broadening of the absorption and emission profiles, which simply is not included in any theory to date. (In practice, this kind of broadening might well be masked by the broadening due to elastic collisions.) The satisfactory resolution of this problem awaits the development of a complete theory of collisional redistribution. If, as is frequently the case, the collisional deexcitation rate is small in comparison to the spontaneous decay rate, the current practice is probably satisfactory.

In addition to the atoms excited by line radiation, we have to consider the effects of collisional excitation or, more generally, atoms excited by any nonresonant or broadband processes, such as recombination or absorption of white light. It may be shown by a variety of arguments (cf Heitler 1954) that such states decay with a frequency distribution identical to $\phi(\nu)$.

We can now write down an expression for the energy emitted per cubic centimeter per second in the frequency interval ν, $\nu+d\nu$ into the element of solid angle $d\Omega$ about the direction $\mathbf{n}$:

$$\mathcal{E}(\nu\mathbf{n})d\nu d\Omega = \left\{ (N_1 B_{12}\bar{J}) \left(\frac{A_{21}}{A_{21} + B_{21}\bar{J} + C_{21}} \right) \right.$$

$$\cdot \left[\frac{\iint d\nu' d\Omega' I(\nu'\mathbf{n}') R(\nu'\mathbf{n}', \nu\mathbf{n})}{\bar{J}} \right]$$

$$\left. + (N_1 C_{12}) \left(\frac{A_{21}}{A_{21} + B_{21}\bar{J} + C_{21}} \right) \frac{\phi(\nu)}{4\pi} \right\} h\nu_0 d\nu d\Omega \qquad 9.$$

The first term, corresponding to radiative excitation, is composed of three factors, which represent, respectively, 1. the rate at which atoms enter the upper level by radiative excitation from the lower, 2. the fraction of atoms leaving the upper level that do so by spontaneous radiative transitions, and 3. a normalized function of frequency and direction depending functionally on the radiation field. In the second term, corresponding to nonresonant excitation, the factors are 1. the rate at which atoms enter the upper level by collisional excitation from the lower, 2. same as 2 above, and 3. the normalized frequency function characteristic of nonresonant excitation. It is interesting that this expression is proportional to the density of lower-state atoms and independent of the density and velocity of the excited-state atoms, at least explicitly.

3.1 THE SOURCE FUNCTION AND ITS APPROXIMATE FORMS

It is convenient to rewrite the transfer equation as

$$(d/ds)I(\nu\mathbf{n}) = - k(\nu)[I(\nu\mathbf{n}) - S(\nu\mathbf{n})] \qquad 10.$$

where

$$S(\nu\mathbf{n}) \equiv \mathcal{E}(\nu\mathbf{n})/k(\nu) \qquad 11.$$

is called the *source function*. If the source function is known throughout the medium, then Eq. 10 can be integrated to obtain $I(\nu\ \mathbf{n})$ (if the boundary conditions are known), so that knowing S is equivalent to solving the problem. To fix ideas, it is helpful to remember that in LTE,

$$S(\nu\mathbf{n}) = B(\nu, T_e) \quad (\text{LTE}) \qquad 12.$$

Substituting Eqs. 9 and 6 for ϵ and k into Eq. 11 and eliminating the atomic populations through the equation of statistical equilibrium 2, we obtain

$$S(\nu\mathbf{n}) = (1 - \epsilon)[4\pi/\phi(\nu)] \int_0^\infty d\nu' \int_{4\pi} d\Omega' R(\nu'\mathbf{n}', \nu\mathbf{n}) I(\nu'\mathbf{n}')$$

$$+ \epsilon B(\nu_0, T_e) \qquad\qquad 13.$$

where

$$\epsilon \equiv C_{21}/\{C_{21} + A_{21}[1 - \exp(-h\nu_0/kT_e)]^{-1}\} \qquad 14.$$

is the probability per scattering that a photon is lost by collisional deexcitation, and $B(\nu_0, T_e)$ is the Planck function at line center for the local electron temperature T_e. It is clear from Eqs. 13 and 14 that if $C_{21} \gg A_{21}$, the LTE limit (Eq. 12) is recovered.

The derivation of $\mathcal{E}(\nu \mathbf{n})$ and the use of the statistical equilibrium equation implicitly assume that the scattering process occurs in two stages—absorption followed by emission—and thus cannot correctly treat resonance fluorescence, a single quantum-mechanical process. The consequences of this assumption are not clear.

If the redistribution function appropriate to the situation at hand is available, then $S(\nu \mathbf{n})$ is a known functional of the radiation field and the transfer problem can in principle be solved. Because of the very considerable complexity of this source function, a hierarchy of approximations has been developed that leads to rather simpler transfer problems.

The *isotropic* source function is obtained by replacing $I(\nu'\mathbf{n}')$ in the expression for $S(\nu \mathbf{n}')$ by its mean value

$$J(\nu') = (1/4\pi) \int d\Omega' I(\nu'\mathbf{n}') \qquad\qquad 15.$$

leading to the result

$$S(\nu) = (1 - \epsilon)[1/\phi(\nu)] \int_0^\infty d\nu' R(\nu', \nu) J(\nu') + \epsilon B \qquad 16.$$

which is independent of direction. Here we have introduced

$$R(\nu', \nu) = 4\pi \int d\Omega' R(\nu'\mathbf{n}', \nu\mathbf{n}) \qquad\qquad 17.$$

A further set of approximations leads to the *frequency-independent* source function. If for some reason all correlation between the initial and the final photon vanishes, $R(\nu', \nu)$ is separable, and if the normalization conditions are to be satisfied, $R(\nu', \nu)$ must have the form

$$R(\nu', \nu) = \phi(\nu')\phi(\nu), \quad (\nu, \nu' \text{ uncorrelated}) \qquad 18.$$

in which case

$$S = (1 - \epsilon) \int_0^\infty d\nu \phi(\nu) J(\nu) + \epsilon B \qquad 19.$$

The same frequency-independent result is obtained if $J(\nu)$ is independent of frequency, as is clear from the normalization of $R(\nu', \nu)$,

$$\int_0^\infty d\nu' R(\nu, \nu') = \phi(\nu) \qquad 20.$$

By the same argument, when $J(\nu)$ varies slowly with frequency, S is approximately independent of frequency. In particular, if the line is formed primarily in the Doppler core of the absorption coefficient, the approximation that S is independent of frequency in this region is surprisingly good. Quantitative comparisons are given in Section 3.6. It is perhaps worth remarking that although the expression *complete redistribution* refers strictly to situations described by Eq. 18, it has come to be practically synonymous with *frequency-independent source function*.

The approximation of a frequency-independent source function was apparently first introduced by Houtgast (1942). Further discussions, in some cases independent of earlier work, were given by Spitzer (1944), Sobolev (1944, 1949), Biebermann (1947), Holstein (1947), Zanstra (1949, 1951a), and Thomas (1957). Because of the very substantial simplification in the transfer problem that results from this approximation, almost without exception recent work on line formation has been based on it without regard as to whether it is, in fact, valid. The situation is uncomfortably close to the earlier one in which LTE was assumed without question, simply because calculations could then be made.

The expression for the source function can also be written in another form that is sometimes illuminating. Since emission rate is proportional to $N_2 A_{21}$, regardless of the mode of excitation, we can represent the emissivity formally as

$$\mathcal{E}(\nu \mathbf{n}) = (h\nu_0/4\pi) N_2 A_{21} \psi(\nu) \qquad 21.$$

where $\psi(\nu)$ is a normalized function of frequency (in general, a complicated functional of the radiation field). The source function is then

$$S(\nu, \mathbf{n}) = N_2 A_{21} \psi(\nu) / (N_1 B_{12} \phi(\nu) - N_2 B_{21} \psi(\nu)) \qquad 22.$$

where we have now treated the effects of stimulated emission correctly, at least in a symbolic way. Thus we see that the source function is a measure of N_2/N_1, the degree of excitation and in many situations is proportional to this ratio. The assumption of complete redistribution is simply that $\psi(\nu) = \phi(\nu)$, and the result (Equation 19) can be recovered by the use of the equation of statistical equilibrium.

3.2 The Standard Problem

The transfer problem for a frequency-independent source function in a plane-parallel medium with constant temperature and density is particularly simple, especially if no other opacity source is present, and consequently has received a great deal of attention. Since a number of results, one or two of them exact, have been obtained and are well understood, this "standard problem" is very useful in acquiring insight into the behavior of the radiation field in a spectral line and so will be discussed in some detail here. Subsequently, a number of variants will be discussed in the context of the results of this section.

The appropriate transfer equation is

$$\mu(d/d\tau)I(x\mu\tau) = \phi(x)\left[I(x\mu\tau) - S(\tau)\right] \qquad 23.$$

where $S(\tau)$ is the frequency-independent source function defined by Eq. 19. We have introduced τ, the mean optical depth in the line, which is related to the increment of geometrical depth dz by

$$d\tau = kdz \qquad 24.$$

and μ, the cosine of the angle between the direction n and the outward normal. In Equation 24 the quantity k is that defined in Equation 6. The line-center optical depth is $\tau = \tau\phi(0)$. The atmosphere has free surfaces at $\tau = 0$ and $\tau = T$, and no radiation is assumed to fall on either face. If $T = \infty$, we refer to a semi-infinite atmosphere. It is convenient to use the dimensionless frequency variable

$$x = (\nu - \nu_0)/\Delta \qquad 25.$$

where Δ is some characteristic frequency interval, such as the Doppler width at some representative temperature. The profile function $\phi(x)$ is now normalized to unity on the interval $-\infty < x < \infty$. We regard ϵ and B in the source function and T as parameters and solve for $S(\tau)$ and/or $I(x\mu\tau)$.

If we assume for the moment that $S(\tau)$ is known, we can formally integrate Eq. 23 to obtain

$$I(x\mu\tau) = \begin{cases} [\phi(x)/\mu] \displaystyle\int_{\tau}^{T} d\tau' S(\tau')\, \exp\left[-(\tau' - \tau)\phi(x)/\mu\right] & \mu > 0, \\[4mm] [-\phi(x)/\mu] \displaystyle\int_{0}^{\tau} d\tau' S(\tau')\, \exp\left[-(\tau' - \tau)\phi(x)/\mu\right] & \mu < 0 \end{cases} \qquad 26.$$

from which follows the expression for the mean intensity

$$J(x, \tau) = [\phi(x)/2] \int_{0}^{T} dt' E_1[\phi(x)\,|\,\tau - \tau'\,|\,]S(\tau') \qquad 27.$$

Expressions similar to 26 and 27 are readily obtained for the case when $\phi(x)$ depends upon τ.

Here, $E_1(x)$ is the first exponential-integral function. The emergent intensity at $\tau = 0$ or $\tau = T$ follows immediately from substituting these values into Eq. 26.

There are two alternative ways to proceed. Using $S(\tau)$ expressed in terms of the mean intensity J, according to Eq. 19, we have a differential equation in radiation quantities I and J. Alternatively, we can use Eqs. 27 and 19 to obtain an integral equation for $S(\tau)$:

$$S(\tau) = (1 - \epsilon) \int_0^T d\tau' K_1(|\tau - \tau'|) S(\tau') + \epsilon B \qquad 28.$$

where

$$K_1(\tau) \equiv (1/2) \int_{-\infty}^{\infty} dx \phi^2(x) E_1[\tau \phi(x)] \qquad 29.$$

Both integral and differential forms of the problem can readily be solved by various numerical techniques; analytical work appears to be confined to the integral form. A third basic approach, based on the flux integral, has been introduced by Zanstra (1949, 1951b) and was recently exploited by Kuhn (1966) and Athay & Skumanich (1967).

Let us first summarize the results for a semi-infinite atmosphere, $T = \infty$, keeping in mind that ϵ and B are assumed to be constant. In this case, $S(\tau)$ increases monotonically from

$$S(0) = \sqrt{\epsilon} B \qquad 30.$$

to

$$S(\infty) = B \qquad 31.$$

as τ increases from 0 to ∞. These limiting forms are exact and independent of the form of $\phi(x)$. We have already seen that $S = B$ everywhere in LTE, that is, where collisions are sufficiently strong to ensure that $\epsilon = 1$. Thus, when $\epsilon \ll 1$, deviations from LTE are very substantial near the surface and decrease inward, so that at sufficiently great depth, conditions once more approach those of LTE. Informative graphs of $S(\tau)$ for various values of ϵ and forms of $\phi(x)$ appear in Figures 2 and 3 of Avrett & Hummer (1965) (reproduced in Figures 1 and 2 of Hummer (1968b) and Figures 12–1 and 12–2 of Mihalas (1970)) and Figures 1 and 4 of Avrett (1965). From these graphs it can readily be seen that as the profile function $\phi(x)$ decreases more slowly with x_2 (for example, as the Voigt parameter a increases), $S(\tau)$ increases more slowly with τ. The depth at which $S(\tau)$ approaches B is discussed in the next section.

For finite values of T, no exact results are known. However, as T decreases to 0, the surface value approaches the limit

$$\lim_{T \to 0} S(\tau = 0) = \epsilon B \qquad \qquad 32.$$

and in general the source function as a whole decreases with T. Thus, deviations from LTE *increase* as the atmosphere becomes thinner, since the escape of radiation is facilitated. For fixed τ and T, $S(\tau)$ increases with ϵ. Graphs of $S(\tau)$ for various combinations for the parameters ϵ, T, and a are given in Figures 4 and 5 of Avrett & Hummer (1965) (reproduced in Figures 3 and 4 of Hummer 1968b and Figure 12–3 of Mihalas 1970).

The intensity of radiation emerging from an atmosphere of thickness T is an emission line with no continuum. If T is substantially larger than unity, a significant central reversal will occur, reflecting the condition $S<B$ in the boundary region. As T becomes infinite, the self-reversed emission line becomes infinitely wide with the self-reversal becoming an absorption line. Again, the reader is referred to Avrett & Hummer (1965) where Figure 6 illustrates this progression graphically. Part of this material is reproduced in Figure 5 of Hummer (1968b) and Figure 12–4 of Mihalas (1970). It is important to stress that these absorption features arise only because of departures from LTE.

3.3 Escape Probability, Mean Number of Scatterings, and Thermalization Length

In this section we turn to a number of concepts that have proved to be of considerable utility both in making preliminary estimates of various properties of the radiation field and in interpreting the results of detailed computations.

Escape probability.—Consider an excited atom at optical depth τ in an atmosphere of thickness T. The probability that it will emit a photon of frequency x, $x+dx$ in the element of solid angle $d\Omega$ is

$$(1 - \epsilon)\phi(x)dxd\Omega/4\pi \qquad \qquad 33.$$

and the probability that this photon will escape *without further scattering* is

$$\begin{aligned} \exp\left[-\,\tau\phi(x)/\mu\right], & \qquad \mu > 0 \\ \exp\left[-(T - \tau)\phi(x)/\mu\right], & \qquad \mu < 0 \end{aligned} \qquad 34.$$

Integrating over all frequencies and over all directions, we find for the escape probability

$$P_e(\tau) = \left[(1 - \epsilon)/2\right]\left[K_2(\tau) + K_2(T - \tau)\right] \qquad 35.$$

where

$$K_2(\tau) \equiv \int_{-\infty}^{\infty} dx\phi(x)E_2[\tau\phi(x)] \qquad 36.$$

Capriotti (1965) has made a number of interesting suggestions regarding the calculation and application of escape probabilities.

The ratio of the escape probability to the deexcitation probability can be expressed as

$$u \equiv P_e(\tau)/\epsilon = [(1 - \epsilon)/2\epsilon][K_2(\tau) + K_2(T - \tau)] \qquad 37.$$

For $T = \infty$ and $\epsilon \ll 1$, Ivanov (1965) has shown that, to a good degree of approximation, the source function takes the form

$$S(\tau, \epsilon) = B[1 - F(u)] \qquad 38.$$

where, independent of the form of $\phi(x)$,

$$F(u) \sim \begin{cases} u, & u \ll 1 \\ 1 - 2/(\pi\sqrt{u}), & u \gg 1 \end{cases} \qquad 39.$$

Although this result is too large at $\tau = 0$ by about 25%, it is correct to order ϵ at large depths. Warming (1970) gives further discussion of this similarity principle.

Mean number of scatterings.—The probability that an excited atom is destroyed or that the photon escapes directly is $\epsilon + P_e(\tau)$. Since this probability is clearly a minimum at $\tau = T/2$, an upper bound to the mean number of scatterings $\langle N \rangle$ is the reciprocal of $\epsilon + P_e(T/2)$,

$$\langle N \rangle \leq 1/[\epsilon + (1 - \epsilon)K_2(T/2)] \qquad 40.$$

This relation was derived in another way by Hummer (1964), who showed that for noncoherent scattering this bound actually provides a good estimate of $\langle N \rangle$. Since $K_2(\tau)$ decreases from unity as τ increases from zero, we find for sufficiently large T

$$\lim_{T \to \infty} \langle N \rangle = 1/\epsilon \qquad 41.$$

as is required by the meaning of ϵ. In the other limit in which T is small enough so that $K_2(T/2) \gg \epsilon$, an insignificant fraction of the photons is destroyed, and we can use the asymptotic forms of $K_2(\tau)$ (cf Avrett & Hummer 1965, Appendix I) to obtain

$$\langle N \rangle \lesssim \begin{cases} T[\ln(T/2\sqrt{\pi})]^{1/2}, & \text{Doppler} \\ (3/2\sqrt{2})(T/a)^{1/2}, & \text{Voigt}, \epsilon < a < 0.1 \\ (3/2\sqrt{2})T^{1/2}, & \text{Lorentz} \end{cases} \qquad 42.$$

For coherent scattering, $\langle N \rangle \sim T^2$. Exact asymptotic expressions for $\langle N \rangle$ have been obtained by Nagirner (1967):

$$\langle N \rangle \sim \begin{cases} (T/2)[\ln(T/\sqrt{\pi})]^{1/2}, & \text{Doppler} \\ 1.11\,T^{1/2} & \text{Lorentz} \end{cases} \qquad 43.$$

Thermalization and related length scales.—Obviously the physical conditions for which the quantity u defined by Eq. 37 is much less than or much greater than unity must delineate two different but complementary regimes. Let us first consider the case in which T is infinite, and consider that $\epsilon \ll 1$. Then the value of τ for which $u = 1$, say $\tau = \tau_b$, is the solution of

$$K_2(\tau_b) = 2\epsilon/(1 - \epsilon) \qquad 44.$$

We refer to τ_b as the *boundary relaxation length* (on the optical-depth scale). For $\tau \gg \tau_b$ collisional deexcitation dominates photon escape, and conditions approaching those of LTE are found; thus τ_b provides an estimate of the depth at which $S \sim B$. If ϵ is small, τ_b will be large and we can use the asymptotics of $K_2(\tau)$ to find

$$\tau_b \sim \begin{cases} (C/4)\epsilon^{-1}, & \text{Doppler} \\ (a/9)\epsilon^{-2}, & \text{Voigt, } \epsilon < a < 0.1 \\ (1/9)\epsilon^{-2}, & \text{Lorentz} \end{cases} \qquad 45.$$

where C is a slowly varying function of ϵ of order unity.

For atmospheres of finite thickness T, u achieves its minimum value at $\tau = T/2$, so that we might expect atmospheres for which $[(1-\epsilon)/\epsilon]\,K_2(T/2)$ exceeds unity to be substantially different from those in which it is less than unity. If we define τ_c by

$$K_2(\tau_c/2) = \epsilon/(1 - \epsilon) \qquad 46.$$

then in atmospheres for which $T \ll \tau_c$, substantially all the photons escape, while for $T \gg \tau_c$, only a small fraction of them do so; we refer to these cases as *effectively thin* and *effectively thick*, respectively. In the first case, the radiation field and the degree of excitation are proportional to the photon creation rate, so that frequently a simplified treatment is possible, while for the effectively thick case, the destruction of photons must be treated carefully.

Scale lengths such as τ_b or τ_c are related to the more general idea of the thermalization length Λ, defined as the optical distance from the depth where a typical photon is created from thermal energy to the depth where the photon is destroyed. Thus, Λ provides a measure of the distance over which a perturbation of the thermal structure of the atmosphere will be felt through the radiation field in the line or continuum under discussion. While for the two-level atom the concept is obvious, the generalization to multilevel atoms is not.

The basic idea and the expression "thermalization length" are due to J.

T. Jefferies, although the first derivation of the correct dependence on ϵ was given by Avrett & Hummer (1965), who arrived at the form of Eq. 46. This approach was generalized by Hummer & Stewart (1966). Ivanov (1965a) subsequently proposed Eq. 44 as the definition, and later Ivanov & Nagirner (1966) gave a definition based on a point source in an infinite medium.

None of these procedures was entirely satisfactory, either because the precise nature of the thermalization length was unspecified or because it was defined in terms of processes, like the escape of photons, that are not an intrinsic part of the thermalization process. Rybicki & Hummer (1969) derived and solved an integral equation for the distribution of thermalization distances in an infinite atmosphere with a plane source and defined the thermalization length as the median of this distribution. This definition depended on ϵ in the same way as all those above, i.e., as in Eq. 45 but with different numerical factors. This approach was generalized by Hummer & Rybicki (1970) to include cases in which the frequency dependence of the emissivity could not be specified a priori. This more general view requires that the thermalization process be regarded as a function of the initial frequency of the photon and leads to a thermalization length for each thermalization mechanism.

All the discussion based on escape and destruction probabilities ignores the remaining possibility, that a photon travels some distance and is again scattered. Yet experience has shown that such arguments are generally quite successful for noncoherent scattering but fail for coherent scattering. Rybicki & Hummer (1969) derived the distribution function for pathlengths between scatterings and showed that for scattering with complete redistribution, the photon most often escapes with one very long flight, after making only short excursions about the point where it was created. For coherent scattering, this is not the case. Thus, all the above arguments can be understood on this basis.

3.4 Radiative Equilibrium

Avrett & Hummer (1965) pointed out that for a semi-infinite atmosphere with constant properties the source function could be expressed approximately as

$$S(\tau, \epsilon) \simeq \epsilon^{1/2} f(\tau), \quad 0 \leq \tau \lesssim \tau_\epsilon \qquad 47.$$

where $f(\tau)$ is a function independent of ϵ that increases monotonically from unity at $\tau = 0$, and τ_ϵ is $\sim 10 \, \Lambda^{1/2}$. Although the rate at which photons are created in the line is proportional to ϵ, the source function in the region $(0, \tau_\epsilon)$ is approximately proportional to $\epsilon^{1/2}$. This observation leads us to infer that the source function in this region is insensitive to the local values of ϵ and B and is controlled instead by radiation in the line wings flowing up from deeper layers where $S(\tau) \gg S(0)$. Thus, since as $\epsilon \to 0$ the region controlled by radiation from below becomes infinite in extent, Hummer & Stewart (1966) have argued that one can write

$$\lim_{\epsilon \to 0} [S(\tau, \epsilon)/(\epsilon^{1/2}B)] = S_H(\tau) \qquad 48.$$

where $S_H(\tau)$ is the $\epsilon = 0$ solution of Eq. 28 normalized so that $S_H(0) = 1$. At large depths, $S_H(\tau)$ is unbounded. Ivanov (1965b) obtained the limit (Equation 48) analytically.

The function $S_H(\tau)$ is the source function in a purely scattering atmosphere with an infinitely strong source of radiation situated at infinite depth. Since there are no sources or sinks of energy at any finite depth, the situation is in radiative equilibrium and so is analogous to the familiar Milne problem for a gray opacity.

The expression *radiative equilibrium* is used to indicate that all of the energy is transported radiatively. On the other hand, as the radiation field and the electron gas do not interact in this case, the temperature is not determined. Since $S(\tau, \epsilon) \simeq \epsilon^{1/2} S_H(\tau)$ for a certain distance from the surface, we can conclude that in such regions the flux carried by the line is essentially constant, even if ϵ and B vary moderately. If the profile $\phi(x)$ varies with depth, this conclusion may require modification, as discussed below.

Asymptotic expressions for solution of the Milne problem with complete redistribution are given by Ivanov (1962a,b, 1968), Hummer & Stewart (1966), Abramov, Dykhne & Napartovich (1967), and Warming (1970).

3.5 VARIANTS ON THE STANDARD PROBLEM

We now turn to a number of generalizations of the standard problem in which the assumption of complete redistribution for the line-forming atoms is retained. Each of these introduces some new physical or atmospheric feature that can conveniently be discussed in relation to the standards problem.

Continuous absorption.—The simplest and probably most important variant is that obtained by the addition of some source of continuous absorption and emission. Line photons may be lost in ionizing other atomic species or by extinction on dust grains. Conversely, processes like radiative recombination can create photons in the vicinity of the line in question. This problem has been systematically studied by Hummer (1968c); the following treatment is based on the work reported there.

By defining the source function for continuous absorption,

$$S_c \equiv \mathcal{E}_c/k_c \qquad 49.$$

where $\mathcal{E}_c$ and k_c are the emissivity and absorptivity of the continuous process, respectively, we can write the *total* source function as

$$S(\tau, x) = [\beta S_c(\tau) + \phi(x)S_L(\tau)]/[\beta + \phi(x)] \qquad 50.$$

Here, $S_L(\tau)$ denotes the *line* source function, given by Equation 19, and

$$\beta \equiv k_c/k_L \qquad 51.$$

is the ratio of the continuous opacity per unit frequency interval in the neighborhood of the line to the line opacity. We assume β to be independent of frequency, although this is not in any way essential. Although S_L and S_c are each taken as frequency independent, the total source function has an explicit frequency dependence and, for large values of $|x|$, goes over into S_c, as would be expected.

In general, β and S_c will be determined by a set of transfer and statistical equilibrium equations that may or may not involve the ones for S_L in an important way. For our purpose, it is sufficient to take β and S_c as parameters and discuss their effect on S_L. It turns out that the effective interaction parameter between matter and radiation is

$$\bar{\xi} \equiv \epsilon + (1 - \epsilon)\beta F(\beta) \qquad\qquad 52.$$

where

$$F(\beta) \equiv \int_{-\infty}^{\infty} dx\phi(x)/[\beta + \phi(x)] \qquad\qquad 53.$$

In many respects, $\bar{\xi}$ plays the same role as ϵ in the standard problem. The additional term represents the loss of line radiation to continuous opacity. If $\phi(x)$ has extensive wings, $F(\beta)$ can become very large, so that even if β is much smaller than ϵ, the continuum opacity has a very marked effect.

The magnitude and direction of the changes in the line source function induced by the additional processes depend on the nature of S_c. If $S_c \equiv B \equiv$ constant, S_L increases monotonically with β, whereas if $S_c = 0$, S_L decreases. It may readily occur that the total optical thickness in the continuous opacity, βT, is much less than unity, and yet if $\beta F(\beta) \gg \epsilon$, the line is strongly controlled by the continuous process, so that S_L can be orders of magnitude different from the corresponding situation with $\beta = 0$. Obviously, in this case, the line will be controlled by S_c and will reflect little or nothing about the behavior of ϵ or B. Figures 4 and 5 of Hummer (1968c) illustrate these effects.

The effect of continuous absorption on the emergent intensity is first of all to produce a true continuum. Depending on the circumstances, absorption or emission lines are formed, the latter with varying degrees of self-reversal, including some cases in which the central intensity falls below the level of the continuum. Figure 6 of Hummer (1968a) illustrates all of these possibilities. Thus, the basic transfer process leads naturally to a wide variety of line profiles, with only the simplest of atmospheric configurations.

Nonisotropic scattering.—With one exception, calculations for complete redistribution have also involved the assumption of isotropic scattering, so that both the frequency and the direction of the photon are uncorrelated. Since most strong lines are formed by dipole transitions, the relation between initial and final directions is given by the dipole (Rayleigh) phase function. Because a typical photon is scattered many times, one would not expect the radiation field to be sensitive to the choice of phase function. On the other hand, for depths and frequencies where the *monochromatic* optical depth is small, the radiation field is appreciably anisotropic and one might well expect the emergent radiation to show some effects from a dipole phase function.

Calculations by Hummer (1970) show that the largest effect, obtained for atmospheres with a thickness on the order of unity, was ~5%, with the ratio of the limb ($\mu = 0.2$) to disk center being larger than that predicted by isotropic scattering. For larger optical thickness, the maximum effect was ~2%. Although effects of this magnitude are quite insignificant, this work, along with that reported by Hummer (1969) on general redistribution with the dipole phase function averaged into $R(x',x)$, does suggest that substantial effects from the dipole phase function may be found when the full correlation between frequency and direction is taken into account through the use of the redistribution function $R(x'\mathbf{n}'x\mathbf{n})$.

Overlapping lines.—Normally each line is regarded as isolated. However, for lines high in a series, significant overlap does occur and must be taken into account, for in addition to the opacity contributed to a given line by its neighbors, line photons emitted in one transition can be absorbed by an adjacent one. Van Blerkom & Hummer (1968) have treated this problem by assuming that, in considering the transfer in a given line, the effects of its neighbors can be approximated by replacing them by an infinite array of equally spaced replicas of the line in question. The opacity in this band model is then given by an expression derived by Golden (1967) and Kyle (1967), if the natural width is ignored. Recently, Golden (1969) has published an expression for opacity that includes the natural width.

Van Blerkom & Hummer (1968) solved the transfer problem regarding ϵ, T, and d, the ratio of line separation to Doppler width, as parameters and have discussed the three characteristic length scales that arise. As d decreases from infinity to unity, S increases by a factor of 2 or 3 for $\epsilon = 0.1$ and by an order of magnitude for $\epsilon = 10^{-4}$. For $d \simeq 1$, the opacity is essentially gray. It is then clear that line formation is in general very sensitive to overlap, since overlap increases the opacity in the line wings where photon escape predominately occurs. Where two lines accidentally overlap, large effects are also to be expected, but no simple treatment is now available.

Thermal and density gradients.—Density gradients influence the line-formation process by causing ϵ and the strength of collision broadening to vary through the atmosphere. Thermal gradients can be important in three ways: 1. variations in ϵ, 2. variations in the Planck function, and 3. changes in line and continuous opacity, especially through the Doppler width. Although for any particular line and for a specified atmospheric structure, existing methods can be used to calculate the radiation field including all of these effects, it is difficult to arrive at any general conclusions other than very qualitative ones. To facilitate discussion, it is convenient to consider separately the effects of depth variations in ϵ, $B(T_e)$, and the Doppler width.

1. *Variations in* ϵ: We have already established above that $S(\tau)$ depends only weakly on ϵ near a boundary. Moreover, since ϵ is proportional to electron density and varies slowly with electron temperature, in almost all circumstances in stellar atmospheres ϵ will decrease outward and, for neutral atoms at least, will go to unity at great depths. Roughly speaking, the effective value of ϵ is that found at a depth of one thermalization length corresponding to that value of ϵ; that is, $\epsilon_{eff} = \epsilon(\tau^*)$, where $\tau^* = \Lambda[\epsilon(\tau^*)]$. Some insight can be gained from examination of Figures 2, 3, and 14 of Avrett (1965).

2. *Variations in* $B(T_e)$: Generally, $S(\tau)$ will depend on the values of B over a region of order Λ about τ. If B varies strongly over a distance smaller than Λ, $S(\tau)$ can exceed B in some regions. Examples are given in Figure 7 of Avrett & Hummer (1965) and Figures 11 to 25 of Avrett (1965). Very hot surface layers are responsible for chromospheric emission lines (Jefferies & Thomas 1959, 1960).

3. *Variations in Doppler width*: Two facts indicate the importance of variations in Doppler width. First, we have already stressed that $S(\tau)$ near the surface depends strongly on radiation flowing in the line wings from deep regions. Second, an optically thick radiating gas loses energy primarily through the wings. Quantitative studies of Doppler-width variations have been made by Hummer & Rybicki (1966) and Rybicki & Hummer (1967), who give a number of examples. Some indication of the possible variety of effects can be seen by considering a semi-infinite atmosphere with a surface layer of optical thickness τ_s in which the Doppler width differs from that in the rest of the medium. We regard β and ϵ as constant throughout. First, consider the surface value of the Doppler width to be smaller than that at great depth. If the thickness τ_s of the "cool" layer is much larger than the thermalization length Λ, the region is energetically self-supporting and remains pretty well independent of the rest of the atmosphere. Since the cooling rate is decreased, $S(\tau)$ is larger than for constant Doppler width. If the Doppler width increases rapidly around $\tau = \tau_s$, the transition region acts like an effective boundary for the rest of the atmosphere and a temporary decrease in $S(\tau)$ occurs. On the other hand, if $\Lambda \gg \tau_s$, excitation in the surface region is dependent on radiation from below, and since the narrow wings are not effective in intercepting this radiation, the source function is abnormally low. If the Doppler width is larger for $\tau < \tau_s$ than it is for $\tau > \tau_s$ and $\tau_s \gg \Lambda$, the local energy input dominates, but the extended wings increase radiative loss and $S(\tau)$ is smaller than normal. Conversely, if $\Lambda \gg \tau_s$, the surface layer acts as a partial reflector by impeding the outward flow of radiation, so that $S(\tau)$ is larger than otherwise. As this latter effect can be substantial, we have called it the *reflector effect*.

In reality, non-Doppler contributions to the opacity will probably weaken the effects discussed above. Moreover, the effects of variations in the Planck function and in the Doppler width are going to interact in a complicated way. If $h\nu_0/kT \gg 1$, then variations in the Doppler width will be pretty well washed out, but if $h\nu_0/kT \ll 1$, they may be important. In any event, variations in the Doppler width play an important role in determining the emergent radiation even if the source function is relatively unaffected.

3.6 GENERAL REDISTRIBUTION

Although our knowledge of the redistribution process for various types of collisional broadening is insufficient to justify the solution of the transfer problem with the isotropic or exact forms of the source function in those cases, no such uncertainties exist for combined natural and Doppler broadening (hereafter called nD broadening), or pure Doppler broadening. Resonance lines in low-density media are described well by nD broadening, and since there are good a priori reasons to question the validity of complete redistribution for this situation, this problem has been discussed a number of times in the past two decades. Pure Doppler broadening is of interest primarily as an approximation for cases when the optical thickness is too small for much transfer to occur in the line wings, where natural broadening is important.

The only work involving the exact redistribution function is that of Hearn (1964), which is limited to pure Doppler broadening in atmospheres of moderate optical depth. For this case, it appeared that this isotropic source function deviated by only a few percent from the exact result.

A complete and fairly detailed account of the work to date on the solutions with isotropic source functions has been given by Hummer (1969), who gives the most systematic discussion of Doppler and nD broadening available. For pure Doppler broadening, the isotropic and frequency-independent (hereafter $F\text{-}I$) source functions agree to within a few percent at line center, except for regions very near the surface. At depth τ, the frequencies for which agreement is found are those for which the *monochromatic* optical depth $\tau\phi(x)$ is larger than unity. However, outside this region, the isotropic source function at any depth can differ by orders of magnitude from the $F\text{-}I$ approximation. For this case, it is readily shown that at depth τ in an isothermal, plane-parallel atmosphere,

$$\lim_{|x| \to \infty} S(\tau, x) < (1 + \epsilon)B/2$$

while the $F\text{-}I$ source function takes on all values between $\sqrt{\epsilon}B$ and B. The ratio of the emergent intensity at line center in the $F\text{-}I$ approximation to that in the isotropic approximation lies between 0.9 and 1.2 and in all cases increases approximately linearly away from the line center.

For nD broadening, the agreement of the $F\text{-}I$ and the isotropic source functions is generally much worse than it is for the pure Doppler case. If the Voigt parameter a (ratio of natural to Doppler width) and the atmospheric thickness T are sufficiently small, the results just summarized apply.

When substantial radiation flows in the region of the wings, the region of agreement between the $F-I$ and the isotropic source functions vanishes and order-of-magnitude differences develop in the core. It appears that the pure Doppler $F-I$ source function provides a better approximation to the isotropic source function for nD broadening than does the nD $F-I$ result. Of course, in computing the emergent intensity, the full expression for $\phi(x)$, i.e., the Voigt function, should be used.

Because in the nD case transfer in the wings is dominated by natural broadening, the redistribution function is sharply peaked there. Although the width of this peak is still on the order of a Doppler width, $\phi(x)$ varies so slowly that the frequency change is inconsequential. The escape of photons via the wings is therefore strongly inhibited. Consequently, the source function for large $|x|$ becomes primarily a function of *monochromatic* optical depth and for a semi-infinite isothermal atmosphere, the emergent intensity in the wings drops to a level on the order of $\sqrt{\epsilon}B$, instead of rising to B as the $F-I$ result does.

If scattering is assumed to occur with complete redistribution over a Lorentz profile in the rest frame of the atom, as it does for an idealized case of impact broadening, and if the further effects of Doppler redistribution are included, the absorption coefficient is a Voigt function, as it is in the nD case, but the redistribution function is quite different. That complete redistribution should be a good approximation was suggested by Hummer (1962) and verified by Finn (1967). While this result does support the current practice of assuming the $F-I$ source function for all situations except those involving extremely low densities, remember that impact broadening usually is valid for only part of the line and we do not yet know how well the assumption of complete redistribution holds for that part.

Mean number of scatterings.—Hummer (1969) has shown that the mean number of scatterings for pure Doppler broadening as computed from the isotropic source function is less than 8% smaller than the $F-I$ value. The situation for nD broadening is not entirely clear. For moderate values of ϵ, $\langle N \rangle$ as a function of T computed from the isotropic source function agrees with the values obtained from the $F-I$ source function at large and small values of T, but for intermediate values it can be significantly larger. When ϵ is so small that the atmosphere is effectively thin even for very large thicknesses, Osterbrock (1962), on the basis of an approximate probabilistic calculation, showed that $\langle N \rangle$ varies approximately as T up to some critical value depending on the Voigt parameter a, where the T^2 dependence characteristic of coherent scattering should take over. This question has been investigated further on the basis of Monte Carlo calculations by Auer (1968) and Avery & House (1968). Recently, Adams (1971) calculated $\langle N \rangle$ for still larger optical thicknesses by a direct numerical solution of the transfer equation. For $\log T \lesssim 4$, all the results agree. For larger values of T, the recent calculations find that $\langle N \rangle$ still depends linearly on T, and even at the largest thickness considered by Adams, log $(\sqrt{\pi}T/2) = 7.3$, the T^2 dependence has not yet appeared. At that thickness, Osterbrock's estimate is 2 orders of magnitude too large. Thus, the question of the asymptotic form of $\langle N \rangle$ is at present unsettled.

4. MULTILEVEL ATOMS

Owing to the scarcity of precise analytical results, our understanding of the multilevel case depends mainly on analogies with the two-level case and on physical arguments. We shall here describe, with a minimum of technical detail, various physical concepts that have proved useful in understanding multilevel transfer problems. A number of limiting cases are considered, since these provide a good source of insight and are of great practical value. The linear multilevel problem will be treated in some detail, since it represents one of the few nontrivial, precisely analyzable models and has allowed several important concepts to be sharpened and interrelated. Much of the recent progress on the multilevel problem has been obtained by purely computational means; this is the domain of the specialist and we therefore deal with it only briefly, referring the reader to recent comprehensive reviews (Avrett 1971, Auer 1971). Other general reviews of the multilevel problem have been given in the books of Jefferies (1968) and Mihalas (1970).

4.1 BASIC EQUATIONS

Under the assumption of complete redistribution, the problem of line formation in a multilevel atom is equivalent to the simultaneous solution of the equations of transfer for all lines and continua and of the equations of statistical equilibrium. These latter equations can be written

$$\sum_j N_i P_{ij} = \sum_j N_j P_{ji} \qquad\qquad 54.$$

where the N_i are the populations of the particular energy levels and stages of ionization to be considered. As in the two-level case, each rate coefficient P_{ij} for a transition $i{\to}j$ consists of a radiative part and a collisional part, generally assumed to be due to thermal electrons or to other thermal particles.

The equation of transfer for the line between the lower state j and the upper state i is of the form

$$dI_\nu/ds = k_{ij}[\phi_\nu{}^{ij}(- I_\nu + S_{ij}) + \beta(- I_\nu + S_c)] \qquad\qquad 55.$$

where the line absorption coefficient is

$$k_{ij} = (h\nu_{ij}/4\pi)(N_j B_{ji} - N_j B_{ij}) \qquad\qquad 56.$$

and the line source function is

$$S_{ij} = (2h\nu_{ij}{}^3/c^2)[(N_j g_i/N_i g_j) - 1]^{-1} \qquad\qquad 57.$$

There is one such equation for each line. The equations of transfer for the continua will not be discussed here, as we are primarily concerned with the line-formation problem; however, these continua do affect the overall problem and must be taken into account in a detailed treatment. See, for instance, Mihalas (1970) or Avrett & Loeser (1969). The multilevel problem may be regarded as equivalent to the determination of all source functions, since the radiation fields and populations can easily be found from them.

An important result following from the equation of transfer is the expression for the mean integrated intensity $\bar{J}_{ij}$ in terms of an integral over the line source function:

$$\bar{J}_{ij}(\tau) = \int K_{ij}(\,|\,\tau - \tau'\,|\,)S_{ij}(\tau')d\tau' \qquad 58.$$

If there is a nonzero radiation field incident on the medium or if there is continuum absorption present (which we shall neglect for simplicity), there will also be terms involving these additional quantities. The kernel K_{ij} will be identical with that occurring in two-level theory, but only if the τ scale is the one appropriate to the $i{\leftrightarrow}j$ transition. In a multilevel problem where a standard depth scale has chosen, these kernels will be more complicated; alternatively, the simple forms can be maintained by transforming the depth-dependent quantities between the various optical scales as required.

The theory of line formation in a multilevel atom has invariably been formulated with the assumption of complete redistribution. This is largely because of uncertainties in the physical theory of rest-frame redistribution for natural broadening, and especially for pressure broadening as discussed earlier. But even in the physically clear case of Doppler redistribution with an assumed rest-frame redistribution, the equations are formidable and no one seems to have done much with them. Where impact broadening dominates, it seems physically reasonable to accept complete redistribution in the multilevel case, since arguments similar to those of the two-level case apply. On the other hand, for resonant scattering from a sharp lower level to naturally broadened upper levels, complete redistribution would probably be a bad assumption, as it corresponds to the nD case in the two-level atom. Since it has not been given proper foundation, the assumption of complete redistribution remains one of the weak points of multilevel theory.

The equations of transfer present no real difficulties over the two-level case, but the statistical equilibrium equations now become a set of linear homogeneous equations for all the populations, with coefficients that depend on the radiation fields at a given point in the medium. These can be solved for the ratio of any two populations N_j/N_i in terms of determinants (Rosseland 1936). These solutions can be interpreted in terms of the probabilities of chains of transitions connecting states i and j (Jefferies 1960, White 1961).

A result of utmost importance follows if this solution of the statistical equilibrium equation is substituted into the source function (Eq. 55); this source function will depend linearly on the radiation field J_{ij}, but, in general, nonlinearly on the other radiation fields (Thomas 1957). This result is often simply stated: a source function is linear in its own radiation field. If the radiation fields are prescribed in all lines except $i \leftrightarrow j$, this result can be written

$$S_{ij} = (1 - \epsilon_{ij})\bar{J}_{ij} + \epsilon_{ij}\tilde{B}_{ij} \qquad\qquad 59.$$

where ϵ_{ij} and $\tilde{B}_{ij}$ are now dependent on the given radiation fields but not on J_{ij}. Since this equation has the same form as the two-level equation, it is often called the *equivalent two-level problem* for the transition $i \leftrightarrow j$. The quantity ϵ_{ij} has a simple interpretation—it is the probability that upon absorption the atom will reach the lower level by any means except a direct radiative transition.

The formulation of an equivalent two-level problem for a given transition can be done in other ways, all giving the same solution but involving different values of ϵ_{ij} and $\tilde{B}_{ij}$. A particularly useful formulation of this type is one in which net radiative rates or *net radiative brackets* (Thomas 1960) for the other radiative transitions are prescribed rather than the radiation fields. This leads to faster convergence in many cases for numerical methods based on equivalent two-level problems (Cuny 1965, Cayrel 1965), but the physical meaning of ϵ_{ij} is no longer clear.

4.2 LIMITING CASES

Owing to the difficulty of the exact multilevel problem, simplifications are particularly welcome. There are indeed several types of limiting cases in which the problem simplifies in some substantial way. The most important of these are the optically thin limit, line saturation, and source-function equality. Unfortunately, much is still to be learned about the precise conditions under which the last two hold. The major difficulty seems to center on finding suitable generalizations of the concept of thermalization length for multilevel problems.

Optically thin case.—If the medium is optically thin in a particular radiative transition (line or continuum), the associated radiation field will be equal to the incident field to a good approximation. The radiative rate coefficients are then known, and there is no need for a concurrent solution of that particular transfer equation, although there may be transfer problems in other transitions. A common case is where the incident field is zero; then the spontaneous rate is the only significant radiative rate. Once the populations have been determined under these conditions, the emission in the transition can be found by integrating a simplified transfer equation in which the absorption is neglected. These rules for an optically thin medium follow systematically from a perturbation treatment in which the opacity is treated as being of first order in smallness.

The case of a multilevel atom in which all transitions are optically thin involves no significant transfer problem at all and is, from the present point of view, trivial. A more typical situation is where the medium is optically thin in some, but not all, transitions. Even when the medium is not optically thin in any transitions, the usual vast discrepancies between mean free paths in various lines and continua often make it possible to isolate significant regions of the medium in which some transitions are optically thin. If the radiation field from the rest of the medium in these transitions can be treated as a known incident field, the above simplifications apply. In practice, this requires either that some additional knowledge exist about the radiation field in the rest of the medium or that this radiation be found without solving a full-scale problem; this can often be accomplished by invoking the concept of line saturation.

Line saturation.—The general behavior of any source function is to vary most rapidly near a surface and to vary more slowly deep in the medium. When the source function varies sufficiently slowly, it can, to a good approximation, be removed from under the integral in Eq. 58, setting $\tau' = \tau$. This gives

$$\bar{J}_{ij}(\tau) = N S_{ij}(\tau) \qquad 60.$$

where N is a normalizing factor due to the remaining integral over the kernel. The factor N represents the probability that a photon at τ will be absorbed, so that sufficiently deep in the medium where escape is negligible, it approaches unity and we have

$$\bar{J}_{ij}(\tau) = S_{ij}(\tau) \qquad 61.$$

This is the simplest case of *line saturation*. Because it implies the equality of upward and downward radiative rates, it is also called *radiative detailed balance*. Since the radiation field is expressed directly in terms of the populations, this field can be eliminated from the problem and no significant transfer problem exists in that transition. Line saturation has been discussed by Kalkofen (1966) and by Thomas (1965), who calls it the *locally opaque situation* (LOS). It is also equivalent to the *on-the-spot approximation* of Zanstra (1951a); see Van Blerkom & Hummer (1969).

When a continuum is present, the appropriate generalization of the above is

$$\bar{J}_{ij}(\tau) = N S_{ij}(\tau) + N_c S_c(\tau) \qquad 62.$$

where N and N_c are normalizing factors giving the probabilities of absorbing a photon in the line and continuum, respectively. Deep in the medium where escape is negligible, we have $N + N_c = 1$ and $N_c = \beta F(\beta)$. Unless the continuum source function has the appropriate value, line saturation will not be equivalent to radiative detailed balance, although the transfer problems in the transition can still be eliminated by use of Eq. 62. This has been discussed by Kalkofen (1966).

Saturation can also be defined from the equation of transfer 55 directly, where it corresponds to neglect of the spatial derivative on the left-hand side. The result (Eq. 62) is easily recovered, as is Eq. 61 in cases where the continuum is negligible.

When all transitions are saturated, the condition of LTE will be valid. This need not be due to large dominant collision rates, but simply to the enclosing of the radiation field over a sufficiently large region, so that it comes into equilibrium with the local electron temperature. Thomas (1965) has given this limit the special designation LOS–Σ.

Another limiting case that is equivalent to a saturation approximation is found in the theory of moving atmospheres of Sobolev (1957, 1960). Here the normalization of the kernel is not unity, since photons may escape from great depths owing to the Doppler shifting of the line profile, and we have $N = 1 - \rho$, where ρ is the escape probability. Sobolev (1957) shows that his theory is a result of a saturation approximation and that ρ can be easily determined from local velocity gradients for any line profile when complete redistribution is assumed. The theory of line formation in moving atmospheres has been reviewed by Rybicki (1970). Recently, Lucy (1971) has presented what appears to be a theory equivalent to Sobolev's, but it approaches the line saturation from the differential-equation point of view. It is an essential part of Lucy's theory that the equation of transfer be written in terms of a local frequency variable.

The major difficulty in applying the concept of line saturation is knowing just where in the medium it is valid. One gains an appreciation of this by noting the stringent degree to which a result such as Eq. 61 must hold in order to be useful. It can be deduced from Eq. 59 that $S_{ij} = B_{ij}$, for example, only if Eq. 61 is valid with an error much less than ϵ_{ij}. The problem bears much similarity to the two-level case of thermalization, and it has been common to assume that saturation occurs at an optical depth given by the two-level thermalization length based on the value of ϵ_{ij} (Kalkofen 1966, Thomas 1965). But as we have seen, the value of ϵ_{ij} of an equivalent two-level problem is not unique, but depends on how the information on the other radiation fields is introduced—as total or net rates, for example. There is at present no completely satisfactory resolution of this difficulty, although much light has been shed on it by recent discussions of the linear multiplet problem (see below).

Source-function equality in multiplets.—A limiting case of a different sort occurs when the collisional rate dominates all other rates between two levels i and j that are very closely spaced in energy, as in a multiplet. The detailed balance relation 1 then implies that

$$N_i/g_i = N_j/g_j \qquad\qquad 63.$$

If k is the common level for the multiplet, it follows from this relationship that the source functions of the two multiplet lines are equal:

$$S_{ik} = S_{jk} \qquad 64.$$

This *source-function equality* refers to quantities at the same physical point and therefore may occur at different optical depths when measured in the separate lines. If the energy difference between levels i and j is not small, there still will be set up a more complicated relationship between S_{ik} and S_{jk}; this more general *source-function interlocking* will not be discussed here, as most of the work has been confined to the multiple case.

A question of considerable practical interest is just how strong the collisional rate between i and j must be to ensure the validity of source-function equality. This has been considered by Jefferies (1960, 1965), Waddell (1963), Athay (1964), and Avrett (1966). The discussion of Jefferies (1965) is noteworthy for its clarity and for the introduction of several physical concepts that seem to be worth further development, and we shall give a brief account of it.

Consider a multiplet with a common lower level 1 and two closely spaced upper levels 2 and 3, between which there is no radiative transition. The medium is semi-infinite and is assumed to have constant properties. If the collisional rate C_{23} is zero, there are two independent two-level atoms of the standard type, each approaching the same LTE value B at the thermalization length appropriate to its own optical-depth scale and value of ϵ. The surface values are different in general, so that the two solutions are quite unalike, except for their common asymptotic value. As C_{23} is increased, the photons in the two lines cease to travel independently, since after a certain distance, on the average, a collision between levels 2 and 3 can convert a 2–1 photon to a 3–1 photon, or vice versa. This process is called *photon switching*, and the associated length is termed a *conversion length*. Over distances long compared with the conversion lengths, the upper-level population become thoroughly mixed and come into equilibrium; there is then source-function equality.

Jefferies estimated the conversion lengths using the two-level idea of thermalization with an ϵ that included the *total* collisional depopulation rate. For Doppler profiles, this led to the conversion length in the 2–1 transition,

$$\tau_{21}{}^* \simeq (A_{21} + C_{21} + C_{23})/(C_{21} + C_{23}) \qquad 65.$$

and a similar result for τ^*_{31}. Over a scale less than both of these, the two lines behave independently; one expects therefore that source-function equality will occur only at depths larger than this scale. These conclusions were verified by the direct numerical results of Avrett (1966), who also resolved the controversy between Waddell (1963) and Athay (1964) concerning criteria for source-function equality. Athay (1968) performed a similar analysis to determine the effectively thin boundary layer for a multiplet problem; this was expressed in terms of another length, the *degradation length*.

Another important concept introduced by Jefferies (1965) concerns the transfer process in the region of source-function equality. Each photon is switched into the other one, and back again, many times in such a region, so that it is inconvenient to think of this as successive creation and destruction of photons. Rather one should imagine that the excitation is being carried by a joint *photon pool*, which persists until a true destruction occurs by collisional deexcitation to the ground level. Jefferies (1965) posed the questions of how the concept of a photon pool might be formulated precisely and whether a source function for a photon pool could be defined so as to take advantage of the analogy between complete photon switching and complete frequency redistribution in the single-line case. These questions have been partially answered in the linear multiplet model, to be discussed presently, but not in the generality that the concepts would seem to deserve.

4.3 The Linear Multiplet Model

This model is perhaps the only nontrivial multilevel problem capable of being analyzed rather completely, and it is therefore of considerable interest. Consider a closely spaced set of N upper levels, all having a radiative transition to a common lower level. There are no other radiative transitions, although all levels may be collisionally connected. If stimulated emission is neglected, it is possible to reduce the overall problem to a single-matrix equation identical in form with the two-level problem,

$$S(\tau) = (I - \epsilon) \int K(\tau, \tau')S(\tau')d\tau' + \epsilon B(\tau) \qquad 66.$$

Here S and B are column vectors of the N source functions and Planck functions corresponding to the N transitions to the lower level; I is the unit matrix and K is a diagonal matrix of kernels, each corresponding to transfer in one particular line; the matrix ϵ describes the various collisional couplings between the levels.

By assuming that each kernel function could be approximated by a single exponential, Kalkofen (1965) demonstrated that this equation could be solved exactly for a constant-property, semi-infinite medium. He showed that the overall scale of the solution could be quite different from those of the equivalent two-level atoms that one obtains by solving each transition on the assumption that radiation fields in the others are given. The various scales were related to the eigenvalues of a certain matrix involving ϵ and the various opacities in the different lines. Since the introduction of exponential kernels is tantamount to assuming coherent scattering, these results should not be taken as quantitative.

There is another simplified version of the multiplet model that does not assume exponential kernels, but that does assume all the line opacities are equal. In this case, the kernel is a scalar multiple of the unit matrix: $K = KI$.

The matrix ϵ is not, in general, diagonal, but can be made so by a diagonal-izing transformation; that is, there exists a nonsingular matrix U such that

$$\bar{\epsilon} = U\epsilon U^{-1} \qquad 67.$$

is diagonal.

We now define a new set of "source functions" and "Planck functions,"

$$\bar{S} = US, \quad \bar{B} = UB \qquad 68.$$

These new source functions are linear superpositions of the original source functions, and each independently satisfies its own two-level equation:[1]

$$\bar{S}_i(\tau) = (1 - \bar{\epsilon}_{ii}) \int K(\tau, \tau')\bar{S}_i(\tau')d\tau' + \bar{\epsilon}_{ii}\bar{B}_i(\tau) \qquad 69.$$

This independence leads us to identify these components of $\bar{S}$ as the source functions of the photon pools described in physical terms by Jefferies.

Each $\bar{\epsilon}_{ii}$ is an eigenvalue of the matrix ϵ, and it is this value that deter-mines the scale over which $\bar{S}_i$ saturates to its own modified Planck function $\bar{B}_i$. The smallest eigenvalue of ϵ determines the overall thermalization scale, while the others define scales over which there is saturation in the other pools.

In the case of the three-level multiplet treated above, all matrices reduce to 2×2, and much analytic work can be carried through. When there is a large collisional rate between levels 2 and 3, but small collisional rates from levels 2 and 3 to level 1, it can be shown that there is one small and one large eigenvalue of ϵ. The photon-pool source functions correspond roughly to the sum and difference of the two original source functions. The sum part represents the average source function describing the photon pool that ther-malizes over the long scale given by the small eigenvalue. The difference part saturates over a much shorter scale to a small value. This describes the approach to source-function equality over conversion length. It is interesting that in this simple example, at least, source-function equality appears as a special case of line saturation in one of the photon pools. A similar discussion for the 2×2 exponential kernel case is given by Kalkofen (1965).

The physical question raised by Jefferies of whether complete photon switching is analogous to complete frequency redistribution is answered in part by the introduction of these photon-pool source functions. The photon pool of longest scale is the only one in which there is a significant transfer problem at great depths; therefore, we have a single source function that is independent of a particular line, just as in the simple case we have a single source function that is independent of frequency. However, the analogy is

---

[1] This result of one of us (GR) was incorrectly quoted by Jefferies (1968).

not complete, because for the coupled-line case the region near the surface must be treated in detail for all lines, whereas complete frequency redistribution is usually a good approximation even near the surface.

ACKNOWLEDGMENTS

We are grateful to Dr. John Castor for many informative discussions and for a careful and critical reading of the manuscript.

LITERATURE CITED

Abramov, Yu. Yu., Dykhne, A. M., Napartovich, A. P. 1967. *Astrophysics* 3:215
Adams, T. F. 1971. *Astron. Ap.* In press
Athay, R. G. 1964. *Ap. J.* 140:1579
Athay, R. G. 1968. *Resonance Lines in Astrophysics.* Boulder, Colo: Nat. Cent. Atmos. Res.
Athay, R. G., Skumanich, A. 1967. *Ann. Ap.* 30:669
Auer, L. H. 1968. *Ap. J.* 153:783
Auer, L. H. 1971. *J. Quant. Spectrosc. Radiat. Trans.* In press
Auer, L. H., Mihalas, D. 1971. *Ap. J.* In press
Avery, L. W., House, L. L. 1968. *Ap. J.* 152:493
Avrett, E. H. 1965. *Smithsonian Ap. Obs. Spec. Rep. No. 174,* 101
Avrett, E. H. 1966. *Ap. J.* 144:59
Avrett, E. H. 1971. *J. Quant. Spectrosc. Radiat. Trans.* In press
Avrett, E. H., Hummer, D. G. 1965. *MNRAS* 130:295
Avrett, E. H., Loeser, R. 1969. *Smithsonian Ap. Obs. Spec. Rep. No. 303,* 1
Bely, O., Van Regemorter, H. 1970. *Ann. Rev. Astron. Ap.* 8:329
Biebermann, L. 1947. *J. Exp. Theor. Phys.* 17:416
Capriotti, E. R. 1965. *Ap. J.* 142:1101
Cayrel, R. 1965. *Smithsonian Ap. Obs. Spec. Rep. No. 174,* 295
Cuny, Y. 1965. *Smithsonian Ap. Obs. Spec. Rep. No. 174,* 275
Edmonds, F. N. 1955. *Ap. J.* 121:418.
Enome, S. 1969. *Publ. Astron. Soc. Japan* 21:367
Field, G. B. 1959. *Ap. J.* 129:551
Finn, G. D. 1967. *Ap. J.* 147:1085
Gelinas, R. J., Ott, R. L. 1970. *Ann. Phys. N. Y.* 59:323.
Golden, S. A. 1967. *J. Quant. Spectrosc. Radiat. Trans.* 7:483
Golden, S. A. 1969. *J. Quant. Spectrosc. Radiat. Trans.* 9:1067
Hearn, A. G. 1964. *Proc. Phys. Soc.* 84:11
Heitler, W. 1954. *Quantum Theory of Radiation.* Oxford: Clarendon. 3rd ed.
Henyey, L. G. 1941. *Proc. Nat. Acad. Sci. US* 26:50
Holstein, T. 1947. *Phys. Rev.* 72:1212
Holstein, T. 1950. *Westinghouse Res. Lab. Sci.* Paper No. 1501, Appendix A
House, L. L. 1970a. *J. Quant. Spectrosc. Radiat. Trans.* 10:909
House, L. L. 1970b. *J. Quant. Spectrosc. Radiat. Trans.* 10:1171
Houtgast, J. 1942. Thesis. Univ. Utrecht. Résumé in 1944. *Ap. J.* 99:107

Huber, D. L. 1969. *Phys. Rev.* 178:93
Hummer, D. G. 1962. *MNRAS* 125:21
Hummer, D. G. 1964. *Ap. J.* 140:276
Hummer, D. G. 1965a. *Smithsonian Ap. Obs. Spec. Rep. No. 174,* 13
Hummer, D. G. 1965b. *Smithsonian Ap. Obs. Spec. Rep. No. 174,* 143
Hummer, D. G. 1968a. In *Planetary Nebulae. IAU Symp. No. 34,* ed. D. E. Osterbrock, C. R. O'Dell, 166. Dordrecht-Holland: Reidel
Hummer, D. G. 1968b. *J. Quant. Spectrosc. Radiat. Trans.* 8:193
Hummer, D. G. 1968c. *MNRAS* 138:73
Hummer, D. G. 1969. *MNRAS* 145:95
Hummer, D. G. 1970. *Ap. Lett.* 5:1
Hummer, D. G., Rybicki, G. B. 1966. *J. Quant. Spectrosc. Radiat. Trans.* 6:661
Hummer, D. G., Rybicki, G. B. 1967. *Methods Comput. Phys.* 7:53
Hummer, D. G., Rybicki, G. B. 1970. *MNRAS* 150:419
Hummer, D. G., Stewart, J. C. 1966. *Ap. J.* 146:290
Ivanov, V. V. 1962a. *Astron. Zh.* 39:1020
Ivanov, V. V. 1962b. *Uch. Zap.* LGU No. 307, 52
Ivanov, V. V. 1965a. *Uch. Zap.* LGU No. 328, 44
Ivanov, V. V. 1965b. In *Theory of Stellar Structure,* V. V. Sobelev. NASA transl. TTF-457
Ivanov, V. V. 1968. *Astrophysics* 4:1
Ivanov, V. V. 1971. *Transfer of Line Radiation,* ed. D. G. Hummer. Transl. Washington, DC: US GPO
Ivanov, V. V., Nagirner, D. I. 1966. *Astrophysics* 2:1
Jefferies, J. T. 1960. *Ap. J.* 132:775
Jefferies, J. T. 1965. *Smithsonian Ap. Obs. Spec. Rep. No. 174,* 177
Jefferies, J. T. 1968. *Spectral Line Formation.* Waltham, Mass: Blaisdell. 291 pp.
Jefferies, J. T., Thomas, R. N. 1959. *Ap. J.* 129:401
Jefferies, J. T., Thomas, R. N. 1960. *Ap. J.* 131:429
Kalkofen, W. 1966. *J. Quant. Spectrosc. Radiat. Trans.* 6:633
Kalkofen, W. 1965. *Smithsonian Ap. Obs. Spec. Rep. No. 174,* 187
Kuhn, W. R. 1966. Thesis. Univ. Colorado, Boulder
Kyle, T. G. 1967. *Ap. J.* 148:845
Levich, V. G. 1940. *Zh. Eksp. Theor. Fiz.,* 10, no. 11.
Linsky, J. L., Avrett, E. H. 1970. *PASP* 82:169
Lucy, L. B. 1971. *Ap. J.* 163:95

Mihalas, D. 1970. *Stellar Atmospheres*. San Francisco, Calif: Freeman. 463 pp.

Mihalas, D. 1972. *Ann. Rev. Astron. Ap.* To appear

Nagirner, D. I. 1967. *Astrophysics* 3:133

Osterbrock, D. E. 1962. *Ap. J.* 135:195

Oxenius, J. 1965. *J. Quant. Spectrosc. Radiat. Trans.* 5:771

Rees, D., Reichel, A. 1968. *J. Quant. Spectrosc. Radiat. Trans.* 8:1795

Rosseland, S. 1936. *Theoretical Astrophysics*, p. 143. Oxford Univ. Press

Rybicki, G. B. 1970. In *Spectrum Formation in Stars with Steady-State Extended Atmospheres. NBS Spec. Publ. 332*, 87

Rybicki, G. B., Hummer, D. G. 1967. *Ap. J.* 150:607

Rybicki, G. B., Hummer, D. G. 1969. *MNRAS* 144:313

Sobolev, V. V. 1944. *Astron. Zh.* 21:143

Sobolev, V. V. 1949. *Astron. Zh.* 26:129

Sobolev, V. V. 1957. *Sov. Astron.* 1:678

Sobolev, V. V. 1960. *Moving Envelopes of Stars.* Cambridge, Mass: Harvard Univ. Press

Spitzer, L. 1936. *MNRAS* 96:794

Spitzer, L. 1944. *Ap. J.* 99:1

Thomas, R. N. 1957. *Ap. J.* 125:260

Thomas, R. N. 1960. *Ap. J.* 131:429

Thomas, R. N. 1965. *Some Aspects of Non-Equilibrium Thermodynamics in the Presence of a Radiation Field.* Boulder, Colo: Univ. Colorado Press

Towne, D. H. 1954. Thesis. Harvard Univ., Cambridge, Mass.

Van Blerkom, D., Hummer, D. G. 1968. *Ap. J.* 154:741

Van Blerkom, D., Hummer, D. G. 1969. *J. Quant. Spectrosc. Radiat. Trans.* 9:1567

Waddell, J. H. 1963. *Ap. J.* 138:1147

Warming, R. F. 1970. *J. Quant. Spectrosc. Radiat. Trans.* 10:1219

Weisskopf, V. 1931. *Am. Phys. Leipzig* 9:23

Weisskopf, V. 1933. *Observatory* 56:291

Weymann, R. J. 1970. *Ap. J.* 160:41

White, O. R. 1961. *Ap. J.* 134:85

Woolley, R. v. d. R. 1938. *MNRAS* 98:624

Woolley, R. v. d. R., Stibbs, D. W. N. 1953. *The Outer Layers of a Star.* Oxford: Clarendon

Zanstra, H. 1941. *MNRAS* 101:273

Zanstra, H. 1946. *MNRAS* 106:225

Zanstra, H. 1949. *Bull. Astron. Inst. Neth.* 11:1

Zanstra, H. 1951a. *Bull. Astron. Inst. Neth.* 11:341

Zanstra, H. 1951b. *Bull. Astron. Inst. Neth.* 11:359

FILLED-APERTURE ANTENNAS FOR RADIO ASTRONOMY

J. W. FINDLAY

National Radio Astronomy Observatory,[1] *Green Bank, West Virginia*

INTRODUCTION

This review is restricted to recent developments in the theory and practice of filled-aperture antennas. In radio astronomy these antennas are usually large parabolic reflectors, but other reflector shapes such as sections of spheres or parabolic cylinders are used. The name filled aperture describes the simplest type of radio telescope, one in which the antenna gain and resolution are both determined by the actual size and shape of the antenna aperture. In the unfilled-aperture instruments, such as interferometers, crosses, or synthetic antennas, the resolution is determined by the outer dimensions of the antenna area, but much of this area is not in fact filled with receiving elements.

The general subject of radio telescopes was well treated quite recently in the book by Christiansen & Högbom (1969). There are earlier survey articles by Findlay (1964) and Ko (1964), while the more practical aspects of the subject are discussed in the book by Mar & Liebowitz (1969) and in the report of a conference held in England in 1966 (IEE Conference 1966). The detailed analysis of the behavior of large reflector antennas is also covered in the book of Rusch & Potter (1970).

This article will therefore deal with specific topics in the design and use of filled-aperture antennas where there has been notable progress in the last few years.

RECENTLY COMPLETED LARGE ANTENNAS

Millimeter-wave antennas.—There has been a rapid growth in millimeter-wave observations, and this region of the spectrum has now begun to yield important astronomical results. This progress has been due in part to the steady improvement in the performance of radiometers in the wavelength range below 1 cm and in part to the quite large numbers of antennas capable of work at millimeter wavelengths. A very complete survey of these instruments is given in the article by Cogdell et al (1970). The range of antennas described, and the summary of their performance, is described in Table 1 below, which is reproduced with minor changes from the table in the above paper.

---

[1] Operated by Associated Universities, Inc., under contract with the National Science Foundation.

TABLE 1. Properties of some millimeter-wave antennas

Property	Units	MIT Lincoln Lab.	Crimean RT-22	Aerospace Corp.	University of Texas	AFCRL	Bonn Univ. (MPI)	CRC (DRTE)	NRAO	Univ. of Calif. Berkeley	JPL
Diameter	m	8.53	22.0	4.57	4.87	8.84	10.0	9.14	11.0	6.10	5.49
Optics		Cass	Cass	Cass	Prime focus	Cass	Cass	Cass	Prime focus	Cass	Cass
F/D		0.440		0.300	0.500	0.300	0.433	0.300	0.800	0.420	0.480
Type mount		Az-El	Az-El	Polar	Polar	Az-El	Polar	Az-El	Az-El	Polar	Polar
Computer control		No	Yes	Yes	No	Yes	No	Yes	Yes	Yes	
Pointing accuracy (absolute)	deg	0.01	0.005	0.006	0.006p	0.007	0.02	0.030p	0.003	0.008p	
(tracking)	deg			0.002	0.002		0.001	0.010p	0.001		
Tolerance (mechanical)	mm	0.20	0.12	0.05	0.06	0.15	0.34	0.53	0.10	0.15	0.18
(Ruze method)	mm				0.10				0.15		0.18
Frequency	GHz	35, 90	75	94	35, 97, 134	35	36	35	31.4, 85, 260	38	
Efficiency	%	55, 20	34	54	64, 54, 45	45	43	23	64, 45, 12	>50	
Beam widths	deg	0.072	0.033	0.047	.110, .042, .033	0.067	0.055	0.070	.060, .020, .010		
Sidelobes	dB	24		22	23, 22, 22	18	20	16			
Date began operation		1961	1966	1963	1963	1965	1966	1966	1967	1968	1970
Elevation	km	0.079	0.550	0.047	2.07	0.164	0.435	0.080	1.920	1.050	2.286

Note: The pointing accuracy values are rms unless they are qualified by p, which indicates peak errors.

Some general conclusions can be drawn from this summary. Even the largest of these telescopes, the new 22 m instrument at the Crimean Astrophysical Observatory and the 11 m NRAO telescope on Kitt Peak, have reflector accuracies which allow them to be used at wavelengths around 3 mm. Also, even these two larger instruments can be pointed with precisions commensurate with their very narrow radio beamwidths. None of the antennas is enclosed within a radome, chiefly because it is difficult to build radomes with sufficiently low loss at these short wavelengths, but two of them are located within astrodomes with movable doors of a type similar to that used for optical telescopes. Such protection is very desirable against wind disturbances and the structural temperature differences produced by direct sunlight.

The 130 foot antenna at the Owens Valley Radio Observatory.—The two-element interferometer at the Owens Valley Observatory of the California Institute of Technology (CalTech) is one of the oldest and most productive of the radio telescopes in the USA. In 1968 a new telescope was completed, to be the first of a new synthetic antenna system as well as to serve as a single instrument in its own right. It is an azimuth-elevation telescope mounted on four base trucks which can move along a linear railroad track of 44 foot gauge at speeds up to 60 feet/min. This linear movement is, of course, only required when the telescope is used as part of an array. Mechanical and structural details of the instrument are given by Rule & Gayer (1968), and it is illustrated in Plate 1. The surface and pointing accuracy of the telescope allow of its being used at wavelengths as short as 1.2 cm. The telescope pointing in equatorial coordinates is controlled by a digital computer.

The 210 foot antenna at Goldstone.—This antenna was built to be the first of three large steerable antennas located around the world forming a network for tracking and receiving telemetered data from space vehicles traveling to great distances from the Earth. The antenna is the largest of several instruments located on the desert at Goldstone, California and operated by the Jet Propulsion Laboratory (JPL) for the National Aeronautics and Space Administration. JPL is operated by CalTech and arrangements have been made by CalTech for a part of the observing time on this telescope to be used for radio astronomical research, so that it should be included in any list of major astronomical antennas.

The instrument is an azimuth-elevation telescope built to the quite severe performance specifications required by its space-tracking task (Rocci 1966). It has a surface accuracy and computer-controlled pointing which combine to give good performance at a wavelength of 3 cm. The telescope structure was carefully analyzed before construction and subsequently its deflections have been measured; the results show good agreement with the calculations (Katow & Bathker 1968) and provide a further test of the success of computer techniques in predicting the deflections under gravity of complex structures.

PLATE 1. The Owens Valley 130 foot antenna. (Photograph—G. Stanley)

Much of the work of this antenna has been at fixed frequencies (wavelengths of 13 and 3.6 cm) where the combination of Cassegrain optics and very careful radio engineering has resulted in a very low-noise, high-performance system.

The 100 meter antenna of the Max-Planck-Institut für Radioastronomie.—This antenna, now the largest steerable parabolic reflector in the world, is essentially complete. It is an azimuth-elevation telescope, with computer pointing control. The design of the parabolic reflector support structure was carried out with the intent of minimizing the effects of gravitational deflections on the performance of the instrument, so that the telescope is the first large instrument to be built using the principles of homologous deflection (which is treated in more detail in the next section). Articles by Hachenberg (1968, 1970) and Wielebinski (1970) describe the telescope in some detail and it is illustrated in Plate 2. Since the telescope has not yet been evaluated, it is too early to give its actual performance, but the design and construction suggest that it will behave excellently at wavelengths as short as 3 cm and quite possibly will do even better. The instrument may be used with a Gregorian subreflector or as a prime-focus instrument. For short wavelengths

PLATE 2. The 100 m antenna of the Max-Planck-Institut für Radio astronomie. (Photograph—O. Hachenberg)

the Gregorian mode will be preferred, but the accessibility at all times of the prime focus will allow of good flexibility in the observing program.

The parabolic cylinder antenna in South India.—This antenna has been designed and constructed by the radio astronomy group at the Tata Institute of Fundamental Research for use in specific observational programs; the occultation of radio sources by the moon and studies of pulsars are two where it will be of great value. It has a large physical collecting area (530×30 m) and a good efficiency, which results in an effective aperture of about 9000 m². It is thus the equivalent of a 140 m diameter parabolic dish with an aperture efficiency of 60%. The 12 beams of the antenna can be steered over a range

PLATE 3. The parabolic cylinder antenna near Ootacamund.
(Photograph—G. Swarup)

of $\pm 35°$ in declination and from -4^h to $+5\frac{1}{2}^h$ in hour angle. The operating frequency is 326.5 MHz, with a bandwidth of 4 MHz.

These results have been achieved at relatively low cost by making the antenna in the form of an equatorially mounted parabolic cylinder with its long dimension in the north-south direction. The long dimension is the axis about which the antenna is physically rotated to give the hour-angle coverage; this polar axis is conveniently provided by building the antenna up to the 11°28' slope of a south-facing hillside near Ootacamund in South India (latitude 11°28'N). The declination coverage is provided by phase adjustments along the focal line of dipoles which collect the energy from the reflector. The antenna has been described by Swarup (1968) and Swarup et al (1971) and the feed system by Kapahi et al (1968).

The mechanical construction of the antenna is shown by the photographs in Plates 3 and 4, and Figure 1 is a diagram of one reflector frame and feed support system. There are 24 such reflector frames, rotated in hour-angle in synchronism by being driven through various choices of gear reduction from a long drive shaft.

The line feed consists of 968 dipoles arranged at a spacing of 0.57 wavelengths ($\lambda = 92$ cm) mounted in a 90° corner reflector. The dipoles are colinear and parallel to the reflector surface wires, which themselves run the length of the parabolic cylinder at spacings at 2 cm (near the reflector center) to 4 cm (at the edges).

PLATE 4. The parabolic cylinder antenna near Ootacamund.
(Photograph—G. Swarup)

The dipoles are grouped in 22 blocks of 44 dipoles each. The dipoles in each block feed in series via phase-shifters and directional couplers to a pre-amplifier. It is thus possible to select outputs from these preamplifiers to form 12 beams in the sky, separated by 3 min of arc in declination. Thus, for lunar occultations, the disk of the Moon can be more than covered by the 12 beams. The design of the phase-shifters and directional couplers is interesting but beyond the scope of this review.

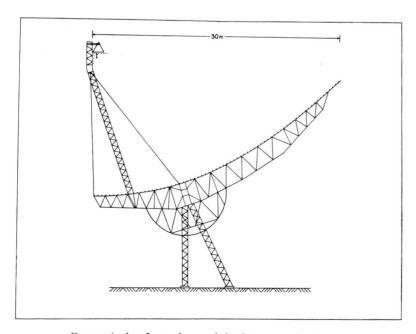

FIGURE 1. A reflector frame of the Ootacamund antenna.

The antenna can be used as a total power system, when its half-power beamwidths (HPBW) are $2° \times 5.5'$, or as a system with the two halves of the antenna phase-switched, when the effective HPBW in declination becomes 3.6'. The antenna is now operational and has made many observations of lunar occultations and pulsars.

Summary table.—The following Table 2[2] summarizes the main filled-aperture telescopes now in use for radio astronomy. The instruments in Table 1 are not included, nor are telescopes of size equal to or smaller than the equivalent of a 90 foot dish.

Gravitational Distortions of Antenna Structures

Most large reflector antennas are made steerable by mounting the reflector support structure on two mutually perpendicular axes. Now that it is easy to make the necessary coordinate conversions rapidly and accurately in a pointing-control computer, it is general to use the azimuth-elevation mounting for big dishes.

The reflector support structure is then required to be stiff enough so that, as the dish is tilted, the surface remains a true paraboloid of revolution to within a small fraction of the shortest operating wavelength (λ_{min}). A root-mean-square (rms) deviation of the surface from the required surface of $\lambda_{min}/16$ is generally considered adequate, although even with this scale of departure from true there is a loss of antenna gain of almost a factor of 2.

TABLE 2. Large filled-aperture radio telescopes

Location	Operating administration	Size
Owens Valley, California, USA	California Institute of Technology	39.6 m (130 foot) steerable paraboloid
Stanford, California, USA	Stanford University	45.7 m (150 foot) steerable paraboloid
Goldstone, California, USA	Jet Propulsion Laboratory	64 m (210 foot) steerable paraboloid
Dixie County, Florida, USA	University of Florida, Gainesville, Florida	640-dipole filled-aperture array for collecting area at 36.3 MHz = 30,000 m²
Vermillion River Observatory, Danville, Illinois, USA	University of Illinois Urbana, Illinois	(a) 183 × 122 m (600 × 400 feet) parabolic cylinder (b) 36.6 m (120 foot) steerable paraboloid

[2] Much of the information in Table 2 has been collected by R. Y. Dow, Secretary of the Committee on Radio Frequencies of the US National Academy of Sciences.

TABLE 2. (*Continued*)

Location	Operating administration	Size
Five College Radio Observatory, New Salem, Massachusetts, USA	University of Massachusetts and others	Two 36.6 m (120 foot) spherical reflectors
Sagamore Hill Radio Observatory, Hamilton, Massachusetts, USA	Air Force Cambridge Research Laboratory	45.7 m (150 foot) steerable paraboloid
Haystack Research Facility Tyngsboro, Massachusetts, USA	MIT Lincoln Laboratories	36.6 m (120 foot) steerable paraboloid
Ohio State-Ohio Wesleyan Radio Observatory, Delaware, Ohio, USA	Ohio State University Columbus, Ohio	103.8×21.4 m (340×70 feet) standing parabola with tilting reflector
Arecibo Observatory, Puerto Rico	Cornell University, Ithaca, New York	305 m (1000 foot) spherical reflector
Green Bank, West Virginia, USA	National Radio Astronomy Observatory	(*a*) 91.4 m (300 foot) paraboloid-meridian transit (*b*) 42.7 m (140 foot) steerable paraboloid
Jodrell Bank, Cheshire, England	Nuffield Radio Astronomy Laboratory, University of Manchester	(*a*) 76.3 m (250 foot) steerable paraboloid (Mark I) (*b*) 38×25 m (125×83 feet) steerable paraboloid (Mark II)
Sugar Grove, West Virginia, USA	Naval Research Laboratory	45.7 m (150 foot) steerable paraboloid
Effelsberg, The Eifel, W. Germany	Max-Planck-Institut für Radioastronomie, Bonn, W. Germany	100 m (328 foot) steerable paraboloid
Lake Traverse, Ontario, Ontario, Canada	Algonquin Radio Observatory National Research Council	45.7 m (150 foot) steerable paraboloid
Nançay, France	The Observatory of Paris, Meudon	300×35 m (984 ×115 feet) standing parabola with tilting reflector
Parkes, NSW, Australia	Australian National Radio Astronomy Observatory (CSIRO)	64 m (210 foot) steerable paraboloid
Ootacamund, South India	Tata Institute of Fundamental Research	530×30 m (1738×98 feet) semisteerable cylindrical paraboloid for 326.5 MHz
Pereyra, Iraola, Argentine	Argentine Radio Astronomy Institute	Two 30.5 m (100 foot) steerable paraboloids

Quite simple considerations (von Hoerner 1967a) suggest that if a material such as steel is used to make the surface support structure and if we accept the $\lambda_{min}/16$ criterion, the greatest diameter (D m) an antenna can have is given by

$$\lambda_{min} = 5.3 \ K \left(\frac{D}{100}\right)^2 \qquad\qquad 1.$$

where λ_{min} is in cm and K is a constant which must be greater than unity, but may lie in the range 1.2–1.8.

Although the reasoning leading to Equation 1 is not precise, a survey of many existing antennas (von Hoerner 1969) shows that the equation well describes their performance. There are other limitations to the accuracy of large reflector surfaces, such as the effects of wind forces and temperature differences across parts of the structure, but von Hoerner (1967a) shows that if it were possible to reduce the gravitational effects then much better telescopes could be built.

In the last few years, two main lines of design have been followed which can lead to this improved performance; they may be classified as "active" and "passive" solutions to the gravitational problem. The active methods involve the use of mechanisms which either sense the deformations of the reflector surface and correct them or apply forces to the reflector surface supports calculated (as a function of elevation angle) to be the values required to keep the surface to a true paraboloidal shape. The passive solutions are those which arrange, by correct design of the reflector support, that as the structure deflects when the elevation angle changes, the surface still remains a paraboloid of revolution. Of course, since the structure is deflecting, the properties of this paraboloid must change with elevation angle, and so both the focal length of the reflector and the direction of its radio axis are permitted to vary as the reflector is tilted. Such deflections, which result in changing a surface of a given type into another surface of the same type, have been described by von Hoerner as homologous deformations. In practice, the small changes in focal length can be allowed for by moving the position of the antenna feed as the elevation angle changes, while the motion of the radio axis (which rotates through angles slightly different from the angles through which the dish has been tilted) results only in a pointing correction which can be calibrated out of the system.

Active control of antenna surfaces.—So far no active control system using a closed-loop method has been employed on a large reflector, but a considerable number of systems have been devised and enough tests made to suggest that the technique can be useful. More than 10 years ago the design of the 600 foot antenna, which was to have been built at Sugar Grove for the US Naval Radio Research Station, included active control of the surface. The required positions of the adjustable panel points were defined by the intersection of light beams projected from a central stable location. Movements of the panel points from these desired positions were sensed, and the movements were automatically corrected by mechanically driven jackscrews.

Such closed-loop active control systems can, of course, in principle correct for more than gravitational effects. Temperature and wind-induced deformations, if their variability in time is within the capabilities of the control-loop response, will also be corrected.

A system of this kind has been built and tested on a 7 m (23 foot) parabolic reflector by Burr (1966). The required position of the surface panels was determined by modulated light beams and the positions were corrected by hydraulic jacks. Despite some mechanical difficulties the tests demonstrated the feasibility of the method.

The method of Swarup & Yang (1961) for measuring small phase-path changes has been tested by several workers as a means of sensing the movements of the reflector surface with the intent of incorporating it into a closed-loop control system. Braude et al (1966) and Bale, Gourlay & Meadows (1966) are two examples of published work.

An interesting and different approach has been suggested and tested on a one-dimensional antenna by Rudge & Davies (1970). A multielement array is used in the reflector focal plane. The first stage of this array gives at its output a spatial Fourier transform of the signals incident on it from the reflector; the array in fact reproduces a set of signals which correspond to the field distribution across the reflector aperture plane. The phase errors in this set of signals can then, in principle, be removed by adaptive antenna techniques, so that the primary reflector phase errors are corrected. The method will work best for those reflector deviations which have a large linear scale size across the reflector surface. Losses in the Fourier transform matrix network are a problem and the system would be complex in practice. Nevertheless, it is a promising and interesting technique which may have useful applications.

The active open-loop method of deflection correction has been applied in the design and construction of the 120 foot Haystack antenna (Weiss et al 1969). The stresses and deflections of the antenna structure were accurately analyzed using good computer programs, and then deflection compensation was introduced by using control cables and counterweights on various parts of the structure. The reflector was also adjusted so that there was some preplanned distortion at the zenith position and thus less average distortion in normal operations. Somewhat more sophisticated control techniques have been discussed by Rothman & Chang (1969) and by Weidlinger (1969).

Passive control of antenna surfaces.—There is now very satisfactory evidence that computations of the stresses and deflections in complex structures yield results which closely agree with practice. It is therefore possible to attempt to design reflector support structures which deflect in a homologous way. Von Hoerner (1965, 1967a) has shown that such designs are possible in principle and that they can be accomplished in a direct manner to yield real structures. The designers of the 100 m telescope for the Max-Planck-Institut für Radioastronomie have arrived at a similar result by trial and success computations. Their method (Hachenberg 1968) analyzed four different basic designs for the dish-support structure. In each analysis, changes were introduced step-by-step into the sizes of the structural members with the intent to bring the surface support points onto a parabolic surface and also to minimize the total reflector weight. Iterations of this process were continued until no further improvement was achieved. Generally, four to six iterations were adequate. The choice of the final structure was made on the basis of the goodness of the fit of the support points to a paraboloid at the zenith and horizon positions, on the size of the focal length and axial position changes, and on the total structural weight.

Von Hoerner (1967b) has described a method for designing a homologous structure which uses a computer to make design modifications in such a way as to guide itself to a homologous solution. In outline, the method works as follows:

(a) A reasonable choice is made of the geometrical shape of the reflector support structure; this defines the undistorted coordinates of all the member joints.

(b) The topology is decided; this says which points are connected by members, which are the surface points which are required to satisfy homology, and which points are to be considered fixed (e.g., the elevation axis support points).

(c) Bar areas are then selected for all members, which in the simplest analysis are to be single pin-jointed members. It is the variation and eventual choice of these bar areas which constitute the most important output of the homology program.

(d) The homology parameters are specified. These define the permitted deformations from one paraboloid to another.

The computer than calculates ΔH, the rms deviation of the surface support points from a paraboloid for two reflector attitudes, zenith and horizon pointing. (These are sufficient for a perfect solution for all pointing directions.) All bar areas are then changed at each iteration, and the computer runs to make $\Delta H \rightarrow 0$. Several iteration steps may be required, and since there can in principle be many homologous solutions, the computer chooses the final structure that is most similar to the initial choice.

The analytic basis for the method is given in von Hoerner (1967b); it relies first on the determination of the stiffness matrix K for the structure and then of the inverse of this matrix K^{-1}. A generalization of Newton's method is used to reach the condition $\Delta H \rightarrow 0$ and this requires that the derivatives with respect to all bar areas of all the elements of the inverse stiffness matrix K^{-1} be computed. Other constraints, necessary to give realistic solutions, are also added. All bar areas selected must be positive and give adequate survival strength to the structure. External loads in addition to the member weights are applied to the structure. The permitted homology conditions are included; the telescope when pointing at the zenith may have a focal length greater than the gravity-free focal length, and, of course, the whole surface may move equally parallel to gravity. With the telescope looking at the horizon the reflector surface may again be permitted to have moved uniformly parallel to gravity and the surface may have rotated (more than 90°) by a small angle.

When the homology program runs, it should converge fairly quickly to a solution where $\Delta H \leq \Delta H_0$, where ΔH_0 is a chosen value considerably smaller than $\lambda_{min}/16$. If it does not converge, generally a new input structure has to be chosen.

The method has been tested and shown to be capable of producing practical homologous structures. These solutions have in turn been checked by more conventional stress and deflection computer calculations and good agreement found.

Measurements of antenna surfaces.—Although it has been generally a fairly straightforward engineering task to measure and set the positions of reflector surfaces, a brief review of methods that can be used where high precision is needed may be of value. Smaller reflectors have often been measured by reference to templates or on large machine tools, but when radial distances of more than 10–15 m are involved other methods are used. Many telescopes have been set by the *tape and theodolite* system illustrated in Figure 2; an accurate theodolite is set at a known height h above the reflector vertex. The angle θ is measured: this may be the angle between VF and TP or between the gravity vector (if the dish aperture is horizontal) and TP. One of the linear distances TP or VP or the distance S along the dish surface, is measured by a good surveyor's tape. This technique can give rms measurement accuracies of about 1 mm at distances of 20 m.

An improvement to the optical part of this system is to replace the theodolite by a set of pentaprisms, each of which corresponds to a desired value of θ. A ring of targets at a constant S is measured with each pentaprism. Such a

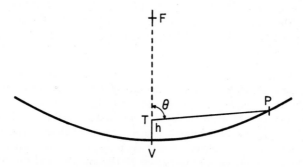

FIGURE 2. Tape and theodolite method for measuring reflector surface profile.

method (Kühne 1966) has been used and shown to give rms measurement precisions of 0.5 mm at 15 m. It is similar to the method used on the Haystack antenna (Weiss et al 1969). An interesting modification by Pearson, North & Kington (1966) sends five laser-light beams from a rotating stack of penta-prisms to pairs of silicon solar cells operated differentially as the rings of targets. A measurement sensitivity of 0.13 mm at 13 m distance is claimed. Other optical measurement methods have been used which have quite satisfactorily given the accuracy required by the instruments on which they were used, for example, Robinson (1966), Minnett et al (1969), and Jeffery (1969).

With the development of high-frequency modulators by which laser-light beams can be modulated at frequencies up to the GHz range, very accurate ranging becomes possible, and angular measurements can be replaced by distance measurements. For example, in Figure 2, the distances FP and VP might be measured for all targets. Justice & Charlton (1966) describe one such system where a modulating frequency of 450 MHz was used. A precision of 0.25 mm was obtained on a 10 foot paraboloid. This can almost certainly be improved, as such ranging instruments as the Mekometer II (Froome & Bradsell 1966) come into use. If very accurate distance measuring were required, true laser-light interferometry, as is used in instruments such as the Hewlett-Packard Model 5525A (Dukes & Gordon 1970) might be adaptable to the tasks, although the precision of such systems is in the nanometer range and may be embarrassingly great.

The modulated radiowave methods have already been mentioned in dealing with the control of surfaces; they obviously can be used for measurement alone.

ATMOSPHERIC EFFECTS ON LARGE ANTENNAS

The effects of the environment of the Earth on large telescopes are of two main kinds. First, there are the limitations imposed on the observations by imperfect transparency of the atmosphere. These limitations arise because the atmosphere absorbs, emits, and refracts radio waves in both a regular and an irregular manner. Second, the local environment of the telescope, particularly the winds and temperature differences to which it may be exposed, limits the precision with which the telescope can be used. This section will treat these two topics.

Atmospheric absorption, radiation, and refraction.—Absorption of radio waves in the atmosphere is an important limitation at wavelengths less than a few centimeters; the general form of the absorption curve is shown in Figure 3. This figure is based partly on calculations of the absorption by oxygen and water vapor and partly on observations. The water vapor contribution to the absorption varies depending on the amount and distribution of water, so the curves are only approximate. The absorptions at the resonance peaks are not well known. The curve has been drawn to conform as well as

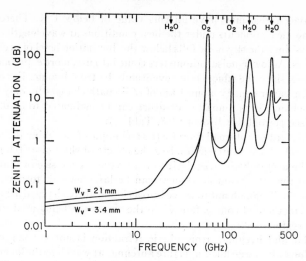

FIGURE 3. Zenith atmospheric attenuation. W_V = total precipitable
water in millimeters.

possible with various recent measurements of total zenith absorption: Ulaby
& Straiton (1969), Haroules & Brown (1968), Shimabukuro & Epstein (1970),
Gautier & Renan (1967), Wulfsberg (1967), Whaley & Fannin (1969), and
Ryadov & Furashov (1966). The calculations of Hall (1967), Liebe (1969),
Ulaby & Straiton (1970), and Reber, Mitchell & Carter (1970) have also been
used. It is evident from the figure that observations are possible through
the atmospheric "windows" of a very dry atmosphere to wavelengths as
short as 1.2 mm. The zenith absorption in these windows may be as low as 1
dB (a power loss of 26%). However, the difficulties of atmospheric radiation
and of the irregular behavior of the absorption add difficulties to observational programs.

A warm, partially absorbing atmosphere will radiate radio signals and
these add to the radio telescope signal in the form of noise. Atmospheric irregularities cause this noise level to fluctuate, and the resultant signal can be
a serious limitation to observations at wavelengths less than a few centimeters.

A number of observations have been made of the magnitude and statistical behavior of the sky noise; the paper of Haroules & Brown (1968) is a good
example of measurements made at 8, 15, 19, and 35 GHz. It is also possible to
calculate the magnitude of the sky noise fluctuations (Stankevich 1968). The
following conclusions can be drawn. Sky noise fluctuations can be troublesome at 10 cm wavelength and get worse at shorter wavelengths. Clouds and
precipitation give very large fluctuations, for example, at 9 mm Foster (1969)
observed antenna temperature fluctuations of many tens of degrees Kelvin.
However, under clear sky and dry air conditions the sky noise, even at milli-

meter wavelengths, may have an rms value of below 1°K. There are, at present, no measurements under the best conditions at wavelengths of a few millimeters since the sky noise falls below the fluctuation level of present-day radiometers. As lower-noise radiometers come into use, more experiments will be desirable, since the effects must eventually be troublesome. For example, if the water vapor density along 1 km of radio path changes by only 1%, the resultant change in antenna temperature can be estimated to be 0.12°K at 95 GHz (see Haroules & Brown 1968, Table 1).

For observations of small sources, the technique of switching the telescope beam on and off the source minimizes the effects of sky noise (Baars 1970), because the atmospheric irregularities may have a scale size of hundreds of meters, so that the beam movement can be large enough to move off the source yet small enough not to move into a significantly changed atmospheric path. The method, of course, fails when the antenna is being used to map an extended region of sky.

Regular and irregular atmospheric refraction is not yet a problem in observations with large filled-aperture antennas at wavelengths longer than a few centimeters. Regular refraction affects the apparent positions of radio sources, but adequate corrections for centimeter wavelengths can be made using the radio refraction derived from surface atmospheric properties. In the millimeter-wave region the narrower beamwidths and the greater influence of water vapor make the regular refraction more difficult to determine. Davis & Cogdell (1970) have studied the difference between radio and optical refraction at wavelengths of 8.6, 4.3, 3.1, and 2.2 mm using a 16 foot (4.9 m) radio telescope. A simple formula (Bean 1962) for radio refractive index (N)

$$N = (77.6 \, P/T) \, 1 + \frac{4810}{T} \frac{e}{P} \qquad\qquad 2.$$

where T is the temperature in °K, P the pressure in mbar, and e the water vapor partial pressure in mbar, fits the results up to about 100 GHz but there is evidence that at 134 GHz the radio refraction approaches more closely to the optical value.

The atmospheric irregularities show also as variations in refractive index, and this should affect radio telescopes much as the small-scale irregularities affect optical telescopes. However, as interferometer observations show (Hinder 1970, Basart, Miley & Clark 1970), rms path differences of about a centimeter occur between the phase pathlengths of zenith rays entering the atmosphere 1 km apart. This suggests that there will be an rms positional fluctuation with such an instrument of about 10^{-5} radians (2 arc sec). A direct check was made by Unger (1966) with a 70 foot (20.6 m) horn antenna, of the positional fluctuations at 4 GHz due to atmospheric irregularities. He observed the Early Bird satellite at 24.5° elevation and found random position errors of between 2 and 10 sec of arc. These results suggest that, since these angle fluctuations are well behaved and normally distributed, large filled-aperture antennas are not likely to be limited by this effect in their performance under the best atmospheric conditions even at millimeter wavelengths.

Environmental effects on performance.—In general, conventional engineering design methods can ensure that an exposed radio telescope will withstand winds, precipitation, and temperature changes. However, as greater precision becomes desirable, two effects need further study for exposed antennas: the effects of temperature differences across the structure and of irregular winds on the structure and on the pointing precision. Note that when a radome can be used these effects are much reduced. The precision with which a precise radome-enclosed antenna can be pointed is well illustrated in the experimental results reported by Meeks, Ball & Hull (1968) for the Haystack antenna. However, although radomes can be built with adequate transparency for wavelengths down to about 1 cm it becomes difficult to get good performance in the millimeter-wave region.

The effects of temperature differences on different members of an antenna support structure can be estimated. Various temperature differences can be assumed or these can be based on measurements on typical structures. The corresponding changes in member lengths are then included in the computer run to give the shape of the reflector surface. Katow (1969) gives an example of such calculations.

Wind gusts acting on the antenna will cause errors in the servocontrolled pointing system. The gusts provide varying torques on the structure; to counteract these torques the servodrive must develop a positional error signal at its input. The calculation of the magnitude of this positional error requires that the magnitude and the spectrum of the wind torques be known.

Measurements of the actual wind torques on antennas have been reported by Titus (1960) and Basistov (1969). Both experiments showed that the spectrum of the wind torques could be represented by

$$P_t(\omega) = \frac{K}{1 + (\omega/\omega_0)^2} \qquad\qquad 3.$$

Here $P_t(\omega)$ is (torque)^2 per radian/sec, K is a constant depending on antenna size, wind velocity and gust factor, and ω_0 is a measure of the angular frequency at which the spectrum cuts off. The measurements made on a 60 foot and a 22 m antenna showed that a mean value of ω_0 was about 0.15 radians per sec. The form of Equation 3 is shown in Figure 4. Stallard (1964) shows how it is possible to derive the rms pointing errors for an antenna using this knowledge of the wind torque spectrum combined with detailed information of the properties of the control servomechanism and of the antenna dynamics.

FEEDS FOR ANTENNAS

Good feeds for reflector antennas must meet two main requirements; they should give as large an aperture efficiency as possible and should also

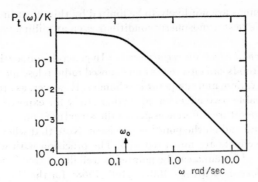

FIGURE 4. Spectral density of the wind torque on a parabolic antenna.

not permit unwanted radio noise to enter the radiometer system. There are other factors of lesser importance to be considered, such as achieving a desired level of sidelobes in the antenna far-field pattern.

For a prime-focus parabolic reflector antenna, it is possible to envisage a feed which satisfies these main needs. The feed should ideally collect all the radio energy in a plane wave incident normally on the dish and collect no energy from elsewhere. Of course, in practice, no feed can achieve this; if it did, the directivity pattern of the feed itself would have to follow the rectangular shape of Figure 5a. A typical horn feed pattern which gives quite good performance is sketched in Figure 5b. It is possible to improve on such patterns in various ways. The scalar feed (Simmons & Kay 1966) has been much used; it gives a small improvement in aperture efficiency and a reduction in antenna noise temperature. These results are obtained by surrounding the waveguide aperture with a conical shield, whose inner surface has corrugated rings. This gives almost similar performance for both E- and H-plane illumination.

The idea of making a feed which collects energy in the correct phase and amplitude from the concentric zones of the circular diffraction pattern in the focal plane of the telescope[3] has been used by Koch (1966). A sketch of the illumination pattern achieved with one of his experimental feeds is reproduced in Figure 5c.

The primary feed pattern can also be shaped by the use of dielectric material (Bartlett & Moseley 1966). The dielectric provides a low-loss guiding structure between the feed and the reflector—the authors refer to it as a *dielguide*. This technique can be used in both prime-focus and subreflector telescopes. When a subreflector is used, further methods are possible for controlling the main dish illumination. For example, both reflectors can, in

[3] For a study of the fields in the image space of a reflecting telescope see Minnett & Thomas (1968). The same authors (1966) report the synthesis of feeds based on this work.

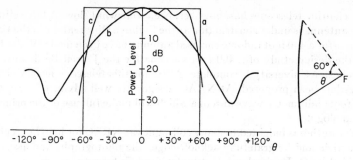

FIGURE 5. Illumination patterns for a parabolic reflector.

principle, be of an arbitrary shape (Galindo 1964) and this freedom can be used to improve the aperture illumination (Williams 1965). This has been used in a number of antennas employed for satellite communications but not so far for radio astronomy.

Some of the best-documented work in which the subreflector shape and the horn illumination have been carefully designed for the best low-noise system is that done on the Jet Propulsion Laboratory (JPL) antennas at Goldstone. Levy et al (1968) describe the installation on an 85 foot Cassegrain antenna which results in a total system noise of 16°K. The subject is treated further in the book by Rusch & Potter (1970).

The largest of all reflector antennas is the Arecibo 1000 foot spherical dish (Gordon & LaLonde 1961) and the potential value of such instruments has emphasized the need for good phase-correcting low-noise feeds. The problem has proved to be difficult in practice, although in principle the fact that the feed is an extended object—it often is a length of slotted waveguide—allows of independent control of the phase corrections and of the amplitude directivity. There have been a number of papers on the design of feeds for correcting spherical aberration; the one by Rumsey (1970) may be used as a survey. Measurements of the fields in the focal region of a spherical dish are described by Sen, Chasse & Rouillard (1970). The most successful feed so far described is one used at Arecibo at 318 MHz (LaLonde & Harris 1970). Excellent aperture efficiency and phase correction was achieved by using a slotted waveguide line feed 40 feet in length.

Correction for spherical aberration by the use of shaped subreflectors is also possible and again there have been a number of papers giving designs of which Burrows & Ricardi (1967) and Clarricoats & Lim (1969) are examples. No such feeds have yet been used on large spherical dishes for radio astronomy.

CONCLUSION

Over the past few years the growth of communication satellites and space research and the advances in radio astronomy have produced an excellent impetus to the development of new antenna technology. Progress in building new scientific telescopes has, however, been somewhat slow. A large tiltable-plate antenna is under construction at the Pulkova Observatory in the USSR. Designs for a 440 foot radome enclosed antenna have been made (Weiss 1969) and the design study of a 400 foot antenna for the Jodrell Bank station of Manchester University is complete. A site for this telescope near Meifod in Wales has been proposed. At NRAO designs are well advanced for a 65 m (213 foot) millimeter-wave antenna which will make full use of the principles of homology.

The author is indebted to many colleagues at NRAO for help in preparing this article and he gratefully acknowledges the photographs and assistance provided by O. Hachenberg, G. Stanley, and G. Swarup.

BIBLIOGRAPHY

Numerous references are made to the following two books, which are listed in the references as follows:

Mar & Liebowitz: 1969. *Structures Technology for Large Radio and Radar Telescope Systems*, ed. James W. Mar, Harold Liebowitz. Cambridge, Mass: MIT Press

IEE Conf. 1966. *Design and Construction of Large Steerable Aerials. Inst. Elec. Eng. (G.B.) Conf. Publ. No. 21*

Baars, J. W. M. 1970. *Dual-Beam Parabolic Antennae in Radio Astronomy*, 94–120. Groningen: Wolters-Noordhoff

Bale, F. V., Gourlay, J. A., Meadows, R. W. 1966. *Electron. Lett.* 2:252–53

Bartlett, H. E., Moseley, R. E. 1966. *Microwave J.* Dec., 53–58

Basart, J. P., Miley, G. K., Clark, B. G. 1970. *IEEE Trans. Antennas Propagat.* AP-18:375–79

Basistov, G. G. 1969. *Wideband Cruciform Radio Telescope Research. Proc. (Trudy) P. N. Lebedev Phys. Inst.*, ed. D. V. Skobel'tsyn, 38:106–14. Transl. Consultants Bureau, New York

Bean, B. R. 1962. *Proc. IRE* 50:260–73

Braude, B. V., Yesepkina, N. A., Petrun'kin, V. Yu., Khaykin, S. E., Umetski, V. N. 1966. *Radiotekh. Elektron.* 11:1499–502. English transl. in *Radio Eng. Electron. Phys.* 11:1302–5

Burr, D. W. 1966. IEE Conf., 84–89

Burrows, M. L., Ricardi, L. J. 1967. *IEEE Trans. Antennas Propagat.* AP-15:227–30

Christiansen, W. N., Högbom, J. A. 1969. *Radiotelescopes.* Cambridge: Cambridge Univ. Press

Clarricoats, P. J. B., Lim, S. H. 1969. *Electron. Lett.* 5:709–11

Cogdell, J. R. et al 1970. *IEEE Trans. Antennas Propagat.* AP-18:515–29

Davis, J. H., Cogdell, J. R. 1970. *IEEE Trans. Antennas Propagat.* AP-18:490–93

Dukes, J. N., Gordon, G. B. 1970. *Hewlett-Packard J.* Aug., 2–8

Findlay, J. W. 1964. *Advances in Radio Research*, ed. J. A. Saxton, 37–119. New York & London: Academic

Foster, P. R. 1969. *IEEE Trans. Antenna Propagat.* AP-17:684–86

Froome, K. D., Bradsell, R. H. 1966. *J. Sci. Instrum.* 43:129–33

Galindo, V. 1964. *IEEE Trans. Antennas Propagat.* AP-12:403–8

Gautier, D., Renan, J. P. 1967. *Ann. Ap.* 30:739–41

Gordon, W. E., LaLonde, L. M. 1961. *IRE Trans. Antennas Propagat.* AP-9:17–22

Hachenberg, O. 1968. *Beitr. Radioastron. Max-Planck Inst. Radioastron.* 1:31–61

Hachenberg, O. 1970. *Sky Telesc.* 40:338–43

Hall, J. T. 1967. *Appl. Opt.* 6:1391–98

Haroules, G. G., Brown, W. E. III. 1968. *IEEE Trans. Microwave Theory Tech.* MTT-16:611–20

Hinder, R. A. 1970. *Nature* 225:614–17

Jeffery, M. H. 1969. Mar & Liebowitz, 219–40

Justice, R., Charlton, T. 1966. IEE Conf., 177–81

Kapahi, V. K., Damle, S. H., Balasubramanian, V. 1968. *A Steerable Array of 968 Dipoles for Illuminating a Parabolic Cylindrical Antenna.* In *Proc. Symp. Antennas* Suppl. 3, ix-xii. New Delhi: Radio Telecommun. Res. Comm., CSIR

Katow, M. S. 1969. Mar & Liebowitz, 185–200

Katow, M. S., Bathker, D. A. 1968. *Radiofrequency Performance of a 210–foot Ground Antenna*, 383–90. Presented at *Int. Antenna Propagat. Symp.*, Boston, New York: IEEE

Ko, H. C. 1964. *Microwave Scanning Antennas.* ed. R. C. Hansen, 263–334. New York & London: Academic

Koch, G. F. 1966. IEE Conf., 163–67

Kühne, C. 1966. IEE Conf., 187–98

LaLonde, L. M., Harris, D. E. 1970. *IEEE Trans. Antennas Propagat.* AP-18:41–48

Levy, G. S., Bathker, D. A., Higa, W., Stelzried, C. T. 1968. *IEEE Trans. Microwave Theory Tech.* MTT-16:596–602

Liebe, H. J. 1969. *IEEE Trans. Antennas Propagat.* AP-17:621–27

Meeks, M. L., Ball, J. A., Hull, A. B. 1968. *IEEE Trans. Antennas Propagat.* AP-16:746–51

Minnett, H. C., Thomas, B. MacA. 1966. IEE Conf., 262–66

Minnett, H. C., Thomas, B. MacA. 1968. *Proc. Inst. Elec. Eng.* 115:1419–30

Minnett, H. C., Yabsley, D. E., Puttock, M. J. 1969. Mar & Liebowitz, 135–49

Pearson, H. E., North, J. C., Kington, C. N. 1966. IEE Conf., 277–82

Reber, E. E., Mitchell, R. L., Carter, C. J. 1970. *IEEE Trans. Antennas Propagat.* AP-18:472–79

Robinson, A. V. 1966. IEE Conf., 75–79

Rocci, S. A. 1966. *The 210-ft Parabolic Fully Steerable Tracking Antenna for a Deep Space Instrumentation Facility (JPL).* In *Deep Space and Missle Tracking Antennas: An Aviation and Space Symp.* Held in Conjunction with *ASME Winter Ann. Meet.* New York, 50–70

Rothman, H., Chang, F. K. 1969. Mar & Liebowitz, 273–86

Rudge, A. W., Davies, D. E. N. 1970. *Proc. Inst. Elec. Eng.* 117:351–58

Rule, B., Gayer, G. F. 1968. *Westinghouse Eng.* 28:No. 6, 162–67

Rumsey, V. H. 1970. *IEEE Trans. Antennas Propagat.* AP-18:343–51

Rusch, W. V. T., Potter, P. D. 1970. *Analysis of Reflector Antennas.* New York & London: Academic

Ryadov, V. Ya., Furashov, N. I. 1966. *I. Ucheb. Zaved. Radiofiz.* 9:859–66. English transl. in *Sov. Radiophys.* 9:504–7

Sen, A. K., Chasse, Y., Rouillard, M. 1970. *IEEE Trans. Antennas Propagat.* AP-18: 426–30

Shimabukuro, F. I., Epstein, E. E. 1970. *IEEE Trans. Antennas Propagat.* AP-18:485–90

Simmons, A. J., Kay, A. F. 1966. IEE Conf., 213–17

Stallard, D. V. 1964. *IEEE Trans. Appl. Ind* 83:105–114

Stankevich, K. S. 1968. *Radiotekh. Elektron.* 13:1570–76. English transl. in *Radio Eng. Electron. Phys.* 13:1366–72

Swarup, G. 1968. *A Large Cylindrical Radio Telescope at Ootacamund for Radio Astronomy Observations.* In *Proc. Symp. Antennas,* i–iii. Suppl. 1. New Delhi: Radio Telecommun. Res. Comm. CSIR

Swarup, G. et al 1971. *Nature, Phys. Sci.* 230:185–88

Swarup, G., Yang, K. S. 1961. *IRE Trans. Antennas Propagat.* AP-9:75–81

Titus, J. W. 1960. *Wind Induced Torques Measured on Large Antenna. NRL Rep. 5549,* US Naval Res. Lab., Washington, DC

Ulaby, F. T., Straiton, A. W. 1969. *IEEE Trans. Antennas Propagat.* AP-17:337–42

Ibid 1970. AP-18:479–85

Unger, J. H. W. 1966. *Bell Syst. Tech. J.* 45:1439–74

von Hoerner, S. 1965. *LFST Rep. No. 4.* Nat. Radio Astron. Obs., Green Bank, W. Va.

von Hoerner, S. 1967a. *Astron. J.* 72:35–47

von Hoerner, S. 1967b. *J. Struct. Div. Proc. Am. Soc. Civil Eng.* 93:461–85

von Hoerner, S. 1969. Mar & Liebowitz, 311–33

Weidlinger, P. 1969. Mar & Liebowitz, 287–309

Weiss, H. G. 1969. Mar & Liebowitz, 29–54

Weiss, H. G., Fanning, W. R., Folino, F. A., Muldoon, R. A. 1969. Mar & Liebowitz, 151–84

Wielebinski, R. 1970. *Nature* 228:507–8

Whaley, T. W. Jr., Fannin, B. M. 1969. *IEEE Trans. Antennas Propagat.* AP-17:682–84

Williams, W. F. 1965. *Microwave J.* 8: July 79–82

Wulfsberg, K. N. 1967. *Radio Sci.* 2:319–24

PHYSICAL CONDITIONS AND CHEMICAL CONSTITUTION OF DARK CLOUDS

CARL HEILES

Department of Astronomy, University of California, Berkeley

This article is primarily concerned with dust clouds, which have gained interest in recent years because they contain molecules in amounts observable by radio astronomy. Clarification of their environment will hopefully result in increased understanding of molecular formation and pumping processes. However, attention should also be directed towards less dense unobscured regions, which have relative molecular abundances comparable to those in the dust clouds (Section 2.8). Molecules are more easily detected in dust clouds mainly because they contain more matter and have larger column densities.

1. INTRODUCTION

Dust particles have long been known to exist in the interstellar medium. They show themselves by their interaction with starlight, both in absorbing it, producing dark regions, and in scattering it, producing bright regions (reflection nebulae, see H. M. Johnson 1968). The absorption manifests itself in two ways: by a general diminution of the brightness of stars over and above that which occurs from the inverse square of the distance, which occurs from dust which occurs everywhere in space along with the gas; and by a large excess of extinction of starlight in small regions as compared with the surroundings. It is the latter regions which are called *dust clouds*. The present review is directed to the densest of the relatively large ($\sim$1/2 degree diam) dust clouds because they contain enough molecules to be observable in the radio region of the spectrum. Dust clouds were first photographed and extensively cataloged by Barnard (1927); the most extensive recent catalog in the Northern sky is by Lynds (1962). The morphological situation has been recently reviewed by Lynds (1968) and Bok, Cardwell & Cromwell (1970), and physical conditions have previously been considered by Hoyle (1957).

Current knowledge of individual dust grains has been reviewed recently by Lynds & Wickramasinghe (1968) and by Greenberg (1968), who also applies the theory of grain optics to the observations in detail. The wavelength behavior of the extinction leads to estimates of the radius of dust grains, about 3×10^{-5} cm. Their albedo is probably about 0.65 (Mattila

1970). However, the grains are nonspherical, because they often show polarization effects due to magnetic alignment (see Jones & Spitzer 1967). Spectra in the infrared, optical, and ultraviolet provide hints as to chemical composition. Favored compositions include graphite, ice, dirty ice, iron, silicates (Woolf & Ney 1969, Gilman 1969, Knacke et al 1969, Donn, Krishna Swamy & Hunter 1970), and even diamond (see Landau 1970), or a mixture (Gilra 1971). In addition to broadband absorption and scattering, 26 fairly narrow spectral features called *diffuse interstellar lines* are thought to be produced by grains; these are discussed in the aforementioned review papers and also by Herbig (1966). Recently Johnson & Castro (1970) showed that the molecule bispyridylmagnesiumtetrabenzoporphine has absorption bands which match the wavelengths and approximate relative intensities of all of these diffuse lines; this molecule might therefore represent an important component responsible for extinction, either in gaseous or solid form, but this is not yet generally accepted. For the purposes of some of the discussion below, we need only assume that the grains are roughly spherical with radius 3×10^{-5} cm and are composed of a mix of elements such that most of their mass results from heavy elements; we take the mass density as 1 g/cm^3.

The elemental composition of the interstellar medium has been determined from studies of ionized regions and of newly formed stars (Aller 1961, Allen 1963). The ratio by mass of H:He:all heavier elements is approximately 63:36:1.4. Dust grains, being composed primarily of heavy elements, cannot possibly be responsible for more than 1.4% of the total mass in a given volume of space; dust will be responsible for this much only if *all* of the heavy elements exist in the grains. However, we know that some heavies remain in the gas, because only they can cool the gas to its observed temperature. This factor, 0.014, is therefore a stringent upper limit on the relative mass of dust; the remainder will be mainly hydrogen, observable in atomic form by virtue of its 21 cm line emission.

The optical extinction of a dust cloud can be measured by comparison of the histogram of brightness of stars behind the cloud with that of nearby stars; this method is reliable unless the cloud is so opaque that no background stars are visible behind it. Since this is often the case, values for extinction are often only lower limits. The extinction yields directly the column density of dust grains, and thus the mass of dust and a lower limit on the total mass. A number of interesting clouds have optical (visual) extinction of $\gtrsim 8$ mag (representing an attenuation $\gtrsim e^{-8/1.086} = e^{-7.37}$ (Bok 1956, Heiles 1968); these clouds might be regarded as typical of the ones in which molecules are visible in the radio region. For some clouds, new lower limits on extinction have been obtained by Grasdalen (1971) from the appearance of reflection nebulae within the clouds. These new limits are in some cases an order of magnitude larger than previous ones, so that the visual extinction $\gtrsim 80$ mag and $n_H \gtrsim 10^4$ cm^{-3}.

We measure the angular size of the cloud; its linear size, and hence total

TABLE 1. Properties of typical dust clouds

Δm^a	$\gtrsim 10$ mag
R	0.7 pc
Mass	$\gtrsim 110\ M_\odot$
$n_H + 2\ n_{H2}$	$\gtrsim 2000$ cm^{-3}
$n_{He}{}^b$	$\gtrsim 168$ cm^{-3}
$n_O{}^c$	~ 0.12
$n_C{}^c$	~ 0.06
$N_H + 2\ N_{H_2}$	$\gtrsim 8 \times 10^{21}$ cm^{-2}
n_{grains}	$\sim 4 \times 10^{-10}$ cm^{-3}
T_{gas}	$\sim 10°$K
T_{grains}	$\sim 10°$K
Gravity P.E./K.E.	$\gtrsim 2$

[a] Edge-to-edge through cloud center.

[b] As determined by radio recombination lines (see Dupree & Goldberg 1970; see also Danziger 1970) in H II regions.

[c] Assumes depletion factor of 10 (see text).

mass and density, depends on its distance which is obtainable only indirectly and hence is somewhat uncertain. The parameters given in Table 1 are typical of clouds which have been studied in the radio region, but are definitely not typical of interstellar dust clouds in general since their characteristics vary widely. Dust cloud radii range from <0.03 pc to tens of pc. The small, dense ones are called *globules* (Bok & Reilly 1947, Bok 1948, Bok, Cardwell & Cromwell 1970), and are thought to be future regions of star formation. It is quite probable that globules are similar to the larger clouds discussed in the present paper, but they have not been adequately studied in the radio region because their small size makes detection of lines of low surface brightness difficult.

2. CHEMICAL COMPOSITION OF THE GAS

Direct knowledge of the chemical composition of the gas in dust clouds comes mainly from radio observations, but indirect evidence and recent ultraviolet stellar absorption spectra of background stars in other regions are also important. The probable abundances as discussed below are summarized in Table 2.

2.1 *Hydrogen atoms.*—The most abundant and readily observable constituent of dust clouds would seem to be the atomic hydrogen, by use of the 21 cm line. The quantity measured in radio astronomy is the brightness temperature T_B, which is the temperature a blackbody would need in order to produce the measured surface brightness. The equation of radiation transfer describes the surface brightness of such a cloud; it is (see Kerr 1968)

TABLE 2. Some molecular abundances

Species	Density (cm^{-3})	Relative abundance[a]	Comments and references
H	<5	—	Cloud 2; Heiles 1969a
H$_2$	>1000	—	Cloud 2; Turner & Heiles 1971
OH	1.4×10^{-3}	$[OH]/[O] \sim 10^{-3}$	Cloud 2; $T_{ex} = 5°K$; Turner & Heiles 1971
H$_2$CO	3×10^{-6}	$[H_2CO]/[C][O] \sim 4 \times 10^{-6}$	Cloud 2; T_{ex} taken as 1°K; Palmer et al 1969
"X-ogen"	—	—	Buhl & Snyder 1970
CO	—	—	Solomon et al 1971
H$_2$C^{13}O	$\gtrsim 10^{-5}$	—	Cloud L134; Zuckerman et al 1969
SH	$\gtrsim 10^{-4}$	$[SH]/[S] \gtrsim 3 \times 10^{-3}$	Heiles & Turner 1971

Other upper limits NO (Turner, Heiles & Scharlemann 1970); CO (Penzias, Jefferts & Wilson 1971); HCN (Snyder & Buhl 1971); H$_2$O (Turner et al 1970); NH$_3$ (Cheung et al 1968); HC$_3$N (Turner 1971); H$_2$CO$_2$ (Zuckerman et al 1971); CH$_3$OH (8380 mHz, Turner, private communication).

[a] Values are upper limits for total atomic content (i.e. including those atoms in dust grains); values for only those atoms in the gas depend on depletion factor (see text).

$$\frac{dT_B}{d\tau} = (T_{ex} - T_B) \qquad\qquad 1.$$

where T_{ex} is the temperature describing the relative population of the levels of interest and τ is the optical depth in the line of interest. Of course, τ depends on T_{ex} and is given by

$$d\tau = A_{ul} \frac{c^2}{8\pi\nu} \frac{g_u}{g_l} \frac{h}{kT_{ex}} f(\nu) n_l ds \qquad\qquad 2.$$

where A_{ul} is the spontaneous emission coefficient (2.85×10^{-15} sec^{-1} for the 21 cm line), g_u and g_l are the degeneracies of the upper and lower states, respectively, n_l is the number density in the lower state, and ds represents incremental pathlength in the line of sight; the other symbols have their usual meanings. It is assumed that $h\nu/kT_{ex} \ll 1$. $f(\nu)d\nu$ is the probability that a transition occurs in the range ν to $\nu + d\nu$; under interstellar conditions, this function is always governed by the Doppler shifts due to motions along the line of sight. To obtain total abundances of a given species, the fraction of molecules which exists in the two energy levels producing an observed line must be known; this produces uncertainties in total abundances, particularly for H$_2$CO with its ortho and para ladders (Snyder et al 1969). If the cloud is homogeneous—an assumption usually made for lack of information—we have

$$T_B = T_{ex}(1 - e^{-\tau}) + T_c e^{-\tau} \qquad\qquad 3.$$

where T_c is the brightness temperature incident on the rear of the cloud. For the 21 cm line, a cloud with $n_H = 2000$ cm^{-3} like that in Table 1 with $T_{ex} \simeq 10°$K (see below) and a linewidth of 5 kHz ($\simeq 1$ km/sec) will provide $\tau \sim 400$; such a line should be easily detected.

Observational evidence confirming this general line of argument exists. Lilley (1955) first showed that in the general interstellar medium—that is, *in regions away from dense dust clouds*—the 21 cm line strength varies linearly with the column density of dust. This was confirmed by Heiles (1967), and Sturch (1969). However, in *dust clouds* the correlation vanishes or even becomes weakly negative. This was shown in detailed studies of several regions and in a sampling of 48 clouds (see Heiles 1969a, and the references quoted therein; Mészáros 1968, Quiroge & Varsavsky 1970, Garzoli & Varsavsky 1970, Sancisi & Wesselius 1970). The upper limits on hydrogen content are usually set by the nearby spatial fluctuations in 21 cm line emission, and typically yield column densities $N_{HI} < 3 \times 10^{19}$ atoms/cm^2 (Heiles 1969a). For a cloud of diam 1.5 pc this implies $n_{HI} < 5$ atoms/cm^3 (see Sections 2.2 and 3.2 for theoretical estimates of atomic hydrogen abundance). There is certainly one, and almost certainly two explanations for this.

The first is that the kinetic temperature, and hence T_{ex}, in dust clouds is small; the 21 cm line has been seen in some dust clouds, but when seen it usually appears in self-absorption against the background 21 cm line emission (Heiles 1969a, Sancisi & Wesselius 1970); unpublished, more sensitive measurements by Knapp at University of Maryland (Kerr 1970) show that about 50% of all clouds show weak 21 cm absorption features. [In two cases 21 cm emission occurs (Heiles 1969a) but the amounts are small compared to those expected from Table 1.] The self-absorption observations place stringent upper limits on T_{ex}, and usually imply $T_{ex} < 30°$K or so. For any individual dust cloud showing weak 21 cm absorption, as the majority do, the weakness of the absorption can be rationalized as being consistent with the amount of hydrogen expected from the arguments given in Section 1; one simply requires near equality of T_{ex} and the brightness temperature in the 21 cm line incident on the rear of the cloud, so that T_B does not vary much from the line to adjacent frequencies. However, such accidental situations cannot be invoked to explain weak lines in a large number of clouds. Given the expected amount of hydrogen in Table 1, one expects to see a large variation in 21 cm line appearance, from fairly strong emission lines (which would occur if the 21 cm line radiation from behind the cloud were weak) to almost complete absorption (which would occur if no 21 cm line emission were emitted between the cloud and the observer). However, the observed situation is one in which most clouds show only weak absorption, and given the range of 21 cm brightness temperature observed in the vicinities of these clouds, we can only conclude that they contain many fewer hydrogen atoms than one would expect from Table 1. Of course, the arguments concerning elemental composition in Section 1 might be inapplicable to dust clouds. In particular, the dust could be swept from the gas by radiation pressure of starlight (Whipple 1946); however, it is difficult to see how gas atoms could pass freely through the dust cloud if light photons cannot.

2.2 *Hydrogen molecules.*—We therefore conclude that the hydrogen has become molecular, and is therefore invisible in the 21 cm line. This conclusion is based only on indirect evidence and is not universally accepted (Sancisi & Wesselius 1970); because of its fundamental character we will briefly review the evidence. H_2 can be observed in interstellar space via its infrared or ultraviolet spectra; the role of H_2 in astrophysics has been reviewed by Field, Somerville & Dressler (1966). Infrared H_2 would be expected in emission in rotational transitions near 8 and 12 μ (Field et al 1968), but not in absorption due to the lack of suitable background sources. H_2 has several electronic transitions in the ultraviolet, for example the Lyman band near 1100 Å, but their detection in the spectrum of a star behind a dust cloud is rendered difficult by virtue of the dust itself, which absorbs ultraviolet even more strongly than visible light. Nevertheless, Carruthers (1970) has managed to detect H_2 in the ultraviolet absorption spectrum of a bright star. This star is located behind a dust cloud which *weakly* absorbs at optical wavelengths, and is therefore not of the same class as the clouds in the present discussion, but nevertheless this observation is an important piece of evidence, less indirect than the arguments in Section 1 and below, for the existence of large amounts of H_2 in dust clouds. Note that many hot stars which are not behind dust clouds have also been observed in the ultraviolet, and they do not show H_2 in their spectra.

Theoretical calculations by Solomon & Wickramasinghe (1969), Hollenbach & Salpeter (1971), and Hollenbach, Werner & Salpeter (1971) predict H_2 to predominate over H in sufficiently thick clouds. This is because the dissociating radiation is composed of a number of ultraviolet lines rather than continuum; the lines are all absorbed in dissociating H_2 at the edge of the cloud and cannot penetrate to the interior. H_2 formation occurs on dust grain surfaces. The theoretical results are quantitatively confirmed by the measurements of H_2 and H by Carruthers (1970), mentioned above.

An indirect argument for the existence of large quantities of H_2 comes from data on the 18 cm OH satellite lines. The principal lines (1665 and 1667 mHz) are interpretable in terms of normal population ratios among the levels involved, which are generated by the usual process of thermal equilibrium with collisions. However, the satellite line ratios are anomalous, usually in the sense that the 1720 mHz line is too strong by the same factor by which the 1612 mHz line is too weak (Turner & Heiles 1971). These authors have interpreted the anomaly as arising from either near (3 μ) or far ($\sim$100 μ)

infrared pumping. The sources of infrared are discussed in Section 3.1; a possible additional source of line radiation is shock waves occurring within the dust cloud itself (Litvak 1970). Litvak's (1969) theory of infrared pumping was applied and shows that the expected inversions are much stronger than observed, given the expected infrared intensities, unless collisions occur rapidly so as to quench the inversion. A density $n_{H_2} \gtrsim 100$ cm^{-3} is required. On the other hand, Gwinn & Townes (1968) believe collisional inversion, of the general type responsible for the inversion of formaldehyde, is responsible; however, Turner & Heiles (1971) do not believe the observations are well explained by this theory. This mechanism also requires comparably high gas densities.

Additional indirect evidence for the existence of large quantities of H$_2$ in dust clouds may be present in the mechanism of excitation of the 6 cm radio lines of H$_2$CO. Formaldehyde was first discovered in interstellar space by Snyder et al (1969) and was found in dust clouds by Palmer et al (1969). The H$_2$CO lines appear in absorption rather than emission, which is surprising because there is no excess brightness behind the clouds to provide a background. The line must therefore be absorbing the 2.8°K cosmic background radiation, having $T_{ex} < 2.8$°K. However, this is impossible unless a pumping mechanism exists to invert the level populations. Such a mechanism has been invented by Townes & Cheung (1969); it results from differences between the rates of collision with neutrals and radiative decay, and requires substantial densities ($n_{H_2} \gtrsim 1000$ cm^{-3}) to occur rapidly enough to produce the inversion. However, their calculation of collisional excitation rates was done classically, and is therefore inapplicable at the low temperatures in dust clouds where a quantum-mechanical treatment is in order. Thaddeus (1971), Thaddeus & Solomon (1971), and White (1971) have considered this problem in detail in a series of papers, including the effects of both neutral and charged particles as well as calculating expected temperature and ionization equilibria. They find that the sense of the inversion is predicted incorrectly by the classical theory, and find instead that the inversion is produced by short-wavelength radio radiation. Although another possible mechanism involving large gas densities is the generation of infrared line radiation in shock waves (Litvak 1970), relative velocities in typical dust clouds are probably too small (see Section 3.5) to produce infrared radiation in sufficient quantities.

2.3 *Other atoms.*—The gas densities of other atoms can be estimated only by knowledge of their abundance relative to hydrogen, and by assuming the H$_2$ abundance to be the lower limit outlined in Section 1 above. Helium atoms are certainly abundant in the gas; because of their low polarizability, neither helium nor neon sticks to the grains (Aannestad, personal communication). Heavier atoms which are not noble gases certainly will, however. Results of calculations of this "depletion" have been quoted by Field, Goldsmith & Habing (1969) and Goldsmith, Habing & Field (1969), but these calculations are highly uncertain and do not always agree with observation. In typical regions of interstellar space, where some heavy elements can be observed by the absorption lines they produce in the optical spectra of background stars, depletion factors can be measured. However, they are rendered highly uncertain by the imprecisely known ionization state of the gas, since interstellar optical absorption lines of a given element have the unfortunate tendency to be produced by the ionic species of lowest abundance (see, for example, Spitzer 1968). Goldsmith, Habing & Field find that observations show $\sim$10% of Na remains in the gaseous state, and often none of the Ca remains gaseous. Depletion rates scale as $n^2 T^{1/2}$ and hence are faster in dust clouds, where the densities are possibly greater than 10^4 cm^{-3}. Characteristic depletion times are calculated to be $\sim$10^8 years for densities $\sim$10 cm^{-3}; thus depletion times are $\sim$10^5 years in dust clouds, and depletion is almost certainly very important, but no observational information is available because background stars are invisible behind dark clouds. The gaseous abundances of some of the more abundant elements given in Table 1 are calculated by *assuming* a depletion factor of 10 for all elements (except helium).

2.4 *Electron density.*—At the high densities in dust clouds, ionization from X rays or cosmic rays is insignificant (Solomon & Werner 1971). The primary contribution to the electron density will result from ionization of carbon by starlight, and therefore depends sensitively on the extinction of the cloud. This has been computed by Werner (1970), who finds that the transition from carbon being nearly fully ionized to neutral occurs in a density range of 10^3 to 10^4 hydrogen atoms cm^{-3}. The total number of electrons will depend on the extent to which gaseous carbon has been depleted by the grains; if the depletion factor is 10, $n_e/n_H < 4 \times 10^{-5}$.

2.5 *OH.*—OH was discovered in a few dust clouds by Heiles (1968). This work was extended to a reasonable sample of dust clouds in the Galaxy by Cudaback & Heiles (1969), who found OH emission from 25% of the dust clouds they surveyed. However, the lines are weak and their sensitivity was poor. The histogram of number of clouds with a given antenna temperature increases with decreasing antenna temperature down to their sensitivity limit, and it is quite reasonable to conclude that OH would have been detected in every cloud if the sensitivity had been three times better. The column density of OH can be derived directly from the observation of the principal OH lines if no level population inversions occur for these main lines, which is probably the case (Turner & Heiles 1971). For the lower limits to projected hydrogen density given by the argument in Section 1, the ratio [OH]/[H$_2$]$\gtrsim$10^{-6} for a typical cloud. The relative abundance [O]/[H]$\simeq$6$\times$ 10^{-4} in the average interstellar medium, so that [OH]/[O]$\gtrsim$10^{-3}, i.e. less than 0.1% of the O is tied up in OH. Of course, some of the O is almost certainly tied up in the dust grains themselves; the relative amount of gaseous O tied up in OH may be 1% if the numbers in Table 2 are correct—or smaller, depending on the actual density of H_2.

2.6 H_2CO.—Formaldehyde was observed in dust clouds by Palmer et al (1969) in absorption against the cosmic background radiation (Section 2.2). It therefore is subject to ill-understood pumping processes which overpopulate the lower state of the transition relative to the upper state, and the derivation of its column density is somewhat uncertain. Assuming a level population T_{ex} characterized by a temperature of 1°K (an upper limit on T_{ex} for the deepest absorption lines observed), we obtain $N_{H_2CO} \gtrsim 10^{13}$ cm^{-2}. (We regard this value as a lower limit, due to unknown relative abundances in the para and ortho states and possible line saturation effects.) The senses of the inequalities for N_{H_2CO} and N_{H_2} make it impossible to put definite limits on the relative abundances, but given the new lower limits for extinction by Grasdalen (1971) it is probably correct to say $[H_2CO]/[H_2] \gtrsim 2 \times 10^{-9}$. It is interesting that $[H_2CO]/[OH] \sim 2 \times 10^{-3}$; this is surprisingly large and implies an efficient formation mechanism for H_2CO, with its large number of atoms. However, the relative abundance of H_2CO and OH is not constant within an individual cloud (see Section 3.5).

2.7 *Other molecules.*—Many other molecules have been found in the interstellar medium near the galactic center; however, most have not been seen in dust clouds. This may be a result of lack of sensitivity. A short compendium of some of these is given in Table 2. The absence of sulfur-containing molecules is possibly significant and has been remarked upon by Heiles & Turner (1971). Possibly CO will turn out to be abundant in dust clouds; its relative abundance in some regions of space is so large that most of the carbon atoms exist in CO (Penzias et al 1971).[a] If carbon is atomic it is the main cooling agent in dust clouds, because of several low-lying energy levels which can be excited at low temperatures; CO is probably an equally efficient coolant. The importance of carbon for cooling is curious, and may mean dense clouds are colder now than during the early history of the Galaxy when nucleogenesis had not yet provided significant amounts of carbon.

2.8 *Comparison with molecular abundances in other regions.*—
Sag B₂: This source lies near the galactic center and is rich in many molecules. It is instructive to compare the molecular abundances here with those in dust clouds. The H_2 column density in Sag B₂ can be estimated from the angular diameter of about 8 arc min and a minimum density of about 10^3 H_2 molecules cm^{-3} which is required to collisionally pump the NH₃

---

[a] *Note added in proof:* Solomon et al (1971) have detected CO emission from dust clouds. The lines appear saturated ($\tau \gg 1$ in Equation 3) so that the column density cannot be determined; the excitation temperature is deduced to be 5.7°K, and should equal the kinetic temperature at these densities.

(Cheung et al 1969). Taking the distance as 10 kpc we find a diameter of 22 pc, a column density $N_{H_2} \gtrsim 7 \times 10^{22}$ cm^{-2} [about equal to N_H (Clark 1966)], and a total mass of about $3 \times 10^6 \, M_\odot$. The temperature, as indicated by several lines of NH$_3$, is 20 to 50°K, depending on position. The velocity spread of the NH$_3$ is large, ranging from 32 to 64 km/sec; the velocity spread for other molecules is even larger, for example from -106 to $+62$ km/sec for H$_2$CO (Zuckerman et al 1970).

The column density of NH$_3$ is about 10^{16} cm^{-2}, so the relative abundance [NH$_3$]/[H$_2$]$\simeq 10^{-7}$; if this were the case in dust clouds NH$_3$ would probably not have been detected, given the sensitivity of the search (Knowles & Cheung 1971). The column density of H$_2$CO is about 10^{15} cm^{-2} for most of the individual velocity components and about 3×10^{16} cm^{-2} for all components (due primarily to one component at 26 km/sec), so the ratio [H$_2$CO]/[H$_2$]$\simeq 3 \times 10^{-7}$ (Zuckerman et al 1970). Robinson (1967) estimates [OH]/[H]$\simeq 10^{-4}$; thus [H$_2$CO]/[OH]$\simeq 3 \times 10^{-3}$, comparable to the ratio in dust clouds. The temperature in Sag B$_2$ is larger than the dust clouds, and the proportion of molecules seems to be larger (but estimates of hydrogen content may be much too low). It is likely that the same chemical processes are at work in both cases. The fact that many more different molecules have been seen in Sag B$_2$ than in dust clouds (see the references in Table 2) may simply reflect a very large column density of hydrogen, which provides large column densities of observable molecules even if their relative abundances are small.

Verschuur's clouds: Verschuur (1969b) has discovered several interstellar clouds with 21 cm linewidths narrow enough to show that they are cold. He has successfully examined two of them, clouds A and G, for the presence of OH. Cloud G, an intermediate-velocity cloud (-29.5 km/sec), has kinetic temperature $\lesssim 86$°K, a column density $N_H \simeq 2 \times 10^{20}$ cm^{-2} (Verschuur 1969c), and a relative abundance [OH]/[H]$\simeq 3 \times 10^{-7}$ (Verschuur 1971a); the OH velocity (-30.5 km/sec) is larger in magnitude than the H velocity. This velocity difference is in the correct sense to consider the OH as being formed in a shock wave (Field et al 1968), a not unreasonable conclusion for intermediate-velocity clouds. Cloud A, a low-velocity cloud (3.6 km/sec), has kinetic temperature $\gtrsim 20$°K, $N_H \simeq 5 \times 10^{19}$ cm^{-2}, and [OH]/[H] also $\simeq 3 \times 10^{-7}$; however, there are two OH components placed roughly symmetrically about the H velocity at 0.7 and 7.0 km/sec (Verschuur 1971b).

The relative OH abundances are only an order of magnitude smaller than those in dust clouds (Table 2). However, [OH]/[H] in Table 2 is an upper limit; thus the relative OH abundances may be more nearly equal. It is obviously of interest to search for other molecules in these clouds, particularly H$_2$CO and CO. Although it is just possible that these objects are dust clouds, invisible optically because of the small surface density of stars at high galactic latitudes, the velocity differences between OH and H imply that additional physical processes are at work. Further study of these clouds will be most interesting.

Typical interstellar space: H (Clark 1966), H_2CO (Zuckerman et al 1970), and OH (Goss 1968) have been observed in absorption against a large number of continuum radio sources. In many cases the sources are H II regions or are located behind dusty regions, so that the existence of molecules is perhaps not surprising. However, in 3 cases the source is much more distant than the absorbing gas and there is no dark dust cloud visible so that these observations should represent random samplings of cold, dense regions in the interstellar gas. Optical extinction to Cas A varies from 3 to 5 mag across the source (Minkowski 1968); however, since the distance is 3400 pc the diffuse components of grains in the interstellar medium probably amounts to 3 mag. The nonuniform extinction is probably due to a single dust cloud with maximum extinction 2 mag, but in any case *not more than one* dark cloud (of the kind under consideration here) can lie in front of this source. Visual extinction to Tau A amounts to only 1.1 mag, and is homogeneous in spatial distribution (O'Dell 1962). The extinction to Cyg A is anomalously low, only 1.5 mag in the visual band (2.1 in the photographic band at 4260 Å; Baade & Minkowski 1953). The extinction to these two sources probably is completely due to the diffuse grain component.

The 3 sets of data are summarized in Table 3. The number of molecular velocity components is much too large to result from dense dust clouds. It appears that an H_2CO line usually has an associated OH line, and vice versa. H absorption appears in nearly every source; upper limits at least can be obtained from the interferometric measurements by Clark (1966). In most cases velocity differences can probably be ascribed to blending effects of insufficient sensitivity; exceptions are the 13.2 km/sec OH feature in Tau A, the H_2CO or OH components having no counterparts and perhaps the -40 km/sec features in Cas A. Although there is no real consistency, H_2CO lines tend to be narrower than OH lines and both tend to be narrower than the H lines. In some cases the relative widths can be interpreted in terms of thermal broadening. But they cannot in some others, where we must conclude that the molecules are more concentrated in space than the hydrogen, and often H_2CO more than OH. This point should be kept in mind when comparing relative abundances.

Derivation of the relative abundances from the absorption spectra requires knowledge of the excitation temperatures. The H excitation temperature will equal the kinetic temperature under any reasonable conditions in interstellar space (Field 1958), but this is not true for OH (Goss 1968) or H_2CO, which is affected by pumping even in dense dust clouds. Following Zuckerman et al (1970), we have taken the H_2CO, OH, and H excitation temperatures as 3, 10, and 50°K respectively. This typically produces $[H_2CO]/[H] \approx 10^{-8}$ and $[OH]/[H] \approx 10^{-6}$. Thus $[H_2]/[H] \ll 1$ (presumably), while $[H_2CO]/[OH] \approx 10^{-2}$; this is to be contrasted to the situation in dust clouds, where $[H_2]/[H] \gg 1$ and $[H_2CO]/[OH] \approx 10^{-3}$. This seems somewhat surprising, given the densities which are ~ 10 H atoms per cm³, less than 1% of those in dust clouds; one might have instead expected polyatomic molecules to be

TABLE 3. Absorption in continuum sources[a]

Source	V_H (km/sec)	VO_H (km/sec)	V_{H_2CO}[b] (km/sec)	ΔV_H[c]	ΔVO_H[c]	ΔV_{H_2CO}[c]	n_H[d] (cm^{-3})	N_H[e] (10^{20} cm^{-2})	N_{OH}[e] (10^{13} cm^{-2})	N_{H_2CO}[e] (10^{11} cm^{-2})
Cyg A	3.6	3.5	4.6	3.9	3.2	0.9	9?	1.0	2.0	3

Other H components at 0.3, 10.5, −19.0, and −85.1 km/sec contain a total of 1.7×10^{20} cm^{-2}

Source	V_H	VO_H	V_{H_2CO}	ΔV_H	ΔVO_H	ΔV_{H_2CO}	n_H	N_H	N_{OH}	N_{H_2CO}
Cas A	−47.9	−48.2	−48.1	3.6	1.2	1.5	72?	22	2.4	5
	—	−46.3	−46.4	—	3.4	2.2	—	—	10	9
	−44.8	—	−44.3	3.1	—	2.5	3?	1.0	—	5
	−41.6	−40.9	−39.6	4.2	3.6	3.2	5?	1.6	8.8	15
	−37.9	−37.0	−37.0	5.6	2.5	2.2	60	13	5.3	9
	—	—	−35.3	—	—	1.2	—	—	—	3
	−1.5	−1.6	−1.5	1.8	0.8	0.8	8?	1.3	3.2	6
	−0.4	−0.2	−0.3	2.9	0.9	1.0	25?	4.0	4.0	5

Other H components at −32.7, −11.5, −4.8, +3.4, and +7.9 km/sec contain a total of 2.2×10^{20} cm^{-2}

Source	V_H	VO_H	V_{H_2CO}	ΔV_H	ΔVO_H	ΔV_{H_2CO}	n_H	N_H	N_{OH}	N_{H_2CO}
Tau A	2.3	2.3	—	1.7	3.6	—	1?	0.4	0.55	<0.3
	—	13.2	—	—	2.9	—	—	—	0.9	<0.3

Other H components at −1.2, 3.9, 8.0, and 10.8 km/sec contain a total of 8.1×10^{20} cm^{-2}

[a] Sources: H, Clark (1966); OH, Goss (1968); H₂CO, Zuckerman et al (1970).
[b] Corrected for new determination of rest frequency by Tucker, Tomasevich & Thaddeus (1970).
[c] Full width at half maximum, km/sec.
[d] ?Denotes 10 pc diam assumed; in most cases the density is therefore an upper limit.
[e] Assumed excitation temperatures 50°K for H, 10°K for OH, 3°K for H₂CO.

much less abundant in regions of low density. It does seem that H_2, as opposed to H, is not important in the chemistry.

Other molecules have been detected in absorption against the optical spectra of a few stars. In Zeta Oph, for example, the lines are thought to be produced in a region with gas density $n_H \simeq 10^3$ cm^{-3} and the expected associated grain density; here we have [CH]/[H]$\simeq 10^{-8}$, [CN]/[H]$\simeq 10^{-9}$, while OH and CO$^+\gtrsim 10^{-8}$ and NH, MgH, and SiH$\lesssim 10^{-9}$ (Herbig 1968).

3. Physical Conditions

Some of these have already been discussed.

3.1 *Radiation field.*—The radiation field in dust clouds results from the radiation field which exists in normal regions of interstellar space, as modified by the dust itself and also by the usual environment in which dust clouds find themselves. The known components are summarized in Figure 1. Data are lacking from 100 to 912 Å. Reliable infrared data longward of about 300 μ are nonexistent, and one can only interpolate between the infrared and short radio wavelengths where the 2.8°K cosmic background dominates. Of course, this spectrum represents an underestimate near sources of various kinds; particularly important for the galactic center sources is the infrared contribution from the galactic center (Aumann & Low 1970), while molecules near H II regions are affected by the strong far-infrared radiation field of the H II regions (Low & Aumann 1970).

For sufficiently short wavelengths the radiation is absorbed and scattered by dust, and it will be appreciably weakened in the interior of dust clouds. This is shown in Figure 1, where it has been assumed that the cloud has a visual extinction (edge to edge) of 10 mag. In accounting for extinction the dust has been assumed to completely absorb light; the resulting overestimation of the effective opacity is perhaps balanced by the fact that extinctions quoted for dust clouds are often lower limits. The wavelength dependence of extinction, shown in Figure 2, has been taken from Bless & Savage (1970) in the ultraviolet and from H.L. Johnson (1968) elsewhere. The attenuation of optical and ultraviolet light is certainly important in considering ionization states of atoms and molecules.

The infrared radiation density in dust clouds is likely to be significantly enhanced by the presence of T Tauri–type stars, which are usually associated with dust clouds (Herbig 1962, Herbig & Peimbert 1964, Herbig, 1970) and have large near-infrared excesses (Mendoza 1968) due to the presence of circumstellar dust. Recent observations suggest that these stars also have far-infrared excesses; the flux of T Tauri itself, for example, appears to be increasing slightly with wavelength at 20 μ and is thought to contain abnormally large (4 μ diam) dust particles in its envelope (Low et al 1970). This is consistent with the theoretical model of Low & Smith (1966) from which the spectrum in Figure 1 has been estimated; however, the extrapolation to wavelengths longer than 20 μ is unsupported by any direct observational

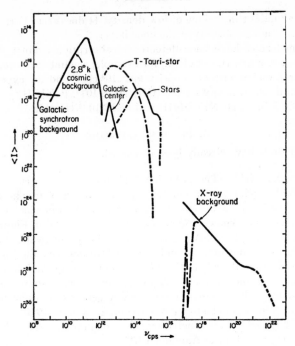

FIGURE 1. $\langle I \rangle$, the specific intensity averaged over the sphere $((1/4\pi)\int Id\Omega)$ in ergs/cm²-sec-Hz-ster, as a function of frequency ν. The solid curve applies to a typical region of space in the solar neighborhood; the dash-dot line to the center of a dust cloud with 5 mag of visual extinction from edge to center (see text). The wavelength dependence of extinction is shown in Figure 2.

X ray: The X-ray portion is taken from the compilations by Silk (1970) and Schwartz, Hudson & Peterson (1970); below frequency 3×10^{15} the neutral interstellar gas, composed primarily of H and He, absorbs X rays and will seriously depress the low-energy X-ray intensity under the extragalactic component in a manner which depends on the degree of clumpiness of the interstellar gas (see Bowyer & Field 1969). Contributions of X-ray sources have been neglected; this approximation is quite good for the higher-energy X rays, but less so for the soft X-ray region (Bunner et al 1969, Werner, Silk & Rees 1970). Some cosmological models predicting low-energy X-ray intensities, which have not been measured, are summarized by Silk & Werner (1969).

Ultraviolet: The ultraviolet portion has been taken as the radiation field in the solar neighborhood given by Habing (1968). The cutoff at 912 Å results from photoionization of hydrogen.

Optical: The optical and near-infrared portion has been taken from the compilation by Allen (1963).

Infrared: The spectrum of the galactic center is given by Aumann & Low (1970); the spectrum of a typical T Tauri star at distance 1 pc is estimated *approximately* (see text). Significant additional radiation will be present near H II regions (Low & Aumann 1970) and far-infrared sources (optically unidentified, Friedlander & Joseph evidence and should be taken with a grain of salt. On the other hand, we have assumed one T Tauri–type star to be located at a distance of 1 pc from the center of the cloud, whereas the more realistic situation is that many such stars are scattered around and within the cloud; for this reason the spectrum shown in Figure 1 may be an underestimate for a typical cloud. We note that the infrared contribution from this type of star completely overwhelms the far-infrared contribution from the galactic center, an important point in considering the energy-level populations in molecules (e.g. for OH, see Turner & Heiles 1971). Grasdalen et al (1971) list T Tauri-type stars associated with many known dust clouds.

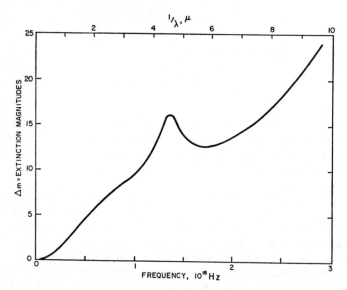

FIGURE 2. Extinction due to dust as a function of frequency. The visual extinction ($\lambda \sim 0.55\mu$) has been taken as 5 mag, which is probably typical of or less than the extinction from space to the center of a dark dust cloud of the kind under consideration in this paper. The ultraviolet portion is from the observations by Bless & Savage (1970), and the visual and infrared from the compilation of observations by H. L. Johnson (1968).

1970); in the solar neighborhood such sources provide as much radiation as the galactic center.

Cosmic background: The strongest contribution at short radio wavelengths is the 2.8°K cosmic background.

Galactic background: At the longer radio wavelengths the galactic background has a spectral index of about −0.9 changing to about −0.5 at 200 mHz (Purton 1966, Penzias & Wilson 1966, Baldwin 1967, Bridle 1967) and turning over at about 3 mHz (Alexander et al 1969, Clark, Brown & Alexander 1970); the intensity of the spectrum was accurately calibrated by Pauliny-Toth & Shakeshaft (1962).

The X-ray background is attenuated both by the gas in the cloud and by the grains. However, the attenuation can be computed as if all the material is gaseous, since absorption occurs primarily by photoionization of inner electrons (see Martin 1970). The material being in grains, rather than in the gas, simply leads to small angle scattering (Overbeck 1965) which is completely unimportant in the present context. Bell & Kingston (1967) have assembled the relevant data and performed calculations of X-ray absorption cross sections for the chemical composition of the interstellar gas. For $\lambda \gtrsim 22$ Å, the effective cross section can be taken as

$$\sigma \simeq 4.6 \left(\frac{\lambda \text{Å}}{22}\right)^{3.2} \times 10^{-22} \, \text{cm}^2/\text{H atom}$$

and for smaller wavelengths we have

$$\sigma \simeq 1.6 \times \left(\frac{\lambda \text{Å}}{22}\right)^{2.15} \times 10^{-21} \, \text{cm}^2/\text{H atom}$$

The X-ray absorption is important only at low energies; it has been accounted for in Figure 1 by assuming a column density of 4×10^{21} hydrogen atoms to the cloud center, a value consistent with the 5 mag of visual extinction used in Figure 1.

3.2 *Cosmic rays and X rays.*—These quantities are important because of their effects on temperature, dissociation of molecules, and ionization of atoms. Cosmic rays exist in interstellar space, as shown by direct measurements on or above the Earth. The energy spectrum is such that the number of particles increases with approximately the inverse 1.5 power of energy (Parker 1968a); since the ionization cross section varies as the inverse of the energy, it is the low-energy cosmic rays which are important in their effects of ionization and heating of the interstellar gas. However, it is in the low-energy region that direct measurements of cosmic ray intensity cannot be made because of the influence of the solar magnetic field and the solar wind (see discussion by Werner, Silk & Rees 1970). Indirect arguments must suffice in this region. These rest exclusively on ionization of normal predominantly neutral interstellar matter by these particles. X rays will produce similar effects.

The ionization can be measured in three ways. Historically, the first was by the low-frequency radio continuum absorption produced by the ionized component of the neutral medium (Ellis & Hamilton 1966, Gould 1969). The interpretation of this measurement is considerably influenced by what temperature is assigned to the gas, however; in H II (ionized hydrogen) regions it is $\simeq 10^4 \, ^\circ \text{K}$, but in the neutral gas it is more likely to be $\sim 100^\circ \text{K}$. The second method is measurement of the frequency dispersion of pulsar radiation, which yields directly the column density of electrons to the pulsar.

The average ionization of "neutral" regions of the interstellar gas has been inferred from these methods to be $n_e/n_H \simeq 10^{-2}$ by Field, Goldsmith & Habing (1969), and $\sim 10^{-1}$ by Hjellming, Gordon & Gordon (1969); here we use the former result. This is more than an order of magnitude higher than that which would occur in the absence of cosmic rays or X rays, in which case the electrons result from ionization of all atoms with ionization potentials less than that of hydrogen (13.6 eV) providing $n_e/n_H \simeq 6 \times 10^{-4}$; this value is determined only by the abundance ratio of ionizable heavy elements to hydrogen.

The third method uses the radio line emission from electrons recombining with hydrogen. These lines are weak in the predominantly neutral gas, but Gottesman & Gordon (1970) were able to detect them. They find that the degree of ionization is apparently enhanced within about 8 kpc of the galactic center to such an extent that $n_e/n_H \sim 0.2$. This result might argue for X-ray rather than cosmic-ray ionization. The Galaxy produces background radio emission from the electron component of the cosmic rays, and if the excess ionization results from cosmic rays we would expect an enhancement in the radio intensity in the interior of the Galaxy. The galactic synchrotron emission is somewhat stronger here (Baldwin 1967), but it is probably not sufficiently strong to produce the requisite enhancement in ionization inferred by Gottesman & Gordon (1970). Also, according to Fowler, Reeves & Silk (1970), spallation in interstellar space limits the maximum allowable ionization rate to $\xi'_H = 6 \times 10^{-17}$ sec^{-1}, from cosmic rays, considerably less than required.

A definitive choice between cosmic rays and X rays as ionization mechanisms must await the test of time; Goldsmith, Habing & Field (1969) and Spitzer & Scott (1969) discuss cosmic rays, while Werner, Silk & Rees (1970) and Sunyaev (1969) discuss X rays. In any case, the important quantity is the ionization rate which must exist to explain the observed value of n_e/n_H, regardless of the mechanism involved. This rate is found to be $\xi'_H = 4 \times 10^{-16}$ sec^{-1} per H atom by Field, Goldsmith & Habing (1969).

The effects of ionization on the ratio [H]/[H$_2$] in a dust cloud have been considered by Hollenbach, Werner & Salpeter (1971) and Solomon & Werner (1971). They find that the number density of atomic hydrogen cannot fall below 50 cm^{-3} with such an ionization rate. However, the observations imply considerably lower limits than this. This argues that cosmic rays and X rays are excluded from dust clouds and that, in dust clouds, $\xi'' \lesssim 10^{-16}$ sec^{-1}. This exclusion must result partially from the absorption in the periphery of the clouds, but possibly not completely because then one could expect a halo of atomic hydrogen around the clouds. It is therefore tempting to think that cosmic rays are stopped even before entering the cloud by an excess magnetic field just outside. Only a small excess of magnetic field is required to exclude protons of energy < 10 MeV, which are the most important for heating and molecular dissociation. An enhanced field is consistent with theoretical expectations, but not necessarily with observations (see Section 3.6).

3.3 Gas temperature.—

Observations: The most direct way to measure temperature would be to observe the intensities of the lines responsible for cooling the cloud. We expect these clouds to be cold, and if suitable molecules exist in the clouds these lines will occur primarily from rotational transitions at very short radio or long infrared wavelengths; however, the relative molecular abundances are unknown. In the absence of such molecules cooling will occur primarily from the fine structure lines of carbon (see below) which occur at long infrared wavelengths and hence are unobservable from the ground. Until such observations can be made with high sensitivity from outside the Earth's atmosphere we must be content with secondary indicators of temperature. See Note[a] added in proof.

At present these include only two: the 21 cm H I line and the 18 cm OH lines. If other molecular lines can be observed in these dust clouds, they will provide valuable additional information as they have in the galactic center regions (see Cheung et al 1969), where the relative intensities of the NH_3 lines arising from transitions within different rotational levels imply temperatures somewhat higher than those which occur in dust clouds.

Only limits on temperature can be estimated from the 21 cm line because the atomic hydrogen in the clouds comprises only a small part of the total in the line of sight. Furthermore, only a few clouds show a usably strong 21 cm line. For clouds exhibiting the 21 cm line in absorption we can drive only an upper limit on the temperature (see Sections 2.1 and 2.2). The only dust cloud for which published information exists, and for which a reliable temperature limit can be determined, is cloud 2 of Heiles (1969a), where $T \lesssim 30°K$. The temperature for this cloud has also been derived from the OH lines (see below). The only known cloud showing strong 21 cm emission is cloud 13 in Table I of Heiles (1969a); for clouds showing emission we can derive only a lower limit for the temperature. For this cloud, $T \gtrsim 60°K$; judging from its high temperature, it is qualitatively different from the typical dark cloud. This cloud was not seen in the OH line survey by Cudaback & Heiles (1969), so no further information is available.

Upper limits on OH excitation temperature can be set if the line appears in absorption. Indeed, Cudaback & Heiles (1969) noticed a trend in their survey: low-latitude clouds appeared in absorption against weak continuum sources having brightness temperatures $> 9°K$ or so. This means the average dust cloud has OH excitation temperature $< 9°K$. The OH excitation temperature can also be derived by comparison of the brightness temperatures in the four OH lines, because the four lines have different spontaneous emission coefficients, and hence different optical depths in Equation 2. The errors which can occur in this procedure are that the excitation temperature may not be the same for the four OH lines, or that inhomogeneities may confuse our interpretation of measurements integrated over volume of space. Indeed, the satellite OH lines do have unequal excitation temperature (Turner & Heiles 1971) which can be explained on the basis of far ($\sim 100 \mu$) or near

($\sim$3 μ) infrared pumping (Section 2.2). Infrared pumping does not affect the principal OH lines, however, for which T_{ex} is determined by an equilibrium between collisional and 18 cm radiative excitation. These have been considered by Heiles (1969b), who found that the excitation temperature is equal to the kinetic temperature as long as the density of molecular hydrogen exceeds 1 cm^{-3}, which is almost certainly the case. The temperature of several clouds has been obtained in this way both by Heiles (1969b) and Turner & Heiles (1971). They find temperatures of about 5°K for three clouds (one of which is cloud 2, discussed above) and a lower limit of 10°K for a third. However, insidious, weak pumping is at work; in one object, the 1665 line is slightly weaker than 5/9 the 1667 line (Heiles & Turner, unpublished); in some objects, then, the excitation temperatures for the two lines are unequal, and in addition may not exactly reflect the gas kinetic temperature.

Theory: The gas temperature in interstellar space results from an equilibrium between heating and cooling processes. Heating processes include photoionization by starlight and ionization by cosmic rays and X rays. Cooling results from collisional excitation of low-energy transitions in atomic and molecular species, which subsequently emit photons which escape from the dust cloud. Equilibrium temperatures have been calculated for normal regions of interstellar space by various authors (see Spitzer & Scott 1969, Hjellming, Gordon & Gordon 1969; Goldsmith, Habing & Field 1969, Werner, Silk & Rees 1970); the theoretically derived temperatures depend on the uncertain cosmic-ray or X-ray densities (Section 3.1) and extent of depletion (Section 2.3). They range from about 15 to 100°K for high-density regions (about 10 atoms/cm^3) and from 10^3 to 10^4°K for low densities (10^{-1} to 10^{-2} atoms/cm^3).

Heiles (1969b) made simple estimates of the conditions which obtain in dust clouds using several questionable assumptions. The calculation assumes no depletion of heavy elements (overestimating the cooling rates), neglects cooling from possible molecular transitions (underestimating the cooling rates), and ignores optical depth effects. The cooling results primarily from neutral carbon, a result confirmed by the more detailed calculations of ionization equilibrium by Werner (1970). The result agrees with temperatures obtained from observations of the 18 cm principal OH line (see above) for $\xi'_{H} \simeq 6 \times 10^{-17}$ sec^{-1}, lending credence to the interpretation that the excitation temperature of the principal 18 cm OH lines actually closely reflects the gas kinetic temperature. Accurate calculations of this sort represent an outstanding theoretical problem at present.

3.4 *Grain temperatures.*—In recent years several authors have concluded that the presence of impurities will lead to the establishment of very low grain temperatures. However, they have violated fundamental physical laws and have overestimated the maximum possible radiation efficiency due to impurities (see Field 1969, Purcell 1969). The problem of grain temperatures in dark clouds, as well as in more typical regions of space, has been considered in detail by Werner & Salpeter (1969). They consider a number of grain compositions, both with and without impurity cooling; here we quote results for grains of 0.15 μ radius composed of a graphite core of 0.05 μ and a dirty ice mantle. Such a grain is a reasonably efficient radiator, does not violate the fundamental laws governing radiation efficiency, and is probably not unrepresentative of the true situation as far as radiation properties of grains are concerned.

The grains are heated by the cosmic background radiation, by far-infrared radiation from nearby grains, and by starlight (which is attenuated and scattered on its way to the inside of a cloud). The adopted radiation field was similar to, but somewhat different from, the one given in Figure 1. They did not include the contribution from T Tauri stars, which are preferentially located near dark clouds; thus their derived temperatures, and particularly the limiting temperatures reached for very dark clouds, will be slightly too small. The equilibrium temperature of these grains was 17°K in normal interstellar space; for a cloud with 5 mag of extinction in the visible band to the center, 10°K; for 20 mag, 6.5°K; and the limiting value, 3.5°K. It is interesting that the gas possibly is colder than the grains; information on the difference between gas and grain temperatures could be obtained from measurements of polarization produced by the grains. Such measurements could only be performed in the infrared, because of the lack of stars in the optical band behind the clouds. However, the magnetic field appears to be small in dust clouds (Section 3.6), which would lead to no appreciable grain alignment at these high densities (Jones & Spitzer 1967).

3.5 *The state of internal motion.*—This information can be inferred from examination of the linewidths of various species having different molecular weights. We have information only from OH and H_2CO. Interpretation of such data is difficult because the angular area covered by an OH line observation is 9 times that in the H_2CO line, since the largest available telescope (the 140 foot at the National Radio Astronomy Observatory at Green Bank, West Virginia) has been used in each case and the beamwidth increases linearly with wavelength due to diffraction. Furthermore, the H_2CO line shape is influenced by the presence of several closely spaced hyperfine components. Published information at present is sparse, which adds to the difficulty.

The OH linewidths in dust clouds are usually about 1.5 km/sec; if due only to thermal motions they imply kinetic temperatures of 800°K, much larger than the upper limits. We must therefore assume that mass motions are present. For most clouds this motion must be small in scale, because the OH lines are single. However, in cloud 2 the OH line is composed of two separate, narrower components (Heiles 1969a; Turner & Heiles 1971), separated by 0.8 km/sec. Cloud 2 is unusual in this regard, and we devote the following to a discussion of this cloud in particular.

Additional information for cloud 2 is available in the H_2CO data, for which one spectrum has been published (Palmer et al 1969). It is surprising that only one velocity component appears in the H_2CO spectrum, not two as for OH. It is tempting to assume that the H_2CO and one of the OH components exist in the same part of the cloud, and a definite decision could be made by comparing the published velocities of the OH component and the H_2CO; they are 5.2 and 4.7 km/sec, respectively, and the difference is much larger than errors. However, the laboratory rest frequencies of both lines are uncertain. Palmer et al (1969) used a rest frequency of 4829.649 mHz in reducing their H_2CO data; Tucker, Tomasevich & Thaddeus (1970) have redetermined their H_2CO rest frequency to be 10.6 kHz higher. This correction raises the velocity derived by Palmer et al to 5.3 km/sec, almost exactly equal to the 5.2 km/sec of the OH component (any difference might result from an error in the rest frequency of OH: see Manchester & Gordon 1970). Thus, we will assume for the present purposes that the H_2CO line and 5.2 km/sec OH component emanate from the same region. The two spectra are shown in Figure 3.

The width of the 5.2 km/sec OH component implies a kinetic temperature of 240°K if the broadening is due to thermal motions alone. The corresponding H_2CO linewidth would then be about 0.6 km/sec, equal to that observed. However, the apparent agreement between the H_2CO and OH linewidth

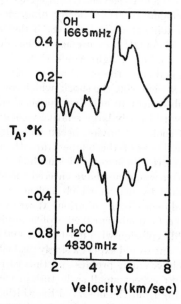

FIGURE 3. OH profile (Turner & Heiles 1971) and H_2CO profile (Palmer et al 1969) taken at nearby positions in cloud 2. The velocity zero of the H_2CO data has been adjusted as described in the text. The OH has two velocity components, the H_2CO only one; the small dip in the H_2CO profile at 6.2 km/sec is simply a hyperfine component associated with the strong feature at 5.2 km/sec.

by assuming thermal broadening alone must be regarded as fortuitous, because of the lower limits on temperature from other data (Section 3.3). It is almost certainly a result of the 9 times smaller area covered by the H_2CO line observation, which implies that the velocity of this material varies with position on the sky. This point should be explored by mapping this portion of the cloud in the H_2CO line; it would be of interest to find systematic velocity variations which could be ascribed, for example, to rotation.

Are the relative motions within dust clouds supersonic, resulting in shock waves with associated higher-temperature regions? This is an important question for several reasons, not the least of which involves pumping of molecular energy levels (Litvak 1970). Although it might appear so, they are not necessarily supersonic. The gas consists mainly of H_2 and He and is expected to have a molecular weight of about 2.3, while the OH has a molecular weight of 17; thus the sound velocity is considerably larger than indicated by the OH line width. However, the question is not simple. For the case of cloud 2 discussed here we consider a symmetrical collision between two clouds at half of the total velocity difference between the clouds. Two cases are possible (in both we arbitrarily assume there is no relative motion perpendicular to the line of sight).

In the first case the 0.8 km/sec velocity difference represents the relative velocity of the two clouds, i.e. the velocity difference *before* entering the shocks. Here the motion is not supersonic even if the gas temperature is the 5°K indicated by the 18 cm OH measurements; the addition of relative motion perpendicular to the line of sight would make the motion only weakly supersonic. In the second case, the measured velocity difference represents the relative velocity *after* entering the shock, which could result from cooling during the motion subsequent to entering the shocked region. In this case one can construct any arbitrarily large relative velocity between the clouds by a proper choice of cooling behavior, because cooling leads to slower velocities relative to the shocks as the temperature drops and the density increases. The first case corresponds to a collision which began recently, because then most of the matter would not have entered the shock. The line radiation would be generated by matter which had not yet entered the shock, with little matter (and hence little radiation) at zero velocity relative to the shocks; one would expect a doubly peaked profile. Similarly, the second case corresponds to an "old" collision with most of the matter at small velocities relative to the shocks; one would expect a singly peaked profile.

Most dust clouds have single profiles and thus correspond to the second picture; here we might expect small-scale turbulence, perhaps supersonic (but weakly so, judging from the observed linewidths). Cloud 2, discussed here, corresponds instead to the first case of a "young" collision; it is the only known cloud showing a double profile. Additional support for this conclusion comes from the 21 cm data. As mentioned in Section 2.1, most clouds show only a very small 21 cm line; cloud 2 has a deep self-absorption feature.

Almost any reasonable interpretation of this situation, as well as the theory of H_2 equilibrium abundances in dust clouds (Section 2.2), predicts $H_2 \rightarrow H$ as time goes by until equilibrium is established, rather than the reverse; we would think of cloud 2 as still being in the process of converting H to H_2, i.e. a young cloud, while the others are sufficiently old for the equilibrium to have been established. To summarize, we believe most clouds have already evolved through the stage we see for cloud 2 now and that cloud 2 is a young cloud; the velocity differences we see for cloud 2 represent the relative velocities of two clouds, and hence the gas velocities *before* entering the shock waves. The fact that the velocity difference between the two components in cloud 2 varies but little over the face of the cloud (Heiles 1970) is also consistent with this picture, as is the 21 cm width from cloud 2 (Heiles 1969a). Thus the motion is subsonic, or only weakly supersonic; this conclusion applies as well to other dust clouds which show a single OH profile.

If shock waves occur in dust clouds they will raise the temperature of the material passing through the shock to not more than about 30°K. However, over sufficiently long time scales the shock cannot be considered adiabatic, because the temperature will return to the equilibrium value. The rate at which this occurs depends on the temperature and density dependence of the cooling rate, and also depends on the strength of the magnetic field (see Field et al 1968). With densities $\sim 10^4$ cm^{-3} and cooling in the fine-structure lines of neutral and ionized carbon, the time required for re-establishment of temperature equilibrium will be ~ 100 years (see Field et al 1968, Bahcall & Wolf 1968), under the assumption of a depletion factor of 10. With a velocity of 1 km/sec the material will then cool in a distance $\sim 10^{-4}$ pc, a small fraction (about 10^{-4}) of the cloud's diameter. Therefore in no case will a large fraction of the material exist in the heated state it may briefly attain behind a shock wave. The OH excitation temperature should come into equilibrium with the kinetic temperature in about 3 years from collisions with H_2 at this density (Rogers & Barrett 1968), so that the OH excitation temperature will reflect the kinetic temperature profile behind a shock wave. Since the OH excitation temperatures are measured to be only 5 degrees or so in many dust clouds, most of the OH producing the 18 cm lines has not just passed through a shock.

The energy source for the internal motions in these clouds is undoubtedly gravity, and they are probably contracting (Section 4). Large rotational velocities due to conservation of angular momentum might therefore be expected. However, the observations (Heiles 1970) show that rotational velocities of the cloud as a whole are generally less than a few tenths of a km/sec, and are therefore much smaller than the internal random motions. However, the comparison of OH and H_2CO lines above suggests that smaller regions within the cloud may show rotation. It is tempting to speculate that during contraction the angular momentum is transmitted to neighboring material by a magnetic field. Magnetic fields play an important role in the dynamics of condensing interstellar matter (see Spitzer 1968 and Burbidge et al 1960).

3.6 *Magnetic field.*—The magnetic field of the Galaxy is important in the large-scale dynamics of interstellar gas. It can be measured in various ways, none of which give certain results (see review by Davis & Berge 1968); a value of 10^{-5} G is probably not far from the average value. The field is *frozen in* to the gas because of its high conductivity and the long distances involved; therefore, the magnetic flux is conserved during gas motion (see Spitzer 1968). In a three-dimensional, isotropic contraction of a cloud we therefore expect $B \propto$ radius^{-2}, or $B \propto n^{+2/3}$. During contraction from the average interstellar density (say, 1 cm^{-3}) to $n = 10^3$ cm^{-3} in dust clouds, we thus expect B to be enhanced by a factor of 100.

The magnitude of strong fields in isolated regions can be measured best by Zeeman splitting of a radio line; the method gives the average, along the line of sight, of the component of magnetic field parallel to the line of sight, and therefore represents a lower limit due to possible cancellation from field reversals along the line of sight. Measurements of Zeeman splitting by Verschuur (1969a) in normal interstellar gas clouds confirm the validity of the above line of argument. In dust clouds we are usually restricted to the 18 cm OH lines. Turner & Verschuur (1970) have attempted such a measurement in cloud 4, whose density is probably about 10^3 cm^{-3}. They obtained only an upper limit equal to 1.7×10^{-4} G. This limit, while somewhat smaller than expected, is not grossly inconsistent with expectation.

However, in cloud 2 there exists a sharp, narrow self-absorption feature in the 21 cm line. Verschuur (1970) has attempted to measure the Zeeman splitting of this line, from which much better limits on the magnetic field can be derived because it is very strong. He finds $B < 5 \times 10^{-6}$ G. This value is so small that it is discomforting to ascribe its weakness to the possible cancellation effects mentioned above. The only possible way to reconcile expectation with observation in this case is to imagine that the hydrogen atoms responsible for the absorption exist in a shell around the periphery of the cloud and do not reflect the physical conditions inside the dust cloud. This has relevance for exclusion of cosmic rays (Section 3.2). It is difficult to accept the interpretation that the field inside dust clouds is so weak, and it is desirable to obtain stricter limits on the Zeeman splitting from the OH line, since the OH does certainly exist in the interior, and not at the outside, of dust clouds (Heiles 1970).

3.7 *Relationship to the surrounding interstellar gas.*—If dust clouds are growing smaller one might expect to see a *hole* in the surrounding gas in the volume previously occupied by the cloud, unless its contraction proceeds slowly enough (i.e., slower than the sound velocity in the surrounding medium). Most of the published maps of dust clouds and their surroundings are not large enough or do not have sufficient angular resolution to show such an effect, which is in any case difficult to disentangle from the normal background fluctuations in the 21 cm line intensity. Although from existing data one can conclude that no large effect exists, one would not expect a large effect because the normal interstellar gas density is quite small, 0.5 to 10 hydrogen atoms per cm³.

4. When Are the Molecules Formed?

If we accept the physical conditions in dust clouds as discussed above, we can calculate the equilibrium abundance of molecules known to exist in the clouds and compare with the observed abundances. Such an exercise for OH was performed by Heiles (1968), who considered various reactions involving H_2 molecules, grain surfaces, and atoms given by Carroll & Salpeter (1966) and Stecher & Williams (1967). None of the reaction rates considered differ greatly. The reactions all required activation energy, and the rates are so small at the low temperatures in dust clouds that equilibrium molecular abundances are minute and the times required for equilibrium are longer than the age of the Universe.

More likely are reactions involving grain surfaces. Rates are given by Stecher & Williams (1967), but are derived on the basis of perfect crystalline graphite grains; impurities certainly must abound in grains and on their surfaces, and these may provide centers where enhanced rates of molecular formation can occur which would greatly reduce the temperature dependence of the OH formation rate (Salpeter 1967, Hollenbach & Salpeter 1971). The maximum possible rate of such a mechanism for OH would occur if every oxygen atom colliding with a grain produces a molecule which is freed to the gas. However, even such an optimistic assumption appears just adequate to explain the observed abundances. If we assume that grain surfaces are saturated with hydrogen atoms, this rate would be (see Gould & Salpeter 1963)

$$\frac{dn_{OH}}{dt} = \tfrac{1}{2} J A \quad \text{and} \quad J = \tfrac{1}{4} n_0 \bar{v}$$

where J is the flux of oxygen atoms from the gas incident on the grain surface, $\bar{v}$ is the mean thermal velocity of the atoms, and A is the amount of grain surface area per cm³. Using a grain radius of 3×10^{-5} cm and the densities given in Table 1, we have

$$\frac{dn_{OH}}{dt} \simeq 3 \times 10^{-15} n_O$$

which would produce the observed OH density, about 10^{-3} per cm³, in about 10^6 years if there were no destruction of the OH. This time is probably no more than an order of magnitude less than the age of the cloud. How, then, would H_2CO be formed in the required amount? And how would molecules be formed in unshielded regions of space (Section 2.8)?

Given the difficulty of explaining the observed molecular abundances in terms of chemical equilibria on the basis of present physical conditions, one must not overlook the possibility that the present abundances result from conditions in the past and that the time scale required for equilibrium is too long to consider only the present. Dust clouds are held together by gravity, as indicated by a comparison of their gravitational potential energy with their internal kinetic energy (see Table 1). Under these conditions they will collapse at a rate determined by the rate at which they radiate internal energy by the cooling mechanisms, and it is reasonable to suppose that they are in a state of gradual collapse. Their masses are comparable with the masses of newly formed clusters of new stars, i.e., several hundred solar masses; it is not unreasonable to expect this to be their future. Indeed, the correlation of T Tauri stars, which are young stars, and dust clouds shows that some star formation is already taking place even at these early stages; when a hot O star forms, it will have enough ultraviolet flux to completely ionize the cloud, producing objects not unlike the H II regions, Orion Nebula, M17, W49, etc. The time scale for this is difficult to predict because of the unknown cooling rates. However, if some of the observed OH linewidth results from contraction—say, half—then they will collapse in about 3×10^5 years, this being the shortest possible time scale.

The past, then, should be thought of in reverse: the present clouds were formed from contraction from a larger object which existed in the past. We do not know how one of these clouds gets started, since the influence of self-gravity is smaller when the cloud is physically larger; magnetic (see Parker 1968a, b) and thermal (Field 1965, Goldsmith 1970) instabilities are suspected. But when the cloud is larger the column density of dust and gas is smaller, for a given total amount of mass. Under these conditions the gas temperature will be higher, for example about 30°K for $n = 30$ cm^{-3} (Spitzer & Scott 1969), or perhaps 100°K if depletion is significant in these early stages (Field, Goldsmith & Habing 1969), i.e. typical for that of normal interstellar gas, and may have existed for $10^7–10^9$ years. There may have been a high-temperature ($\sim 10^{4\circ}$K), low-density ($\sim 10^{-1}$ cm^{-3}) phase before the onset of contractional instability. Shock waves may have existed during the initial period of instability, perhaps resulting from cloud collisions, which could have raised the temperature to $10^{4\circ}$K for brief periods of time ($\sim 10^4$ years; Field et al 1968).

It is therefore conceivable that the present molecular abundances in dust clouds reflect such vagaries of past history. This would be disappointing from the standpoint of obtaining a complete understanding of the present situation in dust clouds. However, it could be exceedingly interesting from the standpoint of understanding what conditions are required for the onset of instabilities which lead to the contraction of large, rarefied masses of gas to become dense clouds. This general picture is consistent with the existence of OH and H₂CO in less dense unobscured regions in proportions similar to those in dust clouds (3.8). On the other hand, the lack of chemical homogeneity in dust clouds (3.5) perhaps argues that molecular abundances are determined by present conditions, because the material is expected to be well mixed during contraction of the cloud.

I wish to thank Mr. P. Aannestad, Dr. B. Turner, Dr. G. Field, Mr. G. Grasdalen, Dr. J. Silk, Dr. P. Solomon, Dr. P. Thaddeus, and Dr. M. Werner for interesting and stimulating discussions, and a large number of people for preprints.

LITERATURE CITED

Alexander, J. K., Brown, L. W., Clark, T. A., Stone, R. G., Weber, R. R. 1969. *Ap. J. Lett.* 157:L163

Allen, C. W. 1963. *Astrophysical Quantities*. London: Athlone Press

Aller, L. H. 1961. *The Abundance of the Elements*. New York: Interscience

Aumann, H. H., Low, F. J. 1970. *Ap. J. Lett.* 159:L159

Baade, W., Minkowski, R. 1953. *Ap. J.* 119:206

Bahcall, J. N., Wolf, R. A. 1968. *Ap. J.* 152:701

Baldwin, J. E. 1967. In *Radio Astronomy and the Galactic System*, ed. H. van Woerden, 337. London: Academic

Barnard, E. E. 1927. *Photographic Atlas of Selected Regions of the Milky Way*, ed. E. B. Frost, M. R. Calvert. Washington: Carnegie Inst. Washington

Bell, K. L., Kingston, A. E. 1967. *MNRAS* 136:241

Bless, R. C., Savage, B. D. 1970. In *Ultraviolet Stellar Spectra and Related Ground-Based Observations. Proc. IAU Symp. No. 36*, ed. L. Houziaux, H. E. Butler. Dordrecht-Holland: Reidel. In press

Bok, B. J. 1948. *Centennial Symposia. Harvard Obs. Monogr. No. 7*, 53

Bok, B. J. 1956. *Astron. J.* 61:309

Bok, B. J., Cardwell, C. S., Cromwell, R. H. 1970. *Symposium on Dark Nebulae, Globules, and Protostars*. Tucson: Univ. Arizona Press

Bok, B. J., Reilly, E. F. 1947. *Ap. J.* 105:255

Bowyer, C. S., Field, G. B. 1969. *Nature* 223:573

Bridle, A. H. 1967. *MNRAS* 136:219

Buhl, D., Snyder, L. E. 1970. *Nature* 228:267

Bunner, A. N. et al 1969. *Nature* 223:1222

Burbidge, G. R., Kahn, F. D., Ebert, R., von Hoerner, S., Temesváry, St. 1960. *Die Entstehung von Sternen*. Berlin: Springer-Verlag

Carroll, T. D., Salpeter, E. E. 1966. *Ap. J.* 143:609

Carruthers, G. R. 1970. *Ap. J. Lett.* 161:L81

Cheung, A. C., Rank, D. M., Townes, C. H., Knowles, S. H., Sullivan, W. T. 1969. *Ap. J. Lett.* 157:L13

Cheung, A. C., Rank, D. M., Townes, C. H., Thornton, D. D., Welch, W. J. 1968. *Phys. Rev. Lett.* 21:1701

Clark, B. G. 1966. *Ap. J.* 142:1398

Clark, T. A., Brown, L. W., Alexander, J. K. 1970. *Nature* 228:647

Cudaback, D. D., Heiles, C. 1969. *Ap. J. Lett.* 155:L21

Danziger, I. J. 1970. *Ann. Rev. Astron. Ap.* 8:161

Davis, L., Berge, G. L. 1968. In *Nebulae and Interstellar Matter*, ed. B. M. Middlehurst, L. H. Aller. Chicago: Univ. Chicago Press

Donn, B., Krishna Swamy, K. S., Hunter, C. 1970. *Ap. J.* 160:353

Dupree, A. K., Goldberg, L. 1970. *Ann. Rev. Astron. Ap.* 8:231

Ellis, G. R. A., Hamilton, P. A. 1966. *Ap. J.* 146:78

Field, G. B. 1958. *Proc. Inst. Radio Eng.* 46:240

Field, G. B. 1965. *Ap. J.* 142:531

Field, G. B. 1969. *MNRAS* 144:411

Field, G. B., Goldsmith, D. W., Habing, H. J. 1969. *Ap. J. Lett.* 155:L149.

Field, G. B., Rather, J. D. G., Aannestad, P. A., Orszag, S. A. 1968. *Ap. J.* 151:953

Field, G. B., Somerville, W. B., Dressler, K. 1966. *Ann. Rev. Astron. Ap.* 4:207

Fowler, W. A., Reeves, H., Silk, J. 1970. *Ap. J.* 162:49

Friedlander, M. W., Joseph, R. D. 1970. *Ap. J. Lett.* 162:L87

Garzoli, S. L., Varsavsky, C. M. 1970. *Ap. J.* 160:83

Gilman, R. C. 1969. *Ap. J. Lett.* 155:L185

Gilra, D. P. 1971. *Nature* 229:237

Goldsmith, D. W. 1970. *Ap. J.* 161:41

Goldsmith, D. W., Habing, H. J., Field, G. B. 1969. *Ap. J.* 158:173

Goss, W. M. 1968. *Ap. J. Suppl.* 15:131

Gottesman, S. T., Gordon, M. A. 1970. *Ap. J. Lett.* 162:L93

Gould, R. J. 1969. *Aust. J. Phys.* 22:189

Gould, R. J., Salpeter, E. E. 1963. *Ap. J.* 138:393

Grasdalen, G. L. 1971. In preparation

Grasdalen, G. L., Kuhi, L. V., Harlan, E. A. 1971. Submitted to *PASP*

Greenberg, J. M. 1968. In *Nebulae and Interstellar Matter*, ed. B. M. Middlehurst, L. H. Aller. Chicago: Univ. Chicago Press

Gwinn, W. D., Townes, C. H. 1968. *Quantum Electronics Conf., Miami*

Habing, H. J. 1968. *Bull. Astron. Inst. Neth.* 19:421

Herbig, G. H. 1962. *Advan. Astron. Ap.* 1:47

Herbig, G. H. 1966. In *IAU Symp. No. 31*, ed. H. van Woerden

Herbig, G. H. 1968. *Z. Ap.* 68:243

Herbig, G. H. 1970. *Proc. XVIth Liège Symp.* In press

Herbig, G. H., Peimbert, M. 1964. *Trans. IAU* 12:412

Heiles, C. 1967. *Ap. J.* 148:299

Heiles, C. 1968. *Ap. J.* 151:919
Heiles, C. 1969a. *Ap. J.* 156:493
Heiles, C. 1969b. *Ap. J.* 157:123
Heiles, C. 1970. *Ap. J.* 160:51
Heiles, C., Turner, B. E. 1971. Submitted to *Ap. Lett.*
Hjellming, R. M., Gordon, C. P., Gordon, K. J. 1969. *Astron. Ap.* 2:202
Hollenbach, D. J., Salpeter, E. E. 1971. *Ap. J.* 163:155
Hollenbach, D. J., Werner, M. W., Salpeter, E. E. 1971. *Ap. J.* 163:165
Hoyle, F. 1957. *The Black Cloud.* New York: Harper
Johnson, F. M., Castro, C. E. 1970. Preprint
Johnson, H. L. 1968. In *Nebulae and Interstellar Matter*, ed. B. M. Middlehurst, L. H. Aller. Chicago: Univ. Chicago Press
Johnson, H. M. 1968. In *Nebulae and Interstellar Matter*, ed. B. M. Middlehurst, L. H. Aller. Chicago: Univ. Chicago Press
Jones, R. V., Spitzer, L. 1967. *Ap. J.* 147:943
Kerr, F. J. 1968. In *Nebulae and Interstellar Matter*, ed. B. M. Middlehurst, L. H. Aller. Chicago: Univ. Chicago Press
Kerr, F. J. 1970. Paper presented at IAU Gen. Assembly, Brighton, England
Knacke, R. F., Gaustad, J. E., Gillett, F. C., Stein, W. A. 1969. *Ap. J. Lett.* 155:L189
Knowles, S. H., Cheung, A. C. 1971. *Ap. J. Lett.* 164:L19
Landau, R. L. 1970, *Nature* 226:924
Lilley, A. E. 1955. *Ap. J.* 121:559
Litvak, M. M. 1969. *Ap. J.* 156:471
Litvak, M. M. 1970. *Ap. J. Lett.* 160:L133
Low, F. J., Smith, B. J. 1966. *Nature* 212:675
Low, F. J., Aumann, H. H. 1970. *Ap. J. Lett.* 162:L79
Low, F. J., Johnson, H. L., Kleinman, D. E., Latham, A. S., Geisel, S. L. 1970. *Ap. J.* 160:531
Lynds, B. T. 1962. *Ap. J. Suppl.* 7:No. 64, 1
Lynds, B. T. 1968. In *Nebulae and Interstellar Matter*, ed. B. M. Middlehurst, L. H. Aller. Chicago: Univ. Chicago Press
Lynds, B. T., Wickramasinghe, N. C. 1968. In *Ann. Rev. Astron. Ap.* 6:215
Manchester, R. N., Gordon, M. A. 1970. *Ap. Lett.* 6:243
Martin, P. G. 1970. *MNRAS* 149:221
Mattila, K. 1970. *Astron. Ap.* 9:53
Mendoza, E. E. 1968. *Ap. J.* 151:977
Mészáros, P. 1968. *Ap. Space Sci.* 2:510
Minkowski, R. 1968. In *Nebulae and Interstellar Matter.* ed. B. M. Middlehurst,

L. H. Aller. Chicago: Univ. Chicago Press
Odell, C. R. 1962. *Ap. J.* 136:809
Overbeck, J. W. 1965. *Ap. J.* 141:864
Palmer, P., Zuckerman, B., Buhl, D., Snyder, L. E. 1969. *Ap. J. Lett.* 156:L147
Parker, E. N. 1968a. In *Nebulae and Interstellar Matter*, ed. B. M. Middlehurst, L. H. Aller, 18. Chicago: Univ. Chicago Press
Parker, E. N. 1968b. *Ap. J.* 154:875
Pauliny-Toth, I. I. K., Shakeshaft, J. R. 1962. *MNRAS* 124:61
Penzias, A. A., Wilson, R. W. 1966. *Ap. J.* 146:666
Penzias, A. A., Jefferts, K. B., Wilson, R. W. 1971. Submitted to *Ap. J. Lett.*
Purcell, E. M. 1969. *Ap. J.* 158:433
Purton, C. R. 1966. *MNRAS* 133:463
Quiroge, R. J., Varsavsky, C. M. 1970. *Ap. J.* 160:83
Robinson, B. J. 1967. In *Radio Astronomy and the Galactic System*, ed. H. van Woerden, 49. London: Academic
Rogers, A. E., Barrett, A. H. 1968. *Ap. J.* 151:163
Salpeter, E. E. 1967. In *Radio Astronomy and the Galactic System*, ed. H. van Woerden, 65. London: Academic
Sancisi, P. R., Wesselius, P. R. 1970. *Astron. Ap.* 7:341
Schwartz, D. A., Hudson, H. S., Peterson, L. E. 1970. *Ap. J.* 162:431
Silk, J. 1970. *Space Sci. Rev.* 11:671
Silk, J., Werner, M. W. 1969. *Ap. J.* 158:185
Snyder, L. E., Buhl, D. 1971. *Ap. J. Lett.* 163:L47
Snyder, L. E., Buhl, D., Zuckerman, B., Palmer, P. 1969. *Phys Rev. Lett.* 22:679
Solomon, P. M., Werner, M. W. 1971. *Ap. J.* 165:41
Solomon, P. M., Wickramasinghe, N. C. 1969. *Ap. J.* 158:449
Solomon, P. M., Wilson, R. W., Jefferts, K. B., Penzias, A. A. 1971. Submitted to *Ap. J. Lett.*
Spitzer, L. 1968. *Diffuse Matter in Space.* New York: Interscience
Spitzer, L., Scott, E. H. 1969. *Ap. J.* 158:161
Stecher, T. P., Williams, D. A. 1967. *Ap. J. Lett.* 149:L29
Sturch, C. 1969. *Astron. J.* 74:82
Sunyaev, R. A. 1969. *Astron. Zh.* 46:929
Thaddeus, P. 1971. Submitted to *Ap. J.*
Thaddeus, P., Solomon, P. 1971. Submitted to *Ap. J.*
Townes, C. H., Cheung, A. C. 1969. *Ap. J. Lett.* 157:103

Tucker, K. D., Tomasevich, G. R., Thaddeus, P. 1970. *Ap. J. Lett.* 161:L153

Turner, B. E. 1971. *Ap. J. Lett.* 163:L35

Turner, B. E., Buhl, D., Churchwell, E. B., Mezger, P. G., Snyder, L. E. 1970. *Astron. Ap.* 4:165

Turner, B. E., Heiles, C. 1971. Submitted to *Ap. J.*

Turner, B. E., Heiles, C., Scharlemann, E. 1970. *Ap. Lett.* 5:197

Turner, B. E., Verschuur, G. L. 1970. *Ap. J.* 162:341

Verschuur, G. L. 1969a. *Ap. J. Lett.* 155:L155.

Verschuur, G. L. 1969b. *Ap. Lett.* 4:85

Verschuur, G. L. 1969c. *Astron. Ap.* 1:169

Verschuur, G. L. 1970. *Ap. J.* 161:867

Verschuur, G. L. 1971a. *Ap. Lett.* 7

Verschuur, G. L. 1971b. Submitted to *Astron. J.*

Werner, M. W. 1970. *Ap. Lett.* 6:81

Werner, M. W., Salpeter, E. E. 1969. *MNRAS* 145:249

Werner, M. W., Silk, J., Rees, M. J. 1970. *Ap. J.* 161:965

Whipple, F. 1946. *Ap. J.* 104:1

White, R. E. 1971. In preparation

Woolf, N. J., Ney, E. P. 1969. *Ap. J. Lett.* 155:L181

Zuckerman, B., Ball, G., Gottlieb, C. A. 1971. *Ap. J. Lett.* 163:L41

Zuckerman, B., Buhl, D., Palmer, P., Snyder, L. E. 1970. *Ap. J.* 160:485

Zuckerman, B., Palmer, P., Snyder, L. E., Buhl, D. 1969. *Ap. J. Lett.* 157:L167

CONVECTION IN STARS[1]

I. Basic Boussinesq Convection[2]

E. A. SPIEGEL

Astronomy Department, Columbia University

Convection occurs somewhere in most stars, yet our lack of understanding of convection has not seemed a major impediment to progress in stellar structure in recent years. In part this is true because convection often achieves the idealized adiabatic limit that is expected in convective cores of stars. It has also been true that uncertainties in the other physical processes in stars have been reduced considerably, and this has permitted a better empirical determination of the arbitrary parameters used in stellar convection theory. Of course, there is always the possibility that things are not as satisfactory as one thinks. But if we take the optimistic view that present convective models are qualitatively reasonable, what can one expect of an improved theory? One desirable feature would be the prediction of convective transfer with, in addition, some reasonable estimate of the accuracy of the prediction. For this, a minimal but inadequate test is found in laboratory convection for which some quantitative data are available. Thus, a principal goal of stellar convection theory should be the development of a reasonable deductive theory whose reasonability can be minimally established by laboratory tests.

Having obtained a theory at this level we would next be interested in finer details that characterize stellar convection. That is, we would like to be able to be quantitative about the time dependence and scales of the convection motion and to compare these with solar observations; we would like to know how far convection may penetrate beyond the regions of instability and by large-scale mixing remove chemical inhomogeneities; we would be interested in the precise temperature variations at the tops of convective envelopes to have better input for model atmospheres. And these are only a sample of some of the questions that one would hope to answer at this level of difficulty.

There is, in addition, a series of dynamical questions which raise problems about the interaction of convection with other processes of stellar fluid dynamics. These bring in new instabilities and are probably the most in-

[1] This work was supported in part by the National Science Foundation under NSF GP 18062.

[2] Part II, Special Effects, will appear in Volume 10 (1972) and Part III, Stellar Convection, will appear in Volume 11 (1973).

teresting problems in stellar convection theory at present. Thus we would like to know when convection can stabilize or destabilize pulsation; we would like to understand the role of convection in the rotational history of the Sun; we would like, even in a primitive example, to compute the dynamo effect of rotation and convection from first principles.

Before we can proceed to a discussion of what is known about these various processes, we need to have an outline of the basic theoretical techniques of modern convection theory, some of which involve prodigious calculations. For the most part, these techniques have been developed and tested on the basic problem of convection in a thin-plane layer of fluid. Accordingly, Part I of this review is devoted to the problem of convective transfer in the laboratory situation. It should be stressed, however, that the equations and approximations used are the same as those now used in stellar structure calculations with a common goal—to predict the march of temperature through the convective fluid. It will be seen, for example, that the mixing-length theory as used in stars is not as complete as that now discussed for laboratory convection. Other more difficult, but hopefully more adequate, approaches will be outlined. However, the astrophysicist interested in a simple recipe for calculating stellar models will be disappointed to find that instead, the stellar structure calculation in these approaches constitutes a subroutine in the convection program, and not vice versa. There seems no way around this for the present.

Having outlined these various approaches here, we shall return in the next volume of these *Reviews* to their application to the problem of stellar convection. In Part II the special problems of stellar convection such as large density variation, overshooting, rotation, and radiative transfer will be considered in the context of pure convection theory. It is in these domains that the more involved methods, especially those of Section 8, can be used to advantage, though one would not wish to pursue them without first examining their suitability for the basic problem of convective transfer. Part III is then to be devoted to actual stellar convection and the understanding of it that can be drawn from the discussion of Part II.

It is hardly necessary to add that even the limited subject matter of Part I cannot be treated exhaustively in the space available here. Hence, the discussion is focused on approaches that seem to have a direct bearing on the problems of stellar convection. Since the literature is vast, no attempt is made to cover all the contributions from meteorology (Sutton 1953, Priestley 1959), engineering (Prandtl 1952) and other fields where convection plays an important role. Also, reference is often made to papers which adequately summarize or synthesize previous work, and fundamental papers covered in such discussions are not necessarily cited. Particularly helpful is the thoroughness of existing treatments of linear theory (Chandrasekhar 1961), and the existence of a recent general review of the subject (Brindley 1967), in addition to some less complete ones by the present author (Spiegel

1966, 1967). An interesting discussion of the overall spectral dynamics of convection is also available (Platzman 1965).

1. THE EQUATIONS OF CONVECTION

1.1 *The anelastic and Boussinesq approximations.*—Though it would be out of place here to go into mathematical details of convection theory, it is necessary to discuss the basic equations, since much of the language of the subject stems from them. However, it does not pay to write down the full equations, since no attempt seems to have been made to solve them, except in linear theory. It is more usual to begin at the outset by introducing approximations.

The approximation that seems most appropriate for astrophysical convection is the anelastic approximation familiar to meteorologists (Ogura & Phillips 1962, Gough 1969). The basic idea of this approximation is to filter out high-frequency phenomena such as sound waves since these are thought to be unimportant for transport processes. This approximation is not really valid in the outer layer of convective envelopes or red giants, for example, since the Mach numbers of the convective motions can become appreciable there; but since even the anelastic problem has not been solved for that case, little can be said of this difficulty.

In studying laboratory convection, a further approximation is permitted, namely that the vertical extent of the fluid is much less than its density or pressure scale heights (Spiegel & Veronis 1960). This does not mean that the fluid is incompressible, but it does imply that density variations are very small and permits other such simplifications (Mihaljan 1962, Malkus 1964). This approximation combined with the anelastic approximation leads to the so-called *Boussinesq approximation* which is used in studies of laboratory convection, meteorology [sometimes with other justifications (Dutton & Fichtl 1969)], and even (implicitly) in most calculations of stellar convection. It often is further assumed that material properties such as viscosity and conductivity are insensitive to temperature; that is not an essential part of the Boussinesq approximation, but this "strong" form of the approximation is adequate for many experiments. Further, the principal configuration studied is that of a plane-parallel layer of fluid oriented horizontally in a uniform gravitational field, and that example will serve here.

1.2 *Mean quantities.*—In a convecting fluid, and especially a turbulent one, it is convenient to separate the mean and fluctuating parts of variables such as pressure and temperature. The means should ideally be ensemble averages, but it is computationally more convenient to use means over horizontal surfaces. Thus, one writes for the temperature in the plane-parallel case,

$$T(\mathbf{x}, t) = \overline{T}(z, t) + \theta(\mathbf{x}, t) \qquad 1.1$$

where z is the vertical coordinate and $\bar{\theta}=0$. In the idealized problem of Part I, $\bar{\mathbf{u}}=0$ where $\mathbf{u}$ is the velocity.

When horizontal means are taken of the Boussinesq equations there result (Spiegel 1967)

$$\frac{\partial}{\partial z}\,(\bar{p} + \rho\overline{w^2}) = -\,g\rho \qquad\qquad 1.2$$

and

$$\frac{\partial}{\partial t}\,\overline{T} + \frac{\partial}{\partial z}\,\overline{w\theta} = \kappa\,\frac{\partial^2\overline{T}}{\partial z^2} \qquad\qquad 1.3$$

Here ρ is the density (assumed constant), p is the pressure, g is the acceleration of gravity, w is the vertical component of $\mathbf{u}$, and κ is the thermal diffusivity (molecular or radiative). These two equations are the Boussinesq versions of two of the basic equations of stellar structure theory. The $\rho\overline{w^2}$ is the turbulent pressure and $\overline{w\theta}$ is the convective flux. If these two terms could be simply evaluated in terms of mean quantities, the convective difficulties of stellar structure theory would be essentially overcome, since a reasonably simple set of equations would result. No such possibility is readily found from the equations for $\mathbf{u}$ and θ (which will be displayed below). Indeed, it is clear from looking at the full equations that static structure equations like 1.2 and 1.3 make up one of the least difficult parts of the convection problem.

1.3 *The Boussinesq equations.*—In writing the equations of motion it is advantageous to use natural units appropriate to the problem. Thus we take the vertical extent of the fluid d as the unit of length, d^2/κ as the unit of time, where κ is the thermal diffusivity, ρ as unit of density, and $\Delta T - gd/C_p$ as the unit of temperature, where ΔT is an imposed temperature difference across the fluid, g is the acceleration of gravity, and C_p is the specific heat at constant pressure. The term gd/C_p is the adiabatic temperature change across the layer and meteorologists would call $\Delta T - gd/C_p$ the change in potential temperature (Brunt 1939), but astrophysicists prefer to work with entropy.

In natural units, the equations for the fluctuating quantities are (Malkus & Veronis 1958)

$$\frac{1}{\sigma}\left(\frac{\partial\mathbf{u}}{\partial t} + \mathbf{u}\cdot\nabla\mathbf{u}\right) = -\nabla\varpi + R\hat{\theta}\mathbf{k} + \nabla^2\mathbf{u} \qquad\qquad 1.4$$

$$\frac{\partial\theta}{\partial t} + w\left(\frac{\partial\overline{T}}{\partial z} - \beta_A\right) + \mathbf{u}\cdot\nabla\theta - \overline{\mathbf{u}\cdot\nabla\theta} = \nabla^2\theta \qquad\qquad 1.5$$

$$\nabla \cdot \mathbf{u} = 0 \qquad\qquad 1.6$$

where $\hat{\mathbf{k}}$ is a unit vector in the vertical, $\varpi = (p - \bar{p})/\rho - \overline{w^2}$,

$$R = \frac{g\alpha d^3}{\kappa \nu}\left(\Delta T - \frac{gd}{C_p}\right), \qquad \sigma = \nu/\kappa \qquad 1.7$$

are known as the Rayleigh and Prandtl numbers, and α is the coefficient of thermal expansion.

These equations are completed with the addition of 1.2 and 1.3, but the former is not really essential for the Boussinesq case. Moreover, 1.3 can be simplified by adopting the widely held belief that under stationary external conditions mean quantities are also stationary, at least in turbulent convection. In that case, 1.3 can be written, in natural units, as

$$\overline{w\theta} - \frac{\partial \overline{T}}{\partial z} = N \qquad\qquad 1.8$$

where the Nusselt number N is the (constant) sum of the convective and conductive heat fluxes. If the Boussinesq approximation had not been made, additional terms representing acoustic flux and transport due to viscous stresses would be required, though such effects are usually ignored in stellar convection as well.

To these equations must be added boundary conditions (Chandrasekhar 1961). In experimental studies the attempt is often made to fix the boundary temperatures, hence $\theta = 0$ on the boundaries. On rigid boundaries, $\mathbf{u} = 0$, and on free boundaries $w = 0$ and the tangential stresses vanish. The conditions appropriate to free boundaries are often used in theoretical work even when they do not apply, since they are easier to work with.

2. Stability Theory

Convection as we are considering it here normally arises as an instability which grows on a previously static configuration. In the simplest case, where a perturbation of infinitesimal amplitude does not suffer thermal diffusion or viscous effects, the question of stability can be decided simply. A parcel of fluid displaced vertically and adiabatically suffers a change in energy by an amount $mC_p dT + mg\,dz$ where m is the mass of the parcel. If this change is negative, that is if

$$\frac{dT}{dz} < -g/C_p \qquad\qquad 2.1$$

we can expect instability. This criterion, called the *Schwarzschild criterion*, can be obtained by more rigorous methods (Lebovitz 1966, Kaniel & Kovitz 1967).

When dissipative effects on the perturbation are included, the problem of stability is more difficult but precise treatments are available (Chandrasekhar 1961). An infinitesimal perturbation is applied to a given static configuration. For the initial times at least, the equations of motion are then linear, with constant coefficients in the strong Boussinesq approximation, and they are then separable. In the plane-parallel case solutions of the form

$$w(\mathbf{x}, t) = f(x, y)e^{\eta t}W(z) \qquad\qquad 2.2$$

for w, are found with similar forms for the other variables. Here

$$\nabla_1{}^2 f \equiv \left(\frac{\partial^2}{\partial x^2} + \frac{\partial^2}{\partial y^2}\right)f = -a^2 f \qquad\qquad 2.3$$

and η and a are separation constants representing the growth rate and horizontal wavenumber of the perturbation. Solutions of the resulting eigenvalue problem can be found giving η as a function of R, σ, and a. If for given R and σ there exist solutions with $\mathrm{Re}(\eta) > 0$ for some a, the solution is unstable. If for $\mathrm{Re}(\eta) \geq 0$ we have $\mathrm{Im}(\eta) \neq 0$ the instability is called overstability or vibrational instability. If $\mathrm{Im}(\eta) = 0$ whenever $\mathrm{Re}(\eta) \geq 0$, the principle of the exchange of stabilities is said to hold. For the linearized form of 1.4–1.6 this principle has been established. If we set $\eta = 0$ in the separated linear equations they give a relation defining the condition for marginal stability. As can be seen by inspection, σ drops out of these equations and we are left with a relation between R and a, having the properties $R \to \infty$ for $a \to 0$ and $a \to \infty$. Thus R has a minimum value R_c at a particular $a = a_c$, which implies that for $R > R_c$ there is a band of a for which there exist $\eta > 0$. We conclude that for $R > R_c$ convection occurs. For rigid boundaries the values $R_c = 1708$ and $a_c = 3.1$ are found. A physical discussion of these results suggests that the Rayleigh number may be interpreted as the ratio of buoyancy force to viscous force on the perturbation (Spiegel 1960).

Now stability is much harder to establish than instability, since it is not always possible to test all possible perturbations. For the present elementary example of convection it has also been shown that for $R > R_c$ there are no positive η, but this is not always the case in more complex problems, and overstability may arise if stabilizing forces work against the convection. The overstability then may occur at $R < R_c$ in the sense that R_c is defined here.

Further, some of the more complicated configurations, though stable in the sense used here, may be metastable. That is, if perturbations of sufficient amplitude are introduced, the system does not return to its initial configuration. In fluid dynamics a metastable system is said to exhibit finite amplitude instability. Systems that exhibit overstability often are metastable as well and several examples will arise in Part II.

These various possibilities are discussed here to make clear that the example we are considering is special in that instability occurs only as exponential growth when $R > R_c$ (Sani 1964). But it should also be stressed that when the fluid is unstable there is in general an infinitude of stable modes lying in a continuous band of horizontal wavenumber a. A further degeneracy is that associated with the various possible f's that satisfy 2.2 (Bisshopp 1960).

3. EXPERIMENTAL RESULTS

Most of Part I of this review consists of discussion of various attempts to understand Boussinesq convection and in particular to find how N depends on R and σ. Some astronomers do not consider that this topic is necessarily relevant to stars since the Boussinesq approximation does not hold in stars. However, as noted in Section 1, the Boussinesq equations are a limiting case of the equations governing stellar convection. Thus, any scheme for solving the equations of stellar convection should also work in the Boussinesq approximation or be subject to grave doubts. Whether a method does work in this limit can be tested only by experiment (good numerical experiments may have to suffice) over as wide a range of parameters as possible. Of course, such checks do not by any means guarantee the validity of a stellar convection theory, but they consitute a basic and fairly exacting requirement. Moreover, as we shall see in Section 8, the mixing-length theory now used for stars is a Boussinesq theory.

The onset of convective motions at a critical value of R is well established; the measured critical value is usually within 3% of that given by stability theory or better (Thompson & Sogin 1966) and, as theory predicts, R_c is independent of σ. For R just above R_c, steady cellular motion is observed for $\sigma \geq .7$ and if enough care is taken, the motion is steady and occurs in two-dimensional patterns called *rolls*. The widely quoted remark that hexagonal patterns occur in steady convection is not borne out by modern experiments, except under special circumstances, as when the fluid has properties which are temperature dependent and the strong Boussinesq approximation is not valid (Tippleskirch 1956). The wavenumber of the rolls decreases as R increases (Koschmieder 1966, 1969, Krishnamurti 1970) with a rate which depends on σ but for which no simple experimental relation has as yet been given; this behavior has been attributed to the side boundaries (Davis 1968), but recent numerical experiments without sidewalls show this behavior (Lipps & Somerville 1971). The pattern of the motion changes when R is raised above 22,600 (for $\sigma \geq 1$) and becomes three-dimensional. The preferred form then seems rectangular, and probably consists of crossed rolls (Busse 1970, Busse & Whitehead 1971). Such steady patterns persist to Rayleigh numbers $\sim 5 \times 10^4$ when the motion becomes time dependent, but periodic. Finally at higher values of $R(> 10^6)$ the motion becomes aperiodic, and with even higher R it probably becomes turbulent if σ is not too large. The various transitions in the nature of the flow are currently the object of intense interest (Krishnamurti 1970, Willis & Deardorff 1967a, b, c; Chen & Whitehead 1968).

These results have been obtained for $\sigma \gtrsim 1$. For $\sigma < 1$ the data are too sparse to permit any real conclusions but the various transitions appear to occur at rather reduced values of R (Krishnamurti 1970). (In laboratory fluids typical values of σ are: for silicone oils $\gg 1$, for water $= 7$, for air $= .7$, for mercury $= .025$, for liquid sodium $= .005$.) What is remarkable is the marked change in behavior that occurs in the neighborhood of $\sigma = 1$. In particular for air, transition to time-dependent convection already occurs at $R = 5 \times 10^3$, before the transition to three-dimensional motion (Willis & Deardorff 1970). For mercury, steady motion is not found at all (Rossby 1969).

The kind of experiment that seems easiest to use in checking astrophysically oriented theories is the measurement of the heat transfer, or equivalently the Nusselt number, for different values of R and σ. Of course, the R and σ typical for stellar convection are not really accessible experimentally; in the Sun, for example, R is about 10^{12} to 10^{20} (depending on whether one takes d as a scale height or the depth of the convective zone and on how one chooses the other parameters) and $\sigma \sim 10^{-9}$ because the thermal diffusion is radiative (Ledoux, Schwarzschild & Spiegel 1961). Nevertheless, one should try to push to the highest possible R and the lowest possible σ experimentally to provide data for testing theories as stringently as possible. At present, the data are inadequate to really discriminate among various theories, but let us consider what information is available.

In general, the measured heat fluxes through a convective layer are steady in time. Thus the Nusselt number defined in 1.8 is a constant and one writes

$$J = \kappa \frac{\Delta T}{d} N(R, \sigma) \qquad \qquad 3.1$$

where J is the heat flux divided by ρC_p. Since the only parameters in the equations are R and σ, the assumption that N depends only on them is quite reasonable, though any failure of the theoretical boundary conditions to match the actual ones and any deviations from the strong Boussinesq approximations may spoil this simple functional form.

For $R \leq R_c$, $N = 1$, since convection does not occur. (Exceptions may arise when the fluid is metastable for $R \leq R_c$.) The slope of the curve of N vs R (for fixed σ) breaks from $N = 1$ at $R = R_c$ and N begins to increase with R (e.g. Rossby 1969). A remarkable feature of the experiments is that N is only a piecewise smooth function of R, and a series of breaks in the N-R curves are observed (Malkus 1954b, Krishnamurti 1970, Willis & Deardorff 1967b). The transitions are discontinuities in dN/dR like the one which signals the onset of convection, but the relative jumps are increasingly small as R increases. No transitions have been reported for $R > 5 \times 10^6$. One suggested explanation of the transitions is that they mark the onset of new modes of motion (Malkus 1954a) which become unstable in the convectively altered conditions of fluid. Another possibility is that some of the transitions represent the occurrence of new interactions among the already existing modes.

Most attempts to describe the measurements of N are based on the interpolation formula

$$N = AR^r\sigma^s \qquad\qquad 3.2$$

and experimentalists quote measured values for A, r, and s. Although the quoted uncertainties in several experiments are reasonable, there are some uncomfortably large differences among values given by different experimentalists in A, r, and s, and, less often, in the actual values of N itself. Part of the trouble in the experiments is that heat leaks out of the sides of the apparatus and the correction for this is uncertain. Another experimental difficulty is that to raise R and maintain the strong Boussinesq approximation one must raise d. To keep the same aspect ratio one must then increase the horizontal dimension of the apparatus. The result is a large volume of fluid which is thermally sluggish; such systems take long times to come to equilibrium and are hard to maintain at given boundary temperatures.

Yet another problem relates to data analysis: different workers fit 3.2 to the data in different domains of R, and if r has a weak R dependence, discrepancies are inevitable. Of course, if r depends on R this may imply that 3.2 is inadequate, but there are theoretical approaches suggesting that 3.2 works well in certain well-defined domains of R. If data which spread over more than one of these domains are fitted to 3.2 the results may be misleading. A careful analysis of these points is lacking.

Given these uncertainties we may note some trends in the data. The general impression is that above $R \sim 5 \times 10^5$, for $\sigma > 1$, there appears a marked but continuous increase in the upward variation of N with R. No data for $R > 3 \times 10^9$ seem available and for data in the range 5×10^5 to 3×10^9 the reported values of r range from 0.200 to 0.325 with quoted probable errors $\pm.005$ (Goldstein & Chu 1966, O'Toole & Silveston 1961, Rossby 1969, Sommerscales & Gazda 1969). The values of A associated with these extremes are from about 0.2 to 0.08. The most reliable estimates for r seem to be 0.30–0.33.

These differences may partly be due to Prandtl number dependence, since the value $r = .283$ was found in silicone oil ($\sigma \sim 20$) (Sommerscales & Gazda 1969) and .325 in acetone ($\sigma = 3.7$) (Malkus 1954a). Other data seem consistent with this possibility: for example in one experiment (Rossby 1969) $r = .30 \pm .005$ and $A = .13$ were found for water ($\sigma = 7$) using data down to $R = 4 \times 10^3$. However, limited attempts to find s suggest that it is small for $\sigma \geq 1$ (O'Toole & Silveston 1961), hence one picture that might be considered for $\sigma > 1$ is that $s \sim 0$ but A and r depend on σ in some transcendental way. An alternative interpretation is that the domain of R in which 3.2 is a good representation depends on σ and the apparent variation of A and r with σ is a result of fitting 3.2 to the data in inappropriate domains. This possibility is also implied by data for $R < 5 \times 10^5$ which give lower values of

r (.25–.28) (Rossby 1969), and it may be true that the inclusion of data from lower values of R may depress r. A conclusion widely drawn from these data is that when R becomes large enough, $N \propto R^{1/3}$; this is in keeping with an older suggestion that for $R \lesssim 5 \times 10^5$, $N \propto R^{1/4}$ and for $R \geq 10^5$, $N \propto R^{1/3}$ (Jakob 1946). The modern data do not permit such a simple picture in detail, but support it qualitatively.

For $\sigma \leq 1$, it appears that N drops with σ more discernibly than at higher σ. Data on this point are available only for air ($\sigma = .7$) and mercury ($\sigma = .025$) and a definitive statement on the value of s or on possible σ dependence of N or r cannot yet be made.

The experimental information about N thus seems uncertain in all respects. Even the most recent measurements of N differ between experimentalists by 10% or more at the same R and σ. This level of uncertainty is far greater than is found in calorific measurements in modern physics and there is a definite need to increase the quantity and accuracy of data on convective heat transport. This would seem a useful subject to include in programs of laboratory astrophysics.

Another kind of measurement which has been made in some detail is of $\bar{T}(z)$ (Goldstein 1964; Rossby 1969, Sommerscales & Gazda 1969, Thomas & Townsend 1957, Townsend 1959, Willis & Deardorff 1967c). At large R it is found qualitatively that $\bar{T}$ is nearly constant away from the boundaries; near the boundaries $\bar{T}$ varies rapidly with z. (A variation in $\bar{T} \sim gd/C_p$ would be hard to detect in the laboratory.) From 1.8 and the boundary conditions, it is clear that $-\partial \bar{T}/\partial z = N$ at the boundaries, simply because the convective flux vanishes there. But, as to further details about $\bar{T}$, different experimentalists do not agree. Some results indicate that $\bar{T}$ is not symmetric about the midplane of the layer but others do not. The equations are invariant to reflection through the midplane but this does not necessarily imply that $\bar{T}$ is symmetric, since solutions may occur in asymmetric pairs. The different results may have to do with variations in experimental setup, non-Boussinesq effects, different measuring techniques (which mix some of θ into $\bar{T}$), different averaging times, and differing states of motion.

An interesting feature found in the experiments is that $\bar{T}$ is not always monotonic. In particular, there are often bumps in $\bar{T}$ just inside one or both of the thermal boundary layers (Sommerscales & Gazda 1969). These bumps are not universally accepted, but their existence now seems quite likely.

Other aspects of the motion, such as $\bar{\theta^2}$, are measured and other various external effects, such as rotation, have been studied. Many of these will be discussed in Part II.

4. CALCULATIONS FOR MILDLY SUPERCRITICAL R

If R is just above the critical value R_c for the onset of convection, there are straightforward analytical and numerical techniques for finding solutions of the basic equations (Joseph 1966, Kuo 1959, Kuo & Platzman 1961, Malkus & Veronis 1958, Schlüter, Lortz & Busse 1965, Segel 1966, Stuart

1958). Though this topic (known as *finite amplitude theory*) is the most active and successful branch of convection theory, it is not directly helpful to astrophysicists because in stars $R \gg R_c$ in general. Nevertheless, even in discussing what happens at high R, solutions at low R may give a qualitative lead, and a brief outline of the topic is in order here. The remaining sections on theories will then be devoted to large R.

The usual analytic technique often used when $R - R_c$ is small is a version of degenerate perturbation theory. The basic assumption is that for R slightly above R_c the motions will develop only small amplitude and a perturbation expansion in amplitude is made. Thus one writes $\mathbf{u} = \epsilon \mathbf{u}_0 + \epsilon^2 \mathbf{u}_1 + \ldots$, $\theta = \epsilon \theta_0 + \epsilon^2 \theta_1 + \ldots$, where $\mathbf{u}_0$ and θ_0 are normalized functions and the amplitude ϵ is assumed small. Though in convection problems it is usually assumed that R is prescribed, it is not known in advance which value of R will produce a particular amplitude. It is convenient therefore to consider ϵ as given and to find the R required to produce it; thus R can be thought of as a function of ϵ and we can write $R = R_0 + \epsilon R_1 + \ldots$, with the understanding that this relation can later be inverted.

The leading terms in the expansion give the linear equations of stability theory. These are degenerate since an infinity of choices of the wavenumber a and the horizontal planform f can be made (cf Equation 2.3). The problem is usually restricted by choosing the solution at this stage to be one of the marginally stable solutions of linear theory, though recent work has been directed at time-dependent cases (Matkowsky 1970). Having made this choice one goes on to higher orders using known techniques of perturbation theory to suppress resonances. The expansions have been carried to high order, and are convergent (Lortz 1961); solutions good to $R \sim 10 R_c$ or more are now routinely computable.

In the expansions carried out thus far, the linear solution which is perturbed is taken to have a single, fixed horizontal wavenumber a and fixed planform. Since this linear solution is marginally stable, R_0 is the value of R that produces marginal stability. If R_0 were much greater than R_c, there would exist values of a corresponding to highly unstable modes at $R = R_0$, and these would grow to large amplitude and overwhelm the solution studied. Hence, a must be selected so that R_0 is near to R_c. In the case of the strong Boussinesq approximation, one finds $R_1 = 0$, thus $\epsilon \propto (R - R_c)^{1/2}$, which identifies the perturbation parameter ϵ in terms of the usual known quantities.

Solutions for a variety of planforms having been found, the question arises which is preferred in nature. To answer this, the stability of the solutions has been studied by the same techniques used to find them. The work is most complete for the case $\sigma \to \infty$ (Busse 1967a). In that limit the only known stable solutions are the two-dimensional rolls in a finite band of a, and even these become unstable at $R = 22,600$. This instability represents excellent agreement with experiments which show that the motions do become three-dimensional at a value of R close to 22,600, but the theory does not predict the wavenumbers of the observed rolls. Also the experiments show the existence of steady three-dimensional motion for $R > 22,000$ and a challenge to the theory is to find a corresponding stable three-dimensional nonlinear solution. The observed cells have a rectangular appearance but theoretical rectangular planforms do not seem to be directly related to observed cells (Stuart 1964) and the motion probably consists of two interacting modes (Busse 1970, Busse & Whitehead 1971).

5. DIMENSIONAL ARGUMENTS

A powerful way to test ideas about convective transfer is to combine them with dimensional analysis to obtain the dependence of N on R and of T on z. There are two simple, but divergent, arguments about the form N should take, and they are worth outlining here.

The first argument stems from the observation that in highly developed convection at large R, T follows the adiabatic gradient over the bulk of the fluid. At the boundaries, according to 1.8, $\partial T/\partial z = -N$, where N is a rather large number. For the case of fixed boundary temperatures, T must therefore change by an amount $(\Delta T - gd/C_p)$ in the thermal boundary layers near the walls. The total thickness of the layers is then $\sim d/N$. We may conclude that the ability to conduct heat into the main body of the fluid from the boundary layers is the limiting factor in fixing the heat transport and thus the structure of the boundary layer fixes N.

If this is true, what is the effect of changing d? One line of argument is that for highly active convection, a change in d should not affect the heat flux J, since it would not modify the boundary layers but merely increase the size of the intervening adiabatic region (Priestley 1954). Application of this argument to 1.7, 3.1, and 3.2 gives

$$N = A\sigma^s R^{1/3} \qquad\qquad 5.1$$

which is not incompatible with experiments at their present level of accuracy.

No information on Prandtl number dependence is provided by these arguments. But in stellar cases, where $\sigma \ll 1$, astrophysicists generally agree that the heat flux should not depend on viscosity, which implies that $s = r$ in 3.2 when $\sigma \ll 1$, i.e. $N = N(R\sigma)$. There is no experimental evidence directly supporting this conjecture except for the observed decrease of N with σ. Other theoretical arguments to be mentioned later also support this.

The other dimensional argument for the dependence of N or R comes from the formulation of methods used in stellar structure in the language of laboratory convection (Spiegel 1971a). Convection in stellar cores is treated as if the convective cores were completely adiabatic with no boundary or transition layers intervening between the convective regions and the radiative envelopes. The neglect of such layers implies that for the convective cores no uncertainty exists in the choice of the correct adiabat. It also assumes that whatever the luminosity of the model, the convective flux required will be carried without the limitation implied by the existence of a boundary layer with dominant conductive transport. Of course, stellar models are less constrained than laboratory models, since their dimensions are adjustable; nevertheless such physical requirements placed on the models are exacting and it is of interest to see what they imply. This is simple to do; if the convective zone is purely adiabatic the heat flux must be independent of thermal conductivity. This, taken with the previous conclusion that $N = N(R\sigma)$ for low σ, gives us

$$N = A(R\sigma)^{1/2} \qquad 5.2$$

if we demand that Equations 3.1 and 3.2 imply that J is independent of κ and ν.

The difference between 5.1 and 5.2 (even with $s = 1/3$ in 5.1) is quite striking and its resolution is important to the theory. One possibility is that the stellar arguments should not be applied to the laboratory configuration. If that is true, then an important point of contact is lost between the two cases. However, other lines of argument which permit us to resolve the discrepancy between 5.1 and 5.2 will be given below. What is suggested by these arguments is that at sufficiently high R, turbulent breakdown of the thermal boundary layer occurs and causes a transition from 5.1 to 5.2. No such transition has been detected experimentally, but this is presumably explained by the limitation of the experiments to what in stellar terms are modest Rayleigh numbers ($R \sim 10^9$). The need to confirm (or deny) 5.2, which is intimately connected with basic ideas of stellar structure theory, poses a great challenge to the experimentalist.

Apart from these results, there have been attempts to discuss the structure of convective turbulence on the basis of similarity arguments (Zel'dovich 1932, Priestley 1959). Conclusions drawn about $\overline{T}(z)$ in this way do not seem to agree well with experiment (Townsend 1966).

6. Bounds on the Heat Transport

Integration of 1.8 over z leads to the expression

$$N = 1 + \langle w\theta \rangle \qquad 6.1$$

where the angular brackets denote a volume average. We may then apply the calculus of variations to this functional expression to place bounds on N which, if they are stringent enough, may be a useful guide in selecting among various theoretical results. The value of such bounds will evidently depend on the constraints added to the variational problem.

One set of constraints that has been used are the so-called power integrals (Malkus 1954b, Sorokin 1957, Chandrasekhar 1961). These are obtained by scalar multiplication of 1.4 by $\mathbf{u}$ and of 1.5 by θ followed by averaging over the fluid volume. If mean quantities are steady there result

$$R\langle w\theta \rangle = \langle |\nabla \mathbf{u}|^2 \rangle \qquad 6.2$$

and

$$\left\langle \left(\frac{\partial \overline{T}}{\partial z} + \frac{g}{C_p}\right)w\theta \right\rangle = -\langle |\nabla\theta|^2\rangle \qquad 6.3$$

The first of these expressions balances the rate of buoyant input of energy against the rate of viscous destruction; the second can be similarly thought of in terms of entropy generation. Neither of these conditions involves the Prandtl number.

Subject to 6.2 and 6.3 as constraints, the expression 6.1 for N can be maximized (Howard 1963) and it has been shown that

$$N \le \frac{\sqrt{3}}{8} R^{1/2}$$

This bound shows an intriguing similarity in R-dependence to 5.2, but for $\sigma \ll 1$ it is much higher than 5.2. Attempts to further tighten this bound by the addition of 1.6 as a constraint (Busse 1969) have only modified the numerical coefficient in front of $R^{1/2}$, but the solutions of the Euler-Lagrange equations that then result are probably better representations of the actual flow.

Another bound has been found by replacing the constraint 6.2 by 1.4 in the limit $\sigma \to \infty$ and retaining 6.3 and 1.6 (Chan 1971). This gives $N \le .325$ $R^{1/3}$ and suggests that for large enough σ, 5.2 cannot be correct. Thus, if as demanded by the astrophysical arguments, 5.2 holds for $\sigma \ll 1$, at fixed, large R, N should increase with increasing σ. At some unknown finite value of σ, $N(\sigma)$ should reach a maximum, $O(R^{1/2})$, and then decrease with increasing σ to an asymptotic value $\le .325\ R^{1/3}$.

It seems technically feasible to derive similar bounds (though not with the same rigor) for astrophysical (non-Boussinesq) circumstances. However, the effort does not yet seem warranted. What is needed more is improvement of the bounds at low values of σ. Bounds for this case have been estimated by physical arguments (Spiegel 1971a), which are really dimensional, and which agree with 5.2. Analogous rigorous bounds have not been found, possibly for want of effort. However, there is another conceivable limitation on the closeness with which the bounds may approach the actual fluxes.

Suppose that the equations admit solutions to which correspond inordinately large heat fluxes, but that these solutions are unstable. Then it could happen that fluxes found in practice would be much lower than for these solutions but that rigorous mathematical bounds would not be. This difficulty might be circumvented by the use of ensemble means over many solutions or by an appropriate constraint; but the problem would be difficult. Such knotty problems notwithstanding, the establishment of further bounds would be helpful, especially on other quantities besides flux, such as the potential energy, or kinetic energy, of the fluid. Other less obvious quantities have been considered too (Busse 1967b).

7. Two-Dimensional Solutions

Naturally, one might try to solve the equations numerically, but even with the largest machines, complex flows in three dimensions have not been successfully treated except for R not much larger than R_c. In this region of low R (<10 R_c) the three-dimensional time-dependent solutions evolve slowly to two-dimensional solutions, in confirmation of laboratory results (Chorin 1968). This does not mean that three-dimensional numerical calculations are not useful at low R; it turns out that if a two-dimensional flow wishes to adjust its horizontal scale, it does so through the action of three-dimensional perturbations (Lipps & Somerville 1971).

However, if the flow is two-dimensional, the Boussinesq equations can now be fairly routinely solved for R up to about 10^6 (Deardorff 1964, Fromm 1965, Plows 1968, Schneck & Veronis 1967, Somerville 1970, Veronis 1966), and no doubt these results can be pushed to higher values if adequate spatial resolution of the boundary layer can be achieved.

Such numerical calculations do not avoid the need to introduce a length scale in the horizontal direction since a finite horizontal dimension is required. Another limitation of two-dimensional solutions is that they do not seem to develop full turbulence. For reasons that are probably associated with this, the two-dimensional results show very little dependence on σ and hence are not directly useful to astrophysics except as a possible check on other theoretical approaches. Some σ dependence may be introduced by allowing the horizontal scale of motion to take its preferred value (Lipps & Somerville 1971), but the two-dimensional solutions have no way of determining this scale.

An especially interesting aspect of the two-dimensional problem is that asymptotic solutions for $R \to \infty$ have been obtained. These provide, apart from the static solution without convection, the only accurate solutions of the convection equations for large R. At large enough R, these are doubtless unstable, but they are of theoretical interest. In addition to having the kind of vertical structure in $\bar{T}$ already discussed, they have thin vertical "boundary" layers as well. For example in a horizontal line in the plane of the motion, $\bar{T}$ will be constant except for sharp bumps at the edges of the two-dimensional rolls. The vertical velocity is likewise concentrated in vertical layers.

For $\sigma \gg R^{3/5}$ the result $N \propto R^{1/5}$ is obtained (Roberts 1969) and $N \propto a^{2/5}$ for $a \gtrsim 1$ where a is the inverse horizontal scale that must be supplied to obtain a solution. For $\sigma \ll R^{3/5}$, the result $N \propto R^{1/4}$ is obtained (Wesseling 1969) but the σ dependence has not been given for this case. Also the corresponding flow exhibits separation in cell corners and this probably indicates instability. These results ($R^{1/5}$ and $R^{1/4}$) hold for rigid boundaries. For free boundaries one finds $N \propto R^{1/5}$ with very little σ dependence (Roberts 1969).

The analytic and numerical results (Veronis 1966) coincide in their lack of σ dependence. As we have mentioned, this limits their astrophysical value to tests of other techniques of solution or in exploring qualitative problems such as density variation. It is, however, becoming increasingly reasonable to think of solving three-dimensional problems even in the region of turbulent convection (Orszag 1969). Though this may be premature for the astrophysical case, it is a possibility that should be kept in mind in the coming decade.

8. MIXING-LENGTH THEORY

Most theories of convection consist of attempts to solve the basic equations in some approximation. Little effort has been made to bypass direct solution by constructing models for the flow which may lead to simpler equations (though some very interesting models for time-dependent convection exist (Chang 1957, Howard 1965, Keller 1966, Welander 1967, Elder 1968). A notable exception to this remark is the mixing-length theory which pictures turbulent transport processes in analogy with molecular processes (Taylor 1970, Prandtl 1952). A characteristic mixing length l, analogous to the mean free path of kinetic theory, and a characteristic turbulent velocity u', analogous to the mean molecular velocity, are introduced (e.g. Sutton 1955). Various formulations of the theory are possible, but the simplest is to introduce a turbulent diffusivity, or Austasch coefficient, lu', to be used to describe the turbulent transport processes. This part of the theory is reasonably clear; the choice of l and u' is more difficult. The random velocity u' is often taken to be of the order of some large-scale velocity in the fluid; in convection this is the vertical velocity w. The mixing length is taken to be a characteristic scale which is sometimes a constant such as the size of the system. More usually it is assumed to be a local scale such as the distance to a boundary or the scale of variation of a dynamically important quantity, such as velocity, shear, pressure, or density. These choices imply a slight inconsistency since the transport is described as a diffusion, using the approximation that l is less than any characteristic scale of motion (Spiegel 1963).

The equations describing the mixing-length model are just 1.4, 1.5, and 1.8 without the $\mathbf{u} \cdot \nabla \mathbf{u}$ and $\mathbf{u} \cdot \nabla \theta$ but with terms added to account for the turbulent diffusion. In the simplest case, the same diffusion coefficient wl is used for all quantities such as temperature and velocity, though it may be more correct to consider different values of l for different quantities. This is done in other problems such as vorticity diffusion (Goldstein 1938).

Even the relatively simple mixing-length equations are difficult to solve and, in view of the drastic approximations involved already, perhaps the effort needed to find exact solutions is not called for. Hence further mathematical approximations are usually introduced. In oceanography, for example, one often sets wl constant. In the astrophysical examples, one typically replaces most spatial derivatives of fluctuating quantities by l^{-1}; for example $\nabla^2 \sim l^{-2}$ (Kraichnan 1962). An equivalent way to obtain such results is to be more explicit about the model itself and to picture the turbulent transport as being effected by parcels of fluid of size l before disruption (Vitense 1953, Spiegel 1971b).

If we proceed by the expedient of writing l^{-1} for derivatives, lump together terms that seem to be of comparable magnitude, and drop the pressure and time derivatives, we can write by inspection from 1.4 and 1.5,

$$g\alpha l\theta = \nu(\text{Re} + 1)w \qquad\qquad 8.1$$

and

$$w\left(\frac{\partial \overline{T}}{\partial z} + \frac{g}{C_p}\right)l^2 = \kappa(\text{Pe} + 1)\theta \qquad\qquad 8.2$$

where

$$\text{Pe} = \frac{wl}{\kappa}, \qquad \text{Re} = \frac{wl}{\nu} \qquad\qquad 8.3$$

These equations have been written here in dimensional form since that is the usual practice in the literature, and are to be solved in connection with 1.8.

The dimensionless ratios Pe and Re are known as the Peclet and Reynolds numbers and measure the ratio of turbulent diffusivity to the two molecular diffusivities of the convection problem. Convection differs from other turbulence problems in that no velocities are externally prescribed, hence Pe and Re are derived rather than imposed quantities. It is possible to make estimates for them if we take for w a characteristic free-fall time through the fluid, namely

$$w \sim [g\alpha(\Delta T - gd/C_p)d]^{1/2} \qquad\qquad 8.4$$

Then

$$\text{Pe} \sim (\sigma R)^{1/2}, \qquad \text{Re} \sim \left(\frac{R}{\sigma}\right)^{1/2} \qquad\qquad 8.5$$

(These estimates provide a qualitative guide, and more precise values differ with differing states of motion.) This shows that under astrophysical conditions we expect convection to be turbulent since $\text{Re} \gg 10^3$, but that in stellar envelopes where κ may be large, turbulent transfer does not necessarily dominate radiation transfer everywhere since $(\sigma R)^{1/2}$ is not always large.

A few further steps are needed to put these equations in the usual form used by astrophysicists. We introduce $\alpha^{-1} = \overline{T}$, which holds for a gas. Let

$$\theta = l\frac{\partial \theta}{\partial z} = l\frac{\partial}{\partial z}(T - \overline{T}) \qquad\qquad 8.6$$

and take $\text{Re} \gg 1$ everywhere. Then the equations are readily rearranged into the form used for stellar models with the notation $\Gamma = \text{Pe}$ (Vitense 1953). There are some differences in numerical coefficients, such as those which arise when a more flexible approximation to the radiative diffusion term is used. These matters aside, it is clear that the equations presently used for stellar convection are essentially those used for laboratory convection in the mixing-length approximation.

Leaving for Part III the question of stellar application and related modifications of the theory, we may summarize the results indicated by the mixing-length theory for laboratory convection (Kraichnan 1962). If careful account is taken of the boundary-layer structure, the Nusselt number for σ large enough ($\sigma > 0.1$) and R quite large becomes $N \simeq (R/1500)^{1/3}$, where the constants in this and the succeeding results are estimated from a variety of empirical data and theoretical calculations. For low σ (<0.1), $N \simeq (\sigma R/70)^{1/3}$, where σR must be large (>300).

An important feature of the standard mixing-length theory as described here is that it includes the effect of small-scale motions only through their damping of the large-scale motions. In turn, these large-scale motions are driven only by differential buoyancy forces, and the mixing-length theory chooses at each position a preferred scale of motion l. Near the boundaries, l becomes small and at some distance from the boundary, a local Rayleigh number computed with l instead of d is $\sim 10^3$. For distances to the wall less than this, no motions can be strongly excited by the buoyancy, and conduction becomes the dominant mode of transport. This distance then is the thickness of the thermal boundary layer which we saw in Section 3 is $\sim d/N$. Hence N is found by saying that the Rayleigh number computed for a scale d/N is R/N^3 and this should be $\sim R_c$. From this we find $N \sim (R/R_c)^{1/3}$. A modification must be added in mixing-length theory since the diffusivities entering into R may be turbulent diffusivities. In this model the thermal boundary layer is the region in which molecular (or radiative) conduction dominates over turbulent conduction. However, when the Prandtl number is small the turbulent viscosity may be larger than the molecular viscosity even in the thermal boundary. Hence, in general, the Rayleigh number of the thermal boundary layer, R/N^3, should be corrected for turbulent viscosity. That is, the Rayleigh number of the thermal boundary layer needs a corrective factor $\nu(\nu + wl)^{-1}$ (Spiegel 1967) which is to be evaluated at the edge of the boundary layer; at high Prandtl number this factor becomes unity. If l is proportional to the distance from the boundary layer, we must take $l = d/N$ and in analogy to 8.4 we take $w \sim [g\alpha\Delta T \, d/N]^{1/2}$ as the value for the edge of the thermal layer. (In the laboratory case we may neglect the correction for the adiabatic term gd/NC_p.) We find

$$N \sim \left[\frac{\sigma R/R_c}{\sigma + (\sigma R/N^3)^{1/2}} \right]^{1/3} \qquad 8.7$$

More detailed arguments, allowing for geometrical factors, suggest that in the denominator, R should be replaced by R/R_c. We then find

$$N \sim \left[\frac{\sqrt{1 + 4\sigma} - 1}{2\sigma}\right]^{2/3} \left(\frac{\sigma R}{R_c}\right)^{1/3} \qquad 8.8$$

as the mixing-length prediction for N. This has the limits mentioned; in particular for low σ, $N \sim (\sigma R)^{1/3}$, which corresponds to the astrophysical notion that heat transport should be independent of viscosity for low σ. At very large σ, 8.8 indicates that the slight dependence of N on σ is not well represented by a power law, but rather by a factor like $(1 - \text{const } \sigma^{-1/2})$.

At large σ, these mixing-length results agree reasonably well with experiment, in the sense that the power r in 3.2 seems to be tending toward $1/3$ experimentally. The lack of σ dependence for large σ in the mixing-length results is also not a bad representation of the data. For low σ, data are available only for mercury and these are not adequate for a real test, but the data for σR large do seem consistent with the mixing-length predictions. Attempts to compare mixing-length predictions for $\bar{T}$ and $\bar{\theta^2}$ with experimental results are qualitatively acceptable for large σ but are unsatisfactory for mercury ($\sigma = 0.025$) (Rossby 1969). The values of $(\sigma R)^{1/2}$ studied experimentally for mercury are, however, not very large and the disagreement may not be wholly damning to the theory. This point needs further experimental scrutiny.

The standard mixing-length theory, as we have seen, considers the heat transport by motions driven by differential buoyancy forces. The motions which provide turbulent diffusivity are presumably dynamically excited by the $\mathbf{u} \cdot \nabla \mathbf{u}$ terms in the equations of motion. The effect of these motions is normally included only as a drain on the large-scale motions, but their contribution to the convective transfer should also be included (Kraichnan 1962). In the interior of the fluid such corrections are unimportant since the gradient is already nearly adiabatic, but they can be of great importance in the boundary layer.

The reason for the frequent omission of such corrections is that standard mixing-length theory asserts that at any location a specific scale of motion l (normally = distance from the boundary) is dominant. Thus, the usual mixing-length theory would not have predicted that large eddies from the interior of the fluid strike the boundaries, and set up appreciable horizontal motions there. There is, however, evidence that this occurs in laboratory convection (Malkus 1954a) and it seems to be manifested on the Sun as supergranulation (Noyes 1967; Simon & Weiss 1968). The neglect of the transport at the boundary by these large-scale motions is justified normally since they do not carry large temperature fluctuations, having traveled mostly through nearly adiabatic regions. What cannot always be neglected is the dynamical effect of these motions.

When large eddies hit the boundaries they set up shear layers which, when they are intense enough, can break down into small-scale turbulent motions. This turbulence may be less intense than the turbulence usually included in mixing-length theory, and it will not in general produce as large an eddy conductivity. But if this turbulent conductivity exceeds the molecular conductivity a marked change in boundary-layer thickness occurs. It must be understood, however, that there is still a thermal boundary layer, whose edge now is at the place where the large eddy conductivity of buoyancy-driven turbulence is no longer important and the weaker conductivity of shear turbulence takes over. This boundary layer is a turbulent boundary layer which permits a greater heat transport than the laminar boundary layer. The details of this layer depend on the boundary conditions and the effect of the boundary-layer turbulence has been estimated only for the laboratory case (Kraichnan 1962).

The procedure is to obtain the velocity shear at the wall by assuming a simple model for the eddies arising from the interior. The intensity of the resulting turbulence driven near the wall may be computed in analogy with the results available from work on shear turbulence. Such mechanically driven motions in the presence of a temperature gradient would transport heat (Prandtl 1952) and, as in ordinary mixing-length theory, the additional transport may be estimated. The details of the calculations are too lengthy for inclusion here. They give, for sufficiently large R,

$$N \sim \left(\frac{\sigma R}{(9 \ln R)^3} \right)^{1/2} \qquad 8.9$$

when σ is small. (A more complicated result for large σ is obtained, but the main factor in the expression for N in that case is $R^{1/2}$.) Apart from the logarithmic term, 8.9 is in substantial agreement with 5.2.

A sensitive question is: when does 8.8 give way to 8.9 at small fixed σ, as R increases? Unfortunately, the estimates for this rely heavily on high powers of badly known constants. It appears that for the case $\sigma \sim 10^{-9}$, the transition from 8.8 to 8.9 begins at $R \sim 10^{24}$. Thus, these corrections for boundary-layer turbulence do not seem required under solar conditions. Whether they may be required under any conditions in stars is an open question since these turbulent corrections cannot be confirmed even for the laboratory case with existing data. Nevertheless, such corrections may well be needed for stellar convection (Spiegel 1971a) and the problem will be further discussed in Part III.

Perhaps the most satisfying aspect of the calculations with boundary-layer turbulence is that they permit a rationalization of the difference between 5.1 and 5.2. The suggestion is that for large R, 5.1 is a reasonable representation until the large-scale motions become turbulent in the thermal boundary layers associated with 5.1. The point at which this happens depends on R in a moderately complicated way. Once the boundary-layer turbulence starts, the heat transfer should then depend on d since it is an important factor in the large-scale interior velocities, as 8.4 indicates. Thus at some enormous value of R the heat transport will depend on d and indeed it is roughly found from $J \sim w\theta$ where w is given by 7.4 and $\theta \sim \Delta T$.

In this new regime, the flux no longer depends on κ, except perhaps logarithmically. This comes about because what was previously a laminar thermal layer becomes a turbulent layer and the role of molecular conductivity is taken up by turbulent conductivity. There may be in the new boundary layer a small laminar sublayer which accounts for the logarithmic terms, but without solid boundaries even the sublayer might disappear (Clauser 1961). Of course there are more details involved, particularly for $\sigma > 1$, but the main qualitative features seem to be consistent with the dimensional arguments.

9. Truncated Expansions

A surprisingly effective approach to solving the equations of convection is the use of truncated expansions in terms of an appropriate set of basis functions or modes. Such a procedure is often called a *Galerkin method* (Reiss 1965) though the terminology used varies depending on whether the basis functions depend on one or more of the independent variables. There exist other related approximation methods, especially those relying on variational approaches (Reiss 1965, Finlayson & Scriven 1966), which, in the convection problem, have not proved as simple to use (Roberts 1966). In the Galerkin and related methods one expands in functions of one or more of the independent variables to obtain an infinite set of coupled equations for the amplitudes in the expansion. Expansion in functions of space and time, of only space coordinates, and of only the vertical coordinates have been considered, with varying degrees of success, depending on the choice and number of basis functions, and the techniques used in solving the reduced equations. The greatest effort, however, has gone into the use of basis functions depending on only the horizontal coordinates. Here too a wide choice of basis is possible, but a promising set is comprised of the planform functions of linear theory. Their relevance is reinforced by the theoretical suggestion that cell shape is preserved even for $R > R_c$ (Stuart 1960). The planform functions satisfy 2.3 and contain as a special case the trigonometric functions. They have the useful property that for two different wavenumbers a_i and a_j the corresponding solutions of 2.3, f_i and f_j, satisfy $\overline{f_i f_j} = 0$. To each a_i there corresponds a subspace of f_i, and strictly we should add a second index to indicate the various members of this subspace. But to keep the formulae simple, we shall merely let f_i represent the most general linear combination of basis functions associated with a_i where $\overline{f_i^2} = 1$. Then we can write the expansions for w and θ

$$w(\mathbf{x}, t) = \sum_i f_i(x, y) W_i(z, t); \qquad \theta(\mathbf{x}, t) = \sum_i f_i(x, y)\Theta_i(z, t) \quad 9.1$$

with similar expansions for u, v. Here, the index i is treated as discrete to avoid the difficulty of infinite norm associated with a continuous spectrum.

The expansions may be introduced into 1.4, 1.5, and 1.8 and the expanded equations projected onto the appropriate f_i in the usual way. The resulting equations for the amplitudes in the expansions can be distilled into equations for W_i, Θ_i, and $\overline{T}$. For simplicity we shall here suppress terms describing vertical components of vorticity; these do not alter the general form of the equations, which are (Gough, Spiegel & Toomre 1971),

$$\left[\frac{1}{\sigma}\partial_t - (\partial_z{}^2 - a_i{}^2)\right](\partial_z{}^2 - a_i{}^2)W_i$$

$$= -Ra_i{}^2\Theta_i - \frac{1}{\sigma}\sum_{j,k} C^{ijk}[W_k L_{kij}\partial_z W_j + (\partial_z W_k)M_{kij}W_j] \quad 9.2$$

$$[\partial_t - (\partial_z{}^2 - a_i{}^2)]\Theta_i$$

$$= -\left(\partial_z\overline{T} + \frac{g}{C_p}\right)W_i - \sum_{j,k} C^{ijk}(A_{ijk}\Theta_j\partial_z W_k + (2a_i{}^2 a_k{}^2)W_k\partial_z\Theta_j) \quad 9.3$$

and

$$\partial_t\overline{T} + \sum_j W_j\Theta_j = \partial_z{}^2\overline{T} \quad\quad 9.4$$

where

$$\partial_z = \partial/\partial z, \qquad \partial_t = \partial/\partial t$$

$$2C^{ijk} = \overline{f^i f^j f^k} \quad\quad 9.5$$

$$L_{ijk} = A_{ijk}(\partial_z{}^2 - a_k{}^2), \qquad M_{ijk} = L_{ijk} + \left(\frac{a_j}{a_k}\right)^2 L_{jki} \quad 9.6$$

and

$$A_{ijk} = \frac{1}{a_k{}^2}(a_j{}^2 + a_k{}^2 - a_i{}^2) \quad\quad 9.7$$

The basic feature of these equations is that the amplitudes, or modes, are coupled in two ways. There are the direct or dynamical couplings whose strengths are mediated by the coupling constants C^{ijk}. These interactions give rise to the disorder in the flow that is the hallmark of turbulence. The other form of modal interaction, the so-called *mean-field interaction*, comes from the term $(\partial_z\overline{T})\,W_i$ in 9.3. In turn, $\partial_z\overline{T}$ is given by 9.4, which shows how $\overline{T}$ is affected by the motion. Thus all the modes may interact through the mean temperature.

Consider now the most drastic truncation, in which only one mode is retained. The resulting equations can be integrated numerically, and for a wide range of parameters and initial conditions the solution tends to a stationary state. However, for a given choice of the parameters, there is not a unique stationary solution, there being nonlinear analogues of the fundamental and overtones (in z) of linear theory. For the one-mode case, we shall confine attention to the fundamental mode.

For one steady mode the nature of the solution depends on whether the self-coupling constant, $C^{111} = C$, vanishes or not. If f_1 contains at least three Fourier components whose wavevectors form a triangle, we will have $C \neq 0$. Evidently, for rolls or rectangles we will have $C = 0$, and this case has been studied extensively (Herring 1963, 1964, 1966). For $R \rightarrow \infty$ one finds (Roberts 1966, Stewartson 1966)

$$N_1 \sim 0.28 \left(1 - \frac{a_1^4}{R} \right)^{6/5} [Ra_1^2 \ln (Ra_1^2 - a_1^6)]^{1/5} \qquad 9.8$$

for rigid boundaries and for $a_1 < R^{1/4}$. (For $a > R^{1/4}$ no convective solutions exist.) As a function of a_1 this expression has a maximum at $a_{1\,\mathrm{max}} \sim (R/13)^{1/4}$, and for most purposes we need consider only $a_1 \lesssim a_{1\,\mathrm{max}}$. Hence to good approximation

$$N_1 \simeq A_1 [Ra_1^2 \ln (Ra_1^2)]^{1/5} \qquad 9.9$$

It is interesting that as in the two-dimensional problem, the boundary conditions are very important; for free boundaries $N \propto R^{1/3}$ (Howard 1965, Herring 1966).

For cases when $C \neq 0$ but $a_1 < a_{1\,\mathrm{max}}$, expression 9.9 continues to hold for $R \rightarrow \infty$ except that A_1 is no longer ~ 0.28 but is a function of C and σ (Gough, Spiegel & Toomre 1971). The functional form of A_1 is not known analytically, but for $C/\sigma < 1$ we have $A_1^5 \sim [3/5(2/\pi)^6]$ while for $C/\sigma > 1$, $A_1^5 = 3/5(2/\pi)^4\sigma/C$. Thus, the introduction of the self-interaction terms causes N_1 to depend on σR for small σ.

Given these results we are left with the problem of choosing the a's and C's for one or more modes. Ideally, we would like to use the experiments as a guide, but they do not directly provide scale information at large R. Nevertheless it is possible to let a_1 and C be functions of R and σ and match the observed N over a fairly wide domain of R and σ even with one mode. However, at $a_{1\,\mathrm{max}} = 0$ ($R^{1/4}$), $N_{1\,\mathrm{max}} \propto R^{\cdot 3}(\ln R)^{\cdot 2}$, while for large enough R, the actual N goes up like $R^{1/3}$ (or faster). Hence one mode can never describe the full behavior of N. Further modes will be needed at very high R and no experimental indication of which set of modes to choose is evident.

In the absence of stringent experimental guides, the choice of modes must be dictated by additional assumptions. One that has been used is to maximize N with respect to a and C. For moderate values of R this gives much too high a heat transport. Alternately, one might take over an idea that has been preferred by many astrophysicists and choose that mode which is most unstable according to linear theory. For large R, η, the growth rate of linear theory is maximum when $a_1 \propto R^{1/8}$ and this gives $N_1 \propto R^{1/4} (\ln R)^{1/5}$, which is like the law suggested by some experiments for laminar convection at moderate values of $R(10^5 - 10^6)$.

These two methods of selecting modes suggest that a_1 should increase with R while in fact the experiments (at least for $R < 10^5$) indicate a decrease (Koschmieder 1969). Hence, perhaps the most reasonable procedure is to take $a_1 = 0(1)$ and to represent the more rapid increase of N with R at large R by adding further modes. Of course, this means that the choice of further modes is left open as well, and one must seek an extension of this procedure.

The first possibility to consider when adding higher modes is to add to the basic mode successively more of its horizontal harmonics to obtain an accurate description of cellular motion. If, for example, we start with a roll, $f_1 = \sqrt{2} \sin (a_1 x)$, and its harmonics, we will have a Fourier representation of nonlinear, two-dimensional cellular convection with horizontal periodicity a_1^{-1}. We have seen that solutions of the two-dimensional problem for $\sigma > R^{3/5}$ yield $N \propto R^{1/5}$. Hence the addition of higher harmonics does not produce a drastic change in the heat transport calculated from one mode (at least for rolls at high σ) and this implies that the vertical structure of mean quantities such as $\bar{T}$ as given by one mode is a reasonably close approximation to a solution of the full equations. (The agreement is even better for free boundaries in which case one mode gives $R^{1/3}$.) On the other hand, the horizontal structure implied by the one-mode approximation is generally not at all like that of the full two-dimensional solution, which suggests the existence of rising and descending plumes. Thus, the modal expansion may well be useful for computing mean structures in a convection layer in spite of its gross misrepresentation of the horizontal variations.

If, as is suggested by this comparison, the lower harmonics are not the most important additional modes to introduce, which are? Presumably, the rule used to choose the first mode should be used to choose successive modes. Thus the second mode might be that which is most unstable according to linear theory, based on the conditions existing with one mode. Or the second wavenumber a_2 might be chosen to be of order unity in terms of the vertical-length scale introduced by the first mode; this would imply $a_2 = O(N_1)$ since the boundary layer has thickness d/N_1. We should also note that once two modes are retained, another option is open: the second mode may be used as a perturbation to study the stability of the first mode. In principle this may be a way to help select the preferred first mode, but in practice if the nonlinear stability problem with two modes is studied, no clear choice is indicated. We must therefore adapt one of the ad hoc selection procedures and this is a weakness of the Galerkin procedure in convection. That the difficulty becomes less serious as more modes are introduced may be seen by estimating N_n, the Nusselt number computed from n modes.

Consider first the case of only Fourier modes; these have $C^{iii} = 0$. Then if we add more and more modes, so long as they are not harmonically related, we would obtain an approximation to Equations 1.4–1.8 with the terms $\boldsymbol{u} \cdot \nabla \boldsymbol{u}$ and $\boldsymbol{u} \cdot \nabla \theta - \overline{\boldsymbol{u} \cdot \nabla \theta}$ neglected. The neglect of these terms is often called the *mean-field* or *weak-coupling approximation* (Spiegel 1967). It can be expected to hold only for very large σ as indicated by the agreement of the one-mode roll solution and the two-dimensional theory. In that limit, the $\boldsymbol{u} \cdot \nabla \boldsymbol{u}$ term is negligible, but there seems no obvious reason to omit $\boldsymbol{u} \cdot \nabla \theta - \overline{\boldsymbol{u} \cdot \nabla \theta}$. Of course, this term vanishes for one mode with $C=0$, and since the observed motion is two-dimensional for large σ and $R \lesssim 10 \ R_c$, the approximation should work well in that domain of parameter space. At higher R it should continue to hold approximately since the term $w \partial_z \overline{T}$ acts qualitatively like $\boldsymbol{u} \cdot \nabla \theta$. Part of the reason for this is that for $\overline{T}$ independent of time the statistically steady solutions of the mean-field equations must be stationary (Spiegel 1962a).

To estimate the effect of higher modes we assume that a rule for choosing the first mode has been adopted. In general, for large R, such a rule may be expressed as $a_1 = k_1 R^\mu$, where $\mu \leq 1/4$ and k_1 may depend on σ. (We might also include a factor involving $\ln R$.) Then, for this mode, we have from 9.9,

$$N_1 = A R^\alpha (\ln R)^{1/5} \qquad 9.10$$

where $A \simeq A_1 (5\alpha k_1{}^2)^{1/5}$ and $\alpha = 1/5 \ (2\mu+1)$. For the second wavenumber let us adopt the same prescription, based on the length unit d/N_1; then $a_2 = k_2 N_1 R_{\mathrm{eff}}{}^\mu$, where the effective Rayleigh number R_{eff}, seen by the second mode is, crudely, $\sim R/N_1{}^3$. Here an arbitrary factor of order unity should be included to allow for the deviation from linearity of $\overline{T}$ in the boundary layer and for the lack of a rigid boundary at the edge of the boundary layer. The ratio k_2/k_1 differs from unity for similar reasons. Let us ignore such corrections and take $R_{\mathrm{eff}} = R/N_1{}^3$ and $k_2 = k_1$. Then the effect of the second mode is to increase N by an approximate factor $A \ R_{\mathrm{eff}}{}^\alpha (\ln R_{\mathrm{eff}})^{1/5}$ and for the two modes, the product of this factor with N_1 gives

$$N_2 = A^{2-3\alpha} (1 - 3\alpha)^{1/5} R^{(2-3\alpha)} (\ln R)^{1/5(2-3\alpha)} \qquad 9.11$$

We may readily extend this procedure to n modes with the result

$$N_n = (1 - 3\alpha)^{n/(15\alpha)} \left\{ A^{1/\alpha} (1 - 3\alpha)^{-1/(15\alpha^2)} (\ln R)^{1/(5\alpha)} R \right\}^{[1-(1-3\alpha)^n]/3} \qquad 9.12$$

We see that as n is increased, the chief feature of the limiting N is a variation like $R^{1/3}$, except for logarithmic factors, and that this behavior does not depend on α. However, for fixed R, N_n tends to zero as n tends to infinity, which shows that some cutoff must occur if the result is to be meaningful. That such a cutoff exists is shown by numerical solutions which indicate that if a_n is too large, the nth mode will not develop a detectable amplitude. The numerical results are consistent with the criterion that $R_{\mathrm{eff}} = R/N_n{}^3 \sim R_c$ for the cutoff n. This criterion not only permits us to estimate $n \sim -\ln \ln R/\ln (1-3\alpha)$, but it also shows that $N \sim (R/R_c)^{1/3}$, irrespective of the details of the theory (Malkus 1954b). Another approach, in the case where one wishes to maximize N, is to choose n to maximize N_n for fixed R (Chan 1971). Curiously, this procedure gives the same cutoff for large R, so that for the mean-field approximation $N_{max} \propto R^{1/3}$. For the case of maximum N, accurate asymptotic solutions of the mean-field equations have recently been constructed (Chan 1970).

The recovery of the $R^{1/3}$ law is an encouraging result, and the lack of σ dependence is not surprising since the mean-field equations should hold only at large σ. We must now consider more general modes with $C \neq 0$. These do introduce a σ dependence. If we proceed again with the introduction of a sequence of modes of higher and higher wavenumber, we will again come to the result 9.12 as long as the modes are not harmonically related so that their mutual coupling constants vanish. This time A will depend on σ in a way that depends on our choice of wavenumbers. If at small σ we expect to have N depend on $R\sigma$ we must choose k_1 so that at small σ it varies like σ^μ. In that case, we shall find that $N \propto (R\sigma)^{1/3}$ for large $R\sigma$ and for any $\mu < 1/4$. Thus the truncation procedure gives us the same kind of result for heat transport as standard mixing-length theory. Its advantage over mixing-length theory is that it can accommodate variable density, time dependence, and many of the other features of the problem that must be dealt with in stellar convection. Many of these complications have already been studied and will be discussed in Part II.

Now we must ask: what happens when we deal with dynamically coupled modes in the truncation theory? In particular, do the dynamical couplings excite modes that would not have been excited by mean-field terms and do these excitations lead for example to an $(R\sigma)^{1/2}$ law at large $R\sigma$? The question has not yet been answered. Numerical solutions for R up to 10^9 with three dynamically coupled modes have been found. These do seem to indicate the possibility of dynamical excitation. They also introduce complicated time dependence into the solution (Toomre 1969). However, the value of $R(=10^9)$ achieved to date is not high enough to show a tendency to deviate from an $R^{1/3}$ law; indeed at that value of R the law is still being approached as R increases.

One difficulty with going to higher R is that if we seek steady solutions, there is at any R and σ for a given set of wavenumbers and coupling constants a great wealth of solutions and it is not clear which solution branch to follow. It is just too demanding of computing time to follow them all. On the other hand, even if the difficulty of choice can be alleviated by computing time-dependent solutions, the computing bill also mounts up quickly because of the long transients. The time-dependent problem is reminiscent of computations in stellar pulsation in having one space and one time dimension; however, for three modes we must deal with a system of 20th order in the spatial derivatives. The problem can be done with existing machines, but it is difficult. It may well be possible to extract the main results for dynamically coupled modes analytically as has been done for the modes with only self-coupling and mean-field interactions. This problem has not been attempted and it is clearly of great analytical complexity.

An alternate procedure is to keep just a few (one to three) modes and add terms based on turbulence theory (crude or refined). This would permit a reasonably accurate handling of many of the features of stellar convection.

10. TURBULENCE THEORIES

The development of the statistical theory of turbulence has been proceeding quite rapidly, the recent developments being largely dominated by improved approximations for highly turbulent flows (Kraichnan 1970, Orszag 1970, Saffman 1968). The approximation techniques are usually developed in terms of turbulent spectra and produce equations, which, though much easier to solve than the original equations of motion, pose great calculational difficulties. In particular, one such approximation scheme, the direct interaction, has been applied to the Boussinesq equations for convection (Kraichnan 1964). The numerical solution of these equations has recently been accomplished for the case of free boundaries with $R \leq 10^4$ and $\sigma = \infty$ (Herring 1969). Remarkably enough the solutions are almost identical to those obtained for the mean-field equations with one mode. The extension of these solutions to higher R and lower σ is difficult, but feasible, and it will be of great interest to see how far such extensions can be carried. It should be stressed that this approach leaves no arbitrary parameters.

The use of approximations from turbulence theory has also been applied to the case of very low σ with rather bewildering results: the solutions grow in amplitude without limit (Herring 1970). This may be a peculiarity of the free-boundary conditions used in connection with a particular limiting form of the Boussinesq equations for low σ (Spiegel 1962b), but the matter has yet to be resolved.

There has also been a renaissance of phenomenological theories, many of which are more sophisticated than the standard mixing-length theory (Crow 1968, Lumley 1970, Nee & Kovasznay 1969, Parker 1969, Saffman 1970). The term phenomenological is sometimes used pejoratively in turbulence theory; here it is not. It simply implies that the approach used is not intended to be completely deductive and is based to varying degrees on some physical picture of the inner workings of turbulence. Nor does it follow that the equations in a phenomenological approach are easy to solve, though usually they are easier to solve than the full equations. The disadvantage of these theories for astrophysical purposes is that, like mixing-length theory, they normally contain disposable parameters. Values for these parameters determined in the laboratory may not be applicable to stars. But this is a question to be faced when these new approaches are applied to convection; for the present they simply represent a trend of which the astrophysicist need only be aware.

11. CONCLUSION

Not everyone who works on convection would agree with the assessments made of the various approaches discussed here. But the discussion as given does seem to bring out certain conclusions and these are:

1. The mixing-length theory, for all the criticisms leveled at it, has been qualitatively successful in predicting convective heat transfer. The form used for stellar convection however seems to be incomplete for very strong convection, and additional ingredients will be suggested in Part III. In any case it is difficult to use the theory for many special effects of astrophysical interest.

2. The procedure of maximizing the heat transfer is quite promising and gives Euler equations that seem to represent the flow qualitatively. Hitherto, these Euler equations have not contained dynamical couplings and this is the key to their tractability. If constraints could be added to the maximization problem which bring out the Prandtl-number dependence, the method will have great promise for stellar convection. Whether this can be done without enormous complications in the Euler equations remains to be seen.

3. The truncated modal expansions provide a reasonably accurate and flexible approach to the problem. They can readily accommodate large density variation, time dependence, and other difficulties of the stellar case. As yet, for more than one mode, they must be solved numerically, if the dynamical couplings are to be treated with acceptable accuracy. Whether these couplings will adequately describe boundary-layer turbulence remains moot. To settle this it may yet be necessary to further approximate the dynamical couplings.

4. The statistical theories of turbulence are beginning to provide dividends and will certainly give answers in the next few years. These results are awaited eagerly although they will oblige interested parties to master this very difficult discipline.

ACKNOWLEDGMENTS

Dr. J. Toomre has been kind enough to go over the manuscript with great care and offer detailed comments; any errors that still exist were probably introduced after his last reading. Comments by Drs. G. O. Roberts, R. C. J. Somerville, and J.-P. Zahn on the presentation have been of enormous help.

Several people kindly provided up-to-date information on their own and related work; these were Drs. J. R. Herring, L. N. Howard, G. O. Roberts, and R. C. J. Somerville. Dr. G. Galytsin provided useful information on Soviet work. Mr. L. Baker helped survey the literature. I am grateful to all of them for their aid.

LITERATURE CITED[1]

Bisshopp, F. E. 1960. *J. Math. Anal. Appl.* 1:373

Brindley, J. 1967. *J. Inst. Math. Appl.* 3:313

Brunt, D. 1939. *Physical and Dynamical Meteorology*, 219. Cambridge

Busse, F. H. 1967a. *J. Math. Phys.* 46:140

Busse, F. H. 1967b. *J. Fluid Mech.* 30:6251

Busse, F. H. 1969. *J. Fluid Mech.* 36:457

Busse, F. H. 1970. *GFD*, 70-50, I:71

Busse, F. H., Whitehead, J. A. 1971. *J. Fluid Mech.* In press

Chan, S. K. 1970. Dissertation. MIT

Chan, S. K. 1971. *Stud. Appl. Math.* In press

Chandrasekhar, S. 1961. *Hydrodynamic and Hydromagnetic Stability*. Oxford: Clarendon

Chang, Y. P. 1957. *Trans. ASME* 79:1957

Chen, M. M., Whitehead, J. A. 1968. *J. Fluid Mech.* 31:1

Chorin, A. J. 1968. *Math. Comput.* 23:241

Clauser, F. H. 1961. *Suppl. Nuovo Cimento* 32:354

Crow, S. C. 1968. *J. Fluid Mech.* 33:1

Davis, S. H. 1968. *J. Fluid Mech.* 32:619

Deardorff, J. W. 1964. *J. Atmos. Sci.* 21:

Dutton, J. A., Fichtl, G. H. 1969. *J. Atmos. Sci.* 26:241

Elder, J. W. 1968. *J. Fluid Mech.* 32:69

Finlayson, B. A., Scriven, L. E. 1966. *Non-Equilibrium Thermodynamics and Stability*, 291. Univ. Chicago

Fromm, J. E. 1965. *Phys. Fluids* 8:1757

Goldstein, R. J. 1964. *Chem. Eng. Sci.* 19:997

Goldstein, R. J., Chu, T. Y. 1966. *Final Rep. on NASA Res. grant NSG 714/24-005-036*. Univ. Minnesota, Mech. Eng. Dep.

Goldstein, S. 1938. *Modern Developments in Fluid Mechanics*. Oxford

Gough, D. O. 1969. *J. Atmos. Sci.* 26:448

Gough, D. O., Spiegel, E. A., Toomre, J. 1971. To be published

Herring, J. R. 1963. *J. Atmos. Sci.* 20:325

Herring, J. R. 1964. *J. Atmos. Sci.* 21:277

Herring, J. R. 1966. *J. Atmos. Sci.* 23:672

Herring, J. R. 1969. *Phys. Fluids* 12:39

Herring, J. R. 1970. *GFD*, 70-50, I:115

Howard, L. N. 1963. *J. Fluid Mech.* 17:405

Howard, L. N. 1965. *GFD*, 65-51, I:125

Jakob, M. 1946. *Trans. ASME* 68:189

Joseph, D. D. 1966. *Arch. Rat. Mech. Anal.* 22:163

Kaniel, S., Kovitz, A. 1967. *Phys. Fluids* 10:1186

Keller, J. B. 1966. *J. Fluid Mech.* 26:599

Koschmieder, E. L. 1966. *Beitr. Phys. Atmos.* 39:1

Koschmieder, E. L. 1969. *J. Fluid Mech.* 35:527

Kraichnan, R. H. 1962. *Phys. Fluids* 5:1374

Kraichnan, R. H. 1964. *Phys. Fluids* 7:1048, 1163, 1169

Kraichnan, R. H. 1970. *J. Fluid Mech.* 41:189

Krishnamurti, R. 1970. *J. Fluid Mech.* 42:295

Kuo, H. L. 1959. *J. Fluid Mech.* 5:113

Kuo, H. L., Platzman, G. W. 1961. *Beit. Z. Phys. Atmos.* 33:10

Lebovitz, N. R. 1966. *Non-Equilibrium Thermodynamics, Variational Techniques, and Stability*, 199. Univ. Chicago Press

Ledoux, P., Schwarzschild, M., Spiegel, E. A. 1961. *Ap. J.* 133:184

Lipps, F. B., Somerville, R. C. J. 1971. *Phys. Fluids.* 14:759

Lortz, D. 1961. Dissertation. Univ. Munich

Lumley, J. L. 1970. *J. Fluid Mech.* 41:413

Malkus, W. V. R. 1954a. *Proc. Roy. Soc. London* 225:185

Malkus, W. V. R. 1954b. *Proc. Roy. Soc. London* 225:196

Malkus, W. V. R. 1964. *GFD*, 64-46, I:1

Malkus, W. V. R., Veronis, G. 1958. *J. Fluid Mech.* 4:225

Matkowsky, B. J. 1970. *SIAM J. Appl. Math.* 18:872

Mihaljan, J. M. 1962. *Ap. J.* 136:1126

Nee, V. W., Kovasznay, L. S. G. 1969. *Phys. Fluids* 12:473

Noyes, R. W. 1967. *IAU Symp. No. 28*, 391. Academic

Ogura, Y., Phillips, N. A. 1962. *J. Atmos. Sci.* 19:173

Orszag, S. A. 1969. *Phys. Fluids. Suppl.* 12:250

Orszag, S. A. 1970. *J. Fluid Mech.* 41:363

O'Toole, J. L., Silveston, P. L. 1961. *Chem. Eng. Progr. Symp. Ser.* 51:81

Parker, E. N. 1969. *Phys. Fluids* 12:1592

Platzman, G. 1965. *J. Fluid Mech.* 23:481

Plows, W. 1968. *Phys. Fluids* 11:1593

---

[1] The notation GFD followed by a reference number indicates notes from the Summer Study Program in Geophysical Fluid Dynamics at the Woods Hole Oceanographic Institution. These can be obtained from The Clearinghouse, National Technical Information Service, Operations Division, Springfield, Virginia 22151. They are literally just notes, but they do contain useful information.

Prandtl, L. 1952. *Essentials of Fluid Dynamics*. London: Blackie & Son

Priestley, C. H. B. 1954. *Aust. J. Phys.* 7: 176

Priestley, C. H. B. 1959. *Turbulent Transfer in the Lower Atmosphere*. Univ. Chicago Press

Reiss, E. L. 1965. *Problems of Mathematical Physics*, ed. N. N. Lebedev, I. P. Skalsbayer, Y. W. Uflayand, 391. New York: Prentice-Hall

Roberts, G. O. 1969. Preprint

Roberts, P. H. 1966. *Non-Equilibrium Thermodynamics, Variational Techniques, and Stability*, 125:299. Univ. Chicago Press

Rossby, H. T. 1969. *J. Fluid Mech.* 36:309

Saffman, P. G. 1968. *Topics in Nonlinear Physics*, ed. N. Zabusky, 485. Springer

Saffman, P. G. 1970. *Proc. Roy. Soc. London*

Sani, R. 1964. *J. Fluid Mech.* 20:315

Schneck, P., Veronis, G. 1967. *Phys. Fluids* 10:927

Schlüter, A., Lortz, D., Busse, F. H. 1965. *J. Fluid Mech.* 23:129

Segel, L. A. 1966. *Non-Equilibrium Thermodynamics, Variational Techniques and Stability*, 165. Univ. Chicago Press

Simon, G., Weiss, N. O. 1964. *Z. Ap.* 69: 435

Somerville, R. C. J. 1970. *GFD*, 70–50, I:87

Sommerscales, E. F. C., Gazda, I. W. 1969. *Int. J. Heat Mass Transfer* 12:1491

Sorokin, V. S. 1957. *Appl. Math Mech.* 18: 197

Spiegel, E. A. 1960. *Ap. J.* 132:716

Spiegel, E. A. 1962a. *Mecanique de la Turbulence*, 181. Paris: CNRS

Spiegel, E. A. 1962b. *J. Geophys. Res.* 67: 3063

Spiegel, E. A. 1963. *Ap. J.* 138:216

Spiegel, E. A. 1966. *Stellar Evolution*, ed. Stein & Cameron, vol. 1, 143. Pergamon

Spiegel, E. A. 1967. *IAU Symp No. 28*, 348. Academic

Spiegel, E. A. 1971a. *Comments Ap. Space Sci.* 3:53

Spiegel, E. A. 1971b. *Stellar Evolution*, ed. H.-Y. Chiu and A. Muriel, vol. 2. In press

Spiegel, E. A., Veronis, G. 1960. *Ap. J.* 131: 442

Stewartson, K. 1966. *Non-Equilibrium Thermodynamics, Variational Techniques and Stability*, 158. Univ. Chicago Press

Stuart, J. T. 1958. *J. Fluid Mech.* 4:1

Stuart, J. T. 1960. *Proc. Int. Congr. Appl. Mech.* 63

Stuart, J. T. 1964. *J. Fluid Mech.* 18:481

Sutton, O. G. 1953. *Micrometeorology*, 119. New York: McGraw-Hill

Sutton, O. G. 1955. *Atmospheric Turbulence, Methner's Monogr. Phys. subjects*

Taylor, G. I. 1970. *J. Fluid Mech.* 41:3

Thomas, D. B., Townsend, A. A. 1957. *J. Fluid Mech.* 2:473

Thompson, J. D., Sogin, H. H. 1966. *J. Fluid Mech.* 24:451

Tippleskirch, H. 1956. *Beitr. Phys. Atmos.* 29:37

Toomre, J. 1969. *GFD*, 69–41, 126

Townsend, A. A. 1959. *J. Fluid Mech.* 5: 209

Townsend, A. A. 1966. *Mecanique de la Turbulence*, 169. Paris: CNRS

Veronis. G. 1966. *J. Fluid Mech.* 26:49

Vitense, E. 1953. *Z. Ap.* 32:135

Welander, G. 1967. *J. Fluid Mech.* 29:17

Wesseling, P. 1969. *J. Fluid Mech.* 36:625

Willis, G. E., Deardorff, J. W. 1967a. *Phys. Fluids* 10:931.

Willis, G. E., Deardorff, J. W. 1967b. *Phys. Fluids* 10:1861

Willis, G. E., Deardorff, J. W. 1967c. *J. Fluid Mech.* 28:675

Willis, G. E., Deardorff, J. W. 1970. *J. Fluid Mech.* 44:661

Zel'dovich, Ya. B. 1937. *JETP* 7:1463

RECENT DEVELOPMENTS IN THE THEORY
OF DEGENERATE DWARFS

Jeremiah P. Ostriker[1]

Princeton University Observatory, Princeton, N.J.

1. Introduction and Summary

Stars exist by virtue of a stable balance between the forces of gravity and pressure. Of all the conceivable stars which might be represented by points on the mass-radius plane, only those occurring in four relatively small islands of stability are known (or thought) to exist. The corresponding objects are supported by the pressure provided by the motions of four kinds of particles. Cold nucleon pressure allows *neutron stars* (cf Wheeler 1966), cold electron pressure allows *degenerate dwarfs*, the subject of the present review, thermal ion and electron pressure supports the ordinary stars, and photon radiation pressure the *supermassive stars* (cf Wagoner 1969). While the radius increases by $\sim 10^2$ from class to class, the typical mass in each island (excepting supermassive stars which may not exist) is near one solar mass. Thus ordinary ($\sim 1\ M_\odot$) stars can pass to either of the two condensed states without losing their identity, and their real (vs possible) existence might have been anticipated. Neutron stars were predicted in just this way, but degenerate dwarfs were discovered before our knowledge of physics was advanced enough to understand them.

The original explanation (Chandrasekhar 1931a, b, 1935) of these very common stars has withstood the test of time. From the observed radii ($\sim 10^{-2} R_\odot$) and masses ($\sim 1 M_\odot$), the internal pressure ($\approx GM^2/R^4$) greatly exceeds that available from thermal motions (for $T_{\text{interior}} < 10^{8.5\circ}$K). However, the number density of electrons n_e is so high that the zero-point momentum ($p_e \approx \hbar n_e^{1/3}$) provides enough pressure to support the stars at radii in the observed range. So long as the electron velocities are nonrelativistic, the degenerate pressure $P_e \approx p_e^2 n_e m_e^{-1}$ depends on the density to the (5/3) power or on $M^{5/3} R^{-5}$. Comparing this with the pressure required to support the star against self-gravitation ($\propto M^2 R^{-4}$) we see that degenerate dwarfs supported by nonrelativistic electron motions have a mass-radius relation of the form $R \propto M^{-1/3}$. For sufficiently high mass, the electrons become relativistic, and the electron pressure $P_e \approx p_e c m_e$ depends on the density to the (4/3) power or on $M^{4/3} R^{-4}$; it is no longer certain that a small enough radius can be found to bring the pressure and gravitational forces in balance. The maximum mass allowed for a spherical star—the Chandrasekhar mass limit—can be seen

---

[1] Alfred P. Sloan Fellow, 1970–1971.

from the above argument to correspond to $\sim(\hbar c/Gm_H{}^2)^{3/2}$ nucleons, or more precisely, $M_C = 5.75\,\mu_e{}^{-2}M_\odot$ where μ_e is the mean molecular weight per electron in the star. While corrections to the equations of state and modification of the equation of equilibrium due to general relativistic effects (cf Section 2.1) reduce the limit slightly, the original result—that spherical electron degenerate stars must be less massive than $\sim 1.4 M_\odot (\mu_e = 2)$—has remained valid.

The term *white dwarf* was originally chosen because of the blue-white color of the first observed members of the class. *Degenerate dwarf* is a more general and appropriate name for a star supported by the pressure of a degenerate electron gas, especially since most such stars are not very hot and it is awkward to discuss *yellow white dwarfs*, *red white dwarfs*, or even *black white dwarfs*. While the mechanical properties are determined largely by electrons, ions (because of the much larger mass per particle) are not degenerate. Since the ions have almost all the thermal energy but provide almost none of the pressure, the mechanical and thermal properties of degenerate dwarfs are largely independent of each other. The electrons support the star and the thermal motions of the ions provide a reservoir of heat which, when all nuclear processes have stopped and gravitational contraction has almost ceased, remains to diffuse slowly through the outermost nondegenerate layers of the star and to maintain the surface temperature and luminosity high enough to keep the typical nearby degenerate dwarf observable for billions of years. During the period of slow cooling, the electronic heat conductivity in the interior is sufficiently high to hold the interior nearly isothermal with the heat loss rate determined by transfer through the *thermal bottleneck*—the partially degenerate layers somewhat below the surface within which both electron and photon conduction of heat are inefficient. The total thermal energy available $[\sim 10^{8.3}\,k(M_\odot/m_H) \approx 10^{49.7}$ ergs$]$ is significant; while it is only about 10^{-2} of the energy released by the same stars during the main-sequence hydrogen-burning phase, it is comparable with the visual output emitted by other stars in their final supernova outbursts.

This general theoretical background has been summarized in the excellent reviews of Schwarzschild (1958) and Mestel (1965); the more recent developments can be subdivided into those concerned primarily with the mechanical or thermal properties of the star. First considered are the modifications to the structure and mass-radius relation of spherical stars caused by the contributions of the ions, general relativity, etc. Then nonspherical degenerate stars—rotating and magnetic—are discussed with the surprising result noted that inclusion of nonzero angular momentum can drastically alter the concept of a "mass limit." Finally the problem of mechanical stability is surveyed for both spherical and nonspherical stars. The discussion of thermal properties is focused on an attempt to quantitatively understand the cooling problem. First local properties in the stellar interior (specific heat, thermal conductivity, crystallization temperature, etc) are assessed; then recently derived model atmospheres and envelopes are summarized;

and finally I will try to roughly assemble likely evolutionary sequences. The principal conclusion here is that convection increases the heat flow through the envelope (cf Böhm 1969) as the heat capacity of the crystalline interior declines (as a result of Debye effects, cf Mestel & Ruderman 1967) at late evolutionary stages; because of this accelerated cooling, degenerate dwarfs will fade to near invisibility in less than the age of the Galaxy. A final section contains notes on some potentially profitable directions for future observational programs and suggestions that, among other things, cold stars may contribute significantly to the local mass density in the Galaxy.

2. EQUILIBRIUM AND DYNAMICS

2.1 *Zero-temperature spherical stars.*—The interior pressure for stars in the mass range $0.3 < M/M_\odot < 1.0$ is produced primarily by the kinetic energy of the degenerate electrons. In this mass range, models constructed using the simple equation of state derived by Chandrasekhar (1939) are accurate if the mean molecular weight per electron μ_e is chosen suitably. The Chandrasekhar $P(\rho)$ relation is derived for noninteracting fermions, but for stars of lower mass (and density) the electrostatic interactions among the ions, between the electrons and the ions, and among the electrons all become important. Salpeter (1961) and Salpeter & Zapolsky (1967) have derived the equation of state in zero-temperature stars including effects due to the classical Coulomb energy of the ion lattice (for the ions crystallize as $T \rightarrow 0$), the Thomas-Fermi deviations from uniform charge distribution of electrons (which tend to cluster about the ions), and the exchange energy and spin-spin interactions among the electrons. At a sufficiently low density, degenerate material approaches the state found in common metals, the mass-radius relation becomes $M \propto R^{+3}$ (vs $M \propto R^{-3}$ in the low-density limit of the Chandrasekhar equation of state). Thus a maximum possible radius exists, which, for cold carbon stars, is approximately $3.9 \times 10^{-2} R_\odot$ at $2.2 \times 10^{-3} M_\odot$ (Zapolsky & Salpeter 1969). The calculated corrections to the pressure (for given density), always negative, increase with increasing Z (ionic charge) and decreasing density. Although, for the high-density regime that is relevant for $\sim 1 M_\odot$ stars, the correction is very small (about 1.5 percent for ^{12}C), the effect on the mass-radius relation is not negligible. In this regime the pressure density relation is nearly that in a polytrope of index $n = 3 (P \propto \rho^{4/3})$; for such stars the binding energy is nearly independent of radius, and small corrections to the pressure produce large changes in the radius.

At high densities a more important effect is inverse β decay—the tendency for electrons to tunnel into nuclei if their energy is as great as that liberated by β decay of the created nuclei. In the high-density regime, the electron energy (Fermi-level) is equal to (atomic constants) $\times \rho^{1/3}$ for a given chemical species. Thus a specific density exists, for each species, above which inverse β decay occurs. A chemically homogeneous star with central density equal to this critical value ($10^{10.65}$ g cc^{-1} for ^{12}C) has the maximum mass allowed by equilibrium for a star of the given composition. If more mass were added, it would be impossible for equilibrium to be re-established at a smaller radius and higher pressure, since the inverse β decay induced by the compression would remove electrons and tend to decrease the pressue. The maximum masses range from $1.40 M_\odot$ for carbon to $1.11 M_\odot$ for iron (cf Hamada & Salpeter 1961). General relativistic corrections (cf Chandrasekhar & Tooper 1964) also tend to reduce the maximum mass attainable; this is found to be about $1.42 M_\odot$ for stars with the Chandrasekhar equation of state and $1.36 M_\odot$ for model calculations which include effects due to both GR and particle interactions (Cohen, Lapidus & Cameron 1969).

For every correction so far discussed, the radius of a zero-temperature degenerate dwarf of given mass is decreased from the "classical" value. However, analysis of observations by Greenstein & Trimble (1967) indicates that radii determined from photometric criteria are slightly larger than those given by observations of gravitational shifts and the classical mass radius relation. Possible resolutions to this discrepancy are suggested in Sections 2.2 and 2.4.

2.2 *Finite-temperature spherical stars.*—The equation of state $P(\rho, T)$ has three principal terms. The pressure of a noninteracting finite temperature Fermi gas (Chandrasekhar 1939; cf convenient formulae in Chiu 1968) dominates (by definition for a degenerate star). The ion pressure which ranges from that of a perfect gas to that in a zero-temperature crystal as the temperature decreases is given, in approximate form, by Van Horn (1969a) and the exchange pressure for a finite-temperature nonrelativistic electron gas is given by Hubbard (1969). Models calculated with finite temperatures by Hubbard & Wagner (1970) have radii larger than the Hamada-Salpeter models as expected. Homogeneous models with $Z > 6$ and $L = 10^{-2} L_\odot$ show a radius increase of 1 to 5 percent for masses in the range 1.0–$0.5 M_\odot$. The radius increase can be a factor of 2 to 4 larger for helium stars or heavier-ion stars with hydrogen envelopes, but is always quite small for stars having $M \geq 1 M_\odot$.

2.3 *Rotating stars.*—For a star of fixed angular momentum J, the kinetic energy of rotation T scales as $J^2 R^{-2}$ and the gravitational energy W, as $M^2 R^{-1}$ (where R is the equatorial radius). Thus we should expect rotation to be relatively more important to a star in its condensed final state than it was in its main-sequence phase. Furthermore Mestel (1965) pointed out that the derived mass limits do not apply to nonspherical stars. Consequently James (1964) and others calculated the effect of solid-body rotation on degenerate dwarfs obeying the Chandrasekhar equation of state. Surprisingly, they found that the gross structure of the models was not significantly altered and the mass limit, in particular, was increased by only 4 percent (for given μ_e). The result, while correct, is somewhat misleading. The restriction to uniform rotation in centrally condensed bodies prevents the construction of models with ratios of $|T/W|$ greater than about 0.006 (cf Ostriker 1970), so that, almost by assumption, uniform rotation will not significantly affect a self-gravitating, centrally condensed object. Contraction without angular momentum loss forces the ratio $|T/W|$ to increase as R^{-1} and so it should be possible to construct degenerate dwarfs of arbitrary mass for any (nonzero) angular momentum (Ostriker, Bodenheimer & Lynden-Bell 1966), the star being always able to find a radius small enough for the pressure and rotational energies to balance the gravitational energy. Of course the resulting models will rotate differentially, inevitably showing a Keplerian velocity distribution in the outer parts of rapidly rotating examples, and they may or may not be stable to inverse β decay (collapse), fission (see Section 3.2), rapid evolution through viscous momentum transport (cf Schwartz & Africk 1970, Ostriker & Durisen 1970), and a host of other possibly catastrophic hazards. Detailed models, calculated by Ostriker & Bodenheimer (1968) in the mass range 1.3–$4.1 M_\odot$ with angular momenta corresponding to main-sequence stars of similar mass, have radii which are typically within a factor 2 of 10^9 cm and do not distinguish the models from nonrotating models in the mass range 0.4–$0.9 M_\odot$. Central densities are less than $10^{8.7}$ g cm^{-3}; consequently the models are stable to inverse β decay; and the adopted Chandrasekhar equation of state is a good approximation if the composition is (C, O, Mg, Si). Surface velocities of the massive models are typically in the range 3000 to 7000 km/sec.

Such stars, if they exist, will not be entirely cold. Durisen (1971) has shown that viscous dissipation of energy, while not rapid (time scale typically $>10^9$ years) is sufficient to maintain the luminosity of the massive models at more than $10^{-1} L_\odot$ and keep the internal temperature high enough to prevent crystallization, unless made of a very high-Z material such as iron (cf also Schwartz & Africk 1970).

2.4 *Cold magnetic stars.*—Magnetic energy $\mathfrak{M}$ also will increase for a contracting star if the flux Φ threading a cross-sectional plane is conserved. But since $\mathfrak{M}$ scales only as $\Phi^2 R^{-1}$ and $|\mathfrak{M}/W|$ as $R^0 \Phi^2/M^2$, magnetic forces will be relatively no more important for a degenerate dwarf than they were for the corresponding main-sequence star. However, the magnetic pressure should be observable in a degenerate dwarf by producing a fractional increase of radius of the same order as $|\mathfrak{M}/W|$; a proportionately similar magnetic pressure in a main-sequence star may remain undetectable since the radius and luminosity are determined more by energy production and transport than by pressure and the equation of state. Constructing models with mixtures of poloidal and toroidal internal fields, Ostriker & Hartwick (1968) found that the fractional radius increase can be significant for relatively massive stars going as $\exp(+3.5 |\mathfrak{M}/W|)$ for a $1.05 M_\odot$ model. For the extreme case ($|\mathfrak{M}/W| \sim 0.1$) the central fields approach $10^{12.3}$ G but the surface poloidal fields remain several powers of 10 less. The mass limit can also be increased by inclusion of magnetic forces, but, since $|\mathfrak{M}/W| \propto R^0$, the maximum increase is finite (and in fact quite small); to first order, the fractional increase in the mass limit is $(\Delta M/M_{\mathrm{C}}) = +|3\mathfrak{M}/2W|$.

The increase in radius produced by internal magnetic fields may help to explain the discrepancy found by Greenstein & Trimble between the observed and theoretical radii of degenerate dwarfs.

3. Oscillations and Stability

3.1 *Radial oscillations.*—The frequency σ_R, of the fundamental radial pulsation mode (perturbation $\delta r = f(r)e^{i\sigma_R t}$) of any star is given to a good approximation by the formula

$$\sigma_R{}^2 \doteq \frac{|W|}{I}\left[(3\bar{\gamma} - 4) + \alpha(3\bar{\gamma} - 5)\frac{T}{|W|} - \beta\frac{|W|}{Mc^2}\right]$$

where α and β are positive constants of order unity (cf Ledoux 1945, Chandrasekhar & Lebovitz 1962a, Chandrasekhar 1964). Here, in addition to previously defined symbols, $\bar{\gamma}$ represents the pressure-weighted average of the adiabatic exponent and I the moment of inertia.

The first term in the bracket is, for most stars under most circumstances, the dominant term. The second term shows the stabilizing effect of rotation (on the radial modes) and the third the destabilizing effect of general relativity. While $\bar{\gamma} \doteq 5/3$ in most stars and low-mass degenerate dwarfs, $\bar{\gamma} \doteq 4/3$ for high-density (relativistic) degenerate dwarfs and the normally small "correction" terms become dominant.

The oscillation periods of cold models satisfying the Chandrasekhar equation of state and ignoring relativistic corrections were calculated by Sauvenier-Goffin (1950) and Schatzman (1961). They decrease from about 20 sec (for $\mu_e = 2$) at $0.4 M_\odot$ to about 6 sec at $1.0 M_\odot$. At still higher mass, because of relativistic degeneracy, $(3\bar{\gamma}-4) \propto (m_e c^2/\epsilon_F)^2$, and, since $|W|/I \propto R^{-3}$ as $M \rightarrow M_{\mathrm{C}}$, the product $(|W|/I)(3\bar{\gamma}-4) \propto R^{-1} \rightarrow \infty$; the periods continue to decrease and the models are all stable $(\sigma_R{}^2 > 0)$. However, comparing the first term $(\propto +R^2)$ with the third term $(\propto -1/R)$, it is clear that general relativistic corrections become important as $M \rightarrow M_{\mathrm{C}}$. Chandrasekhar & Tooper (1964) found that $\sigma_R{}^2$ reaches a maximum (period a minimum) and then goes to zero at $(M/M_\odot) = 1.42$ $(\mu_e = 2)$, when the star's radius is still $\sim$250 times the Schwarzschild radius. Several authors (cf Skilling 1968, Wheeler, Hansen & Cox 1968, Faulkner & Gribbin 1968, and most completely, Cohen, Lapidus & Cameron 1969) combined the complications of general relativity with those of a realistic equation of state. The minimum period found depends somewhat on chemical composition but is typically about 2.0 sec and occurs for stars having $\rho_c \approx 10^{10}$ g/cc; the calculated models are unstable to collapse for radii less than about $10^{8.0}$ cm or masses greater than $1.36 M_\odot$ for a homogeneous carbon star to a somewhat lower value (and determined more by $\gamma < 4/3$ when inverse β decay sets in than by GR)

for more realistic models in which carbon-burning products exist at the center of the models. All of the papers so far discussed treat cold degenerate stars. In hot models two complicating effects exist. The ion pressure changes with a γ of 5/3 and acts as a stabilizing influence (cf Baglin 1969, Vila 1970); this effect is small. If the models are hot enough to support thermonuclear reactions, they may be expected to be overstable (cf Ledoux & Sauvenier-Goffin 1950 and Schatzman 1958); then, since moderate-mass degenerate dwarfs are not very centrally condensed, the damping of radial modes in the radiative envelope of the star is overcome by the energy input in the burning core with the result that a quasiadiabatic mode will slowly grow in amplitude. Hansen & Spangenberg (1971) have shown that low-mass ($\doteq 0.3 M_\odot$) helium-burning models are unstable; they have a fairly low luminosity ($\doteq 1 L_\odot$), long e-folding time ($= 10^6$ years), and periods ($\doteq 1$ min) in a range that may be relevant to the observations of old novae.

Since the rotational term in the approximate equation for the radial oscillation frequency is proportional to R^{-2} it may be even more important than general relativity for rotating high-density models. Even models calculated on the basis of a linearized theory of rotation (Durney et al 1968 and Gribbin 1969) show the stabilizing effect clearly; rapidly rotating degenerate dwarfs (cf Ostriker & Tassoul 1968) can have "radial" pulsation periods as low as 0.25 sec.

As a final correction Saslaw (1968) and Axel & Perkins (1971) have studied the destabilizing effect of the contraction that occurs in a degenerate dwarf accreting mass from either the interstellar medium or a companion in orbit.

3.2 *Nonradial modes.*—For spherical degenerate dwarfs the fundamental nonradial (Kelvin) mode of oscillation σ_K has been determined by Ostriker & Tassoul (1969) and, more accurately, by Harper & Rose (1970). As in other stars, the oscillation frequency σ_K of this mode is given to a good approximation by

$$\sigma_K{}^2 \doteq \frac{4}{5} \frac{|W|}{I}$$

The periods range from 42 sec at $0.22 M_\odot$ to 1.9 sec at $1.2 M_\odot$ and tend to zero as ($M \to M_C$, $R \to 0$). From the two approximate formulae for σ_R and σ_K we should expect (cf Chandrasekhar & Lebovitz 1962b) that when $\bar{\gamma} = 8/5$, a degeneracy occurs; $\sigma_R = \sigma_K$. For spherical degenerate dwarfs this phenomenon takes place at $M \doteq 0.45 M_\odot$, $P \doteq 15$ sec.

Of the fundamental five nonradial ($L = 2$) modes which are split by rotation, the two that distort the equatorial plane into a rotating ellipse are of greatest interest because they can become unstable for sufficiently rapid rotation. It is this type of mode that has been thought to lead to fission (cf Ostriker 1970) of rapidly rotating stars. Ostriker & Tassoul (1969) found that, regardless of mass or angular momentum distribution, models having

$|T/W| \geq 0.14$ are secularly unstable; that is unstable on a time scale set by dissipative processes. In the absence of dissipation, the models are dynamically unstable (time scale ≈ 10 sec) if $|T/W| \gtrsim 0.26$, the result being again independent of mass and angular momentum distribution. Viscous dissipation is very slow (10^9–10^{10} year time scale) for degenerate dwarfs, but Chandrasekhar (1970) has shown that the gravitational radiation emitted by a rotating ellipsoid according to general relativity is sufficient to drive the secular instability of a McLaurin spheroid. The time scale for this instability in a degenerate dwarf is short; they must correspondingly be limited to the range $0 \leq |T/W| < 0.14$. This limit, and the restriction on central density imposed by the inverse β-decay instability, probably limits the range of stable degenerate dwarfs to masses less than $\sim 3 M_\odot$.

4. EVOLUTION AND THERMODYNAMICS

4.1 *Local thermal properties.*—The Coulomb energy per ion, E_C, is approximately $(Ze)^2/a$ where a is related to the ion number density n_i as $(4\pi a^3/3)^{-1} \equiv n_i$. This energy is normally small compared to the energy per electron (essentially the Fermi level ϵ_F), but it may be comparable to or larger than the thermal energy per ion, $3/2 \, kT$. If this should happen, the ions will minimize their energy by forming a crystal. Defining

$$\Gamma \equiv \frac{E_C}{kT} = 2.3 \times 10^5 Z^2 (\rho/\mu)^{1/3} T^{-1}$$

crystallization has been variously calculated to occur at $\Gamma = \Gamma_m \doteq 60$, 120, 170 by Mestel & Ruderman (1967), and Brush, Sahlin & Teller (1966), and Van Horn (1969b), respectively. For typical interior conditions it occurs at about $10^{7.5}$ °K.

The specific heat C_v, which is $3/2 \, k$ per ion before crystallization, becomes $3 \, k$ per ion when the potential energy in the Coulomb springs becomes available. However, at still lower temperatures the ions approach their zero-point energy levels—about $E_0 \equiv \hbar(4\pi Z^2 e^2 n_i/\mu m_H)^{1/2}$ per ion. They are then effectively frozen into their zero-point oscillations and very little more thermal energy is available. This occurs at approximately the Debye temperature $\Theta_D (k\Theta_D \equiv E_0)$

$$\Theta_D \doteq 4 \times 10^3 \rho^{1/2} (2/\mu_e)°K$$

and for temperatures considerably below this point the heat capacity declines rapidly, going as $(16\pi^4/5) (T/\Theta_D)^3 \, k$ per ion as $T \to 0$ (Mestel & Ruderman 1967).

While the thermal energy resides principally in the ion gas, it is carried principally by the degenerate electrons. The electron thermal conductivity, which was originally calculated by Marshak (1940) for complete electron degeneracy, has been refined in stages by Lee (1950), by Mestel (1950) and most recently by Hubbard & Lampe (1969), to include effects due to electron-

electron as well as electron-ion collisions and others due to the nonuniformity of the charge distribution. Canuto (1970) has extended the calculations into the high-density regime where the electron motions are relativistic. In addition, the transport coefficients required for a knowledge of the electronic component of the electrical conductivity and viscosity have been calculated by Hubbard (1966) in the nonrelativistic limit. Van Horn (1969a) has estimated the ion viscosity in liquid and crystalline phases. The ionic contribution to the thermal conductivity (phonon transport), although it may be of some significance in the interior of a degenerate dwarf (and perhaps even in low-mass main-sequence dwarfs), does not seem to have been discussed in the literature.

4.2 *Model atmospheres and envelopes.*—Radiative model atmospheres for degenerate dwarfs have been constructed by Weidemann (1960, 1963), Matsushima & Terashita (1969a,b) Wickramasinghe & Strittmatter (1970) and others. This work, whose earlier phases are reviewed by Weidemann (1968), has led to gravity determinations consistent with the theoretical mass-radius relation (for the stars of known mass); with respect to composition, these studies have shown that stars of spectral types DB and DC are extremely deficient in hydrogen (see Weidemann 1970). Atmospheric compositions, however, are still not known with any security.

In recent papers of great significance to the cooling problem, Böhm (1968, 1969) pointed out that low-luminosity degenerate dwarfs can be expected to have convective envelopes for essentially the same reasons as low-luminosity main-sequence dwarfs do (cf also Schatzman 1958). If the convective envelope extends through the *thermal bottleneck* into the degenerate interior, cooling of the star will be rapidly accelerated. Convective-model envelopes have been studied in greater detail by Van Horn (1970a), Grenfell & Böhm (1970), Böhm (1970), and Strittmatter & Wickramasinghe (1970). The phenomenon of convection reaching down to the degenerate interior occurs for luminosities below $10^{-2.5}$ to $10^{-3.5}L_\odot$ depending on whether the envelope composition is primarily helium or carbon. For van Maanen 2, Böhm (1969) finds that a convective envelope of about 10^1 km is so much more efficient at carrying the thermal energy flux from the interior that the core temperature is a factor 4 less than in the corresponding radiative model having the same surface conditions.

4.3 *Early evolution.*—A number of authors have computed the evolutionary paths of stars as they complete their nuclear burning and pass through the regions of the H-R diagram populated successively by the central stars of planetary nebulae and ordinary degenerate dwarfs. Calculations by Chin, Chiu & Stothers (1966), L'Ecuyer (1966), Rose (1966, 1967), and Vila (1966, 1967) showed the importance of gravitational contraction and neutrino losses for such stars—one process representing a source and the other a sink of energy. Initially gravitational energy input dominates, but as contraction proceeds, the central regions become degenerate, further contraction becomes difficult, and the neutrino losses dominate. A maximum core temperature is attained of approximately $10^{8.5}$°K for a $0.75 M_\odot$ model. Surface luminosities also reach a peak and then decline, the maximum being about $10^4 L_\odot$, which is comparable to the peak luminosity of the central stars of planetary nebulae (Seaton 1966). Models without neutrino losses contract more slowly and do not reach surface luminosities as high as those with interior and surface energy loss. Stothers (1966) proposed that the existence of neutrino-pair reactions could be proved or disproved by the presence or absence of *ultraviolet dwarfs*. However, improved calculations by Beaudet & Salpeter (1969) (see also Savedoff, Van Horn & Vila 1969) indicate that the expected effect would be too small to provide a definitive test of the basic neutrino processes.

Unfortunately evolution in these phases is quite dependent on the starting models; Rose and collaborators (see Rose & Smith 1970) and Deinzer & Hansen (1969) adopt starting models with hot outer parts (from late evolutionary shell burning) but nearly degenerate interiors. Such models have less gravitational and thermal energy to lose in the contraction to the degenerate dwarf state. The peak luminosity reached is similar to that found in the sequences beginning with nondegenerate interiors but the evolutionary times are considerably shorter.

Contracting sequences of very low-mass degenerate dwarfs may contain information on the fate of stars which never become hot enough for significant nuclear burning. Stothers (1971) has calculated several evolutionary sequences and finds that $0.1 M_\odot$ models, which have deep convective envelopes before they become quite degenerate, pass through parts of the H-R diagram containing many observed stars in the Hyades cluster.

4.4 *Final stages of evolution—Nonrotating stars.*—Mestel's (1952) theory for the cooling of degenerate dwarfs is now considered to be fairly secure for masses in the range $0.5 < M/M_\odot < 1.0$ and luminosities in the range $10^{-2.5} \gtrsim L/L_\odot \gtrsim 10^{-3.5}$. While neutrino losses and gravitational contraction are no longer important for such stars, convection and declining specific heat have not yet begun to accelerate the cooling. The theory predicts that the time to reach a given bolometric luminosity L is proportional to $L^{-5/7}$. Weidemann (1967) pointed out that a luminosity function ϕ (M_{bol}) of the form log ϕ (M_{bol}) = const + $(2/7) M_{bol}$, should then be observed. He found excellent agreement with theory in the range of luminosities specified above with weak indications of deficits in the high and low luminosity stars as would be expected from the known failures of the simple theory.

In a comprehensive review and extension of the basic theory, Van Horn (1970b) quantitatively estimates the effects of the various complications mentioned above, and includes in addition the heat of crystallization (see also Van Horn 1968) and the residual specific heat of the electrons [$C_V \approx k(kT/\epsilon_F)$ per electron]. The latter effect appears to be quite important for low-mass ($M/M_\odot < 0.6$) stars. He finds that, for high-mass stars with neutrinos, $t \propto L^{-2}$ when $L < 10^{-2.3}L_\odot$, which would imply $\log \phi = \text{const} + (4/5)M_{\text{bol}}$. Convection becomes important at $\sim 10^{-2.5}L_\odot$ for stars with helium envelopes and $\sim 10^{-3.5}L_\odot$ for stars with hydrogen or carbon envelopes. Debye effects in the crystalline core accelerate the cooling of $1.0 M_\odot$ stars at $\sim 10^{-3}L_\odot$ and $0.5 M_\odot$ stars at $\sim 10^{-4}L_\odot$ according to Ostriker & Axel (1969). These authors, who did not include the additional acceleration of cooling due to convection, estimated that all degenerate dwarfs with $(M/M_\odot) \geq 0.7$ cool to invisibility in less than 10^{10} years. Since convective cooling is much more important for low-mass degenerate dwarfs, the two effects complement one another. Preliminary calculations indicate that all degenerate dwarfs born in the first several billion years of stellar activity in the Galaxy will have faded to $M_{\text{bol}} > 18$ where they will be essentially undetectable.

4.5 *Final stages of evolution—Rotating stars.*—If massive rotating, degenerate dwarfs exist, their thermal evolution will be qualitatively unlike that of nonrotating stars. All the mechanisms responsible for thermal energy losses described in Section 4.4 will operate for the more massive stars. However these stars would typically have about 10^{51} ergs of kinetic energy of rotation available in addition to any residual thermal energy. Since they cannot rotate as uniform bodies, viscous dissipation will generate thermal energy from mass motions. The energy loss leads to contraction with a consequent *increase* in rotational kinetic energy, density, and rate of viscous dissipation. Thus, these stars will not cool but can maintain an equilibrium of $L \sim 10^1 - 10^{-1}L_\odot$ (cf Schwartz & Africk 1970 and Durisen 1971). Ultimately (on the viscous dissipation time scale), contraction will bring the central density to the point of inverse β decay and collapse will ensue. This mechanism is similar to those suggested by Finzi & Wolf (1967) and Hansen & Wheeler (1969) for the initiation of Type I supernova. Prior to collapse such stars could make a nonnegligible contribution to the ultraviolet background of the Galaxy if their birth rate is as much as 10^{-1} of the total birth rate for degenerate dwarfs.

There is one further interesting possibility. If these stars exist, and if some have significant magnetic fields (some of the ordinary degenerate dwarfs apparently do—cf Kemp et al 1970), then they can lose rotational energy via magnetic dipole radiation in the same way that rotating neutron stars are thought to lose energy. This process also would lead to collapse ultimately. It is more important than viscous dissipation for strong fields ($\gtrsim 10^7$ G) and would result in the production of $\sim 10^{51}$ ergs emitted as cosmic rays (cf Ostriker 1969) rather than as thermal luminosity.

5. PERSPECTIVES

The theory of cooling degenerate dwarfs has reached a degree of sophistication sufficient to justify extensive observational studies. For example, according to Weidemann (1968) existing data indicate that cold degenerate dwarfs may make a significant contribution to the dynamically determined local mass density (the *Oort limit*). With the accumulating results of parallax, color determinations, spectra, and proper motions as well as the construction of theoretical model atmospheres, it should be possible to divide the degenerate dwarfs into population classes, measure stellar death rates in these classes directly, and derive some reliable information about the mass and composition of stars in final evolutionary states as a function of population class.

The high-mass degenerate dwarfs remain a theoretical curiosity, whose possible existence seems assured even though no clearly defined class members are known. Observational searches (looking for rotationally broadened lines or "compound" spectra) should concentrate on the hottest stars.

Low-mass ($M < 0.1 M_\odot$) degenerate dwarfs may make a significant contribution to the local mass density if the cool subluminous stars found by Eggen (1968, 1969) are interpreted as such. It would be important to have better astrophysical and astronomical information about these very faint but very common neighbors of the Sun.

LITERATURE CITED

Axel, L., Perkins, F. W. 1971. *Ap. J.* 163:29

Baglin, A. 1969. *Ap. Lett.* 3:119

Beaudet, G., Salpeter, E. E. 1969. *Ap. J.* 155:203

Böhm, K. H. 1968. *Ap. Space Sci.* 2:375

Böhm, K. H. 1969. *Low Luminosity Stars,* ed. S. Kumar, 393. New York: Gordon & Breach

Böhm, K. H. 1970. *Ap. J.* 162:919

Brush, S. G., Sahlin, H. L., Teller, E. 1966. *J. Chem. Phys.* 45:2102

Canuto, V. 1970. *Ap. J.* 159:641

Chandrasekhar, S. 1931a. *MNRAS* 91:456

Chandrasekhar, S. 1931b. *Ap. J.* 74:81

Chandrasekhar, S. 1935. *MNRAS* 95:207

Chandrasekhar, S. 1939. *Introduction to Stellar Structure,* Chap. X and XI. Chicago: Univ. Chicago Press

Chandrasekhar, S. 1964. *Phys. Rev. Lett.* 12:114, 437

Chandrasekhar, S. 1970. *Ap. J.* 161:561

Chandrasekhar, S., Lebovitz, N. R. 1962a. *Ap. J.* 135:248

Chandrasekhar, S., Lebovitz, N. R. 1962b. *Ap. J.* 135:305

Chandrasekhar, S., Tooper, R. T. 1964. *Ap. J.* 139:1396

Chin, C. W., Chiu, H.-Y., Stothers, R. 1966. *Ann. Phys.* 39:280

Chiu, H.-Y. 1968. *Stellar Physics,* Chap. 3. Waltham, Mass.: Blaisdell Publ. Co.

Cohen, J. M., Lapidus, A., Cameron, A. G. W. 1969. *Ap. Space Sci.* 5:113

Deinzer, W., Hansen, C. J. 1969. *Astron. Ap.* 3:214

Durisen, R. H. 1971. In preparation

Durney, B. R., Faulkner, J., Gribbin, J. R., Roxburgh, I. W. 1968. *Nature* 219:20

Eggen, O. 1968. *Ap. J.* 153:195

Eggen, O. 1969. *Ap. J. Suppl.* 19:31

Faulkner, J., Gribbin, J. R. 1968. *Nature* 218:734

Finzi, A., Wolf, R. A. 1967. *Ap. J.* 150:115

Greenstein, J. L., Trimble, V. L. 1967. *Ap. J.* 149:283

Grenfell, T. C., Böhm, K. H. 1970. *Ap. J.* 161:1183.

Gribbin, J. R. 1969. *Ap. Lett.* 4:77

Hamada, T., Salpeter, E. E. 1961. *Ap. J.* 134:683

Hansen, C. J., Spangenberg, W. 1971. *Ap. J.* 163:653

Hansen, C. J., Wheeler, J. C. 1969. *Ap. Space Sci.* 3:464

Harper, R., Rose, W. K. 1970. *Ap. J.* 162:963

Hubbard, W. B. 1966. *Ap. J.* 146:858

Hubbard, W. B. 1969. *Ap. J.* 155:333

Hubbard, W. B., Lampe, M. 1969. *Ap. J. Suppl.* 18:297

Hubbard, W. B., Wagner, R. L. 1970. *Ap. J.* 159:93

James, R. A. 1964. *Ap. J.* 140:552

Kemp, J. C., Swedlund, J. B., Landstreet, J. D., Angel, J. R. P. 1970. *Ap. J.* 161:77L

L'Ecuyer, J. 1966. *Ap. J.* 146:845

Ledoux, P. 1945. *Ap. J.* 102:143

Ledoux, P., Sauvenier-Goffin, E. 1950. *Ap. J.* 111:611

Lee, T. D. 1950. *Ap. J.* 111:625

Marshak, R. E. 1940. *Ap. J.* 92:321

Matsushima, S., Terashita, Y. 1969a. *Ap. J.* 156:203

Matsushima, S., Terashita, Y. 1969b. *Ap. J.* 156:219

Mestel, L. 1950. *Proc. Cambridge Phil. Soc.* 46:331

Mestel, L. 1952. *MNRAS* 112:583

Mestel, L. 1965. *Stars and Stellar Systems,* ed. L. H. Aller, D. B. McLaughlin, 8: Chap. 5. Chicago: Univ. of Chicago Press.

Mestel, L., Ruderman, M. A. 1967. *MNRAS* 136:27

Ostriker, J. P. 1969. *Proc. 11th Int. Conf. Cosmic Rays, Budapest. 1970. Acta Phys. Acad. Sci. Hung.* 29:69

Ostriker, J. P. 1970. *Stellar Rotation,* ed. A. Slettebak, 147. Dordrecht-Holland: Reidel

Ostriker, J. P., Axel, L. 1969. *Low Luminosity Stars,* ed. S. Kumar, 357. New York: Gordon & Breach

Ostriker, J. P., Bodenheimer, P. 1968. *Ap. J.* 151:1089

Ostriker, J. P., Bodenheimer, P., Lynden-Bell, D. 1966. *Phys. Rev. Lett.* 17:816

Ostriker, J. P., Durisen, R. H. 1970. Paper Presented at *IAU Symp. No. 42 White Dwarfs,* St. Andrews, Scotland, August

Ostriker, J. P., Hartwick, F. D. A. 1968. *Ap. J.* 153:797

Ostriker, J. P., Tassoul, J. L. 1968. *Nature* 219:577

Ostriker, J. P., Tassoul, J. L. 1969. *Ap. J.* 155:987

Rose, W. K. 1966. *Ap. J.* 146:838

Rose, W. K. 1967. *Ap. J.* 150:195

Rose, W. K., Smith, R. L. 1970. *Ap. J.* 159:903

Salpeter, E. E. 1961. *Ap. J.* 134:669

Salpeter, E. E., Zapolsky, H. S. 1967. *Phys. Rev.* 158:876

Saslaw, W. C. 1968. *MNRAS* 138:337

Sauvenier-Goffin, E. 1950. *Mem. Soc. Roy. Sci. Liège* 10:1

Savedoff, M. P., Van Horn, H. M., Vila, S. C. 1969. *Ap. J.* 155:221

Schatzman, E. 1958. *White Dwarfs.* Amsterdam: North-Holland

Schatzman, E. 1961. *Ann. Ap.* 24:237

Schwartz, R., Africk, S. 1970. *Ap. Lett.* 5:141

Schwarzschild, M. 1958. *Structure and Evolution of the Stars.* Princeton: Princeton Univ. Press

Seaton, M. 1966. *MNRAS* 132:961

Skilling, J. 1968. *Nature* 218:923

Stothers, R. 1966. *Astron. J.* 71:973

Stothers, R. 1971. *Ap. J.* 163:555

Strittmatter, P. A., Wickramasinghe, D. T. 1970. *MNRAS* 150:435

Van Horn, H. M. 1968. *Ap. J.* 151:227

Van Horn, H. M. 1969a. *Low Luminosity Stars,* ed. S. Kumar, 304. New York: Gordon & Breach

Van Horn, H. M. 1969b. *Phys. Lett.* 28A:706

Van Horn, H. M. 1970a. *Ap. J.* 160:L53

Van Horn, H. M. 1970b. Paper presented at *IAU Symp. No. 42 White Dwarfs,* St. Andrews, Scotland, August

Vila, S. C. 1966. *Ap. J.* 146:437

Vila, S. C. 1967. *Ap. J.* 149:613

Vila, S. C. 1970. *Ap. J.* 162:971

Wagoner, R. V. 1969. *Ann. Rev. Astron. Ap.* 7:553

Weidemann, V. 1960. *Ap. J.* 131:638

Weidemann, V. 1963. *Z. Ap.* 57:87

Weidemann, V. 1967. *Z. Ap.* 67:286

Weidemann, V. 1968. *Ann. Rev. Astron. Ap.* 6:351

Weidemann, V. 1970. Paper presented at *IAU Symp. No. 42 White Dwarfs,* St. Andrews, Scotland, August

Wheeler, J. A. 1966. *Ann. Rev. Astron. Ap.* 4:393

Wheeler, J. C., Hansen, C. J., Cox, J. P. 1968. *Ap. Lett.* 2:253

Wickramasinghe, D. T., Strittmatter, P. A. 1970. *MNRAS* 147:123

Zapolsky, H. S., Salpeter, E. E. 1969. *Ap. J.* 158:804

SOME RELATED ARTICLES APPEARING
IN OTHER *ANNUAL REVIEWS*

AUTHOR INDEX

SUBJECT INDEX

A

Absorption
  continuous, in models,
    254
  cross section for X-rays in
    ISM, 308
  line coefficient
    in multilevel atom, 260
    in two-level atom, 243-
      44
  of radio waves
    Cytherean clouds, 156
    terrestrial atmosphere,
      284-85
Absolute magnitude
  determination from parallax,
    107
Abundance
  see Composition
Acceleration
  perspective change in
    proper motion, 119
Active regions--solar
  changes with height, 233
  chromospheric He I and II,
    232-33
  contrast in transition and
    corona, 214, 230
  corona
    density in, 231
    energy input into, 231
    transition zone, 231
    conductive flux, 231
Adiabatic lapse rate
  see Venus, Mars
Algol paradox, 184, 197
Aluminum
  opacity jump at λ 2080,
    215, 217
Amino acids in meteorites,
  21
Anelastic approximation
  in convective theory, 325
Angular momentum
  in binary systems, 186
    spin-orbit exchange, 199-
      200
    variability, 187, 199-200
  in interstellar clouds, 315
Antennas--filled aperture,
  271-92
  atmospheric effects
    absorption of radio waves,
      284-85
    radiation, sky fluctuations,
      285-86
    refraction, index, path
      length variations, 286
  definition, 271

distortions, 278-84
  active control, 280-81
  closed-loop control, 281
  homologous deformations,
    280-83
  limit requirements, 278
  multielement array, 281
  open-loop control, 281
  passive control, 281-83
  size limit, 280
  temperature effects, 280,
    287
  wind effects, 280, 287-
    88
feeds, 287-89
  diffraction pattern, 288
  long slotted waveguide,
    289
  requirements, 287-88
  scalar, 288
  shaped dielectric, 288
  subreflector, 289
housing, 273, 287
measurement of surface,
  283-84
  modulated waves, 284
  pentaprisms, 283
  theodolite, 283
mounting, 278
recently completed large
  Goldstone, 273
  Max-Planck, 274
  millimeter-wave, 271-
    73
  others, 278-79
  Owens Valley, 273
  Tata Institute, 275
Antennas--unfilled aperture,
  271
Astrometry
  limiting factors, 106
  long-focus photographic,
    103, 117
  reflectors vs refractors,
    106
Atmospheres
  degenerate dwarfs, 361
  planetary
    dynamics, 170
    evolution, 175-78
    exosphere, 163
    lower, 157
    mesopause, 163
    scaling from one planet to
      another, 170
    thermosphere, 163
    see also individual planets
  solar
    see Sun
Atmospheric effects

on radio waves
  absorption, 284-85
  radiation, 285-86
  refraction, 286
Atomic recoil broadening,
  241
Attenuation
  of radio waves by Venutian
    clouds, 159
  see also Extinction
Austasch coefficient
  in mixing-length convec-
    tion, 338

B

Barnard's star
  path, 119
  possible companions, 120
  secular perspective accel-
    eration, 119
  velocity, 119
Binary stars
  blue stragglers, 183
  cataclysmic variables as
    binaries, 201-2
    mass loss evidence, 202
    periods, 201
    spectra, 201-2
    which component?, 202
  as checks on stellar struc-
    ture, 183
  close binaries, 183-205
    abundance anomalies,
      200
    Algol paradox, 184, 197
    angular momentum, 186
    angular momentum vari-
      ability, 187, 199-200
    basic concepts, 185-88
    classification, 184, 205
    correlation between excess
      luminosity and mass
      ratios, 197
    critical surface, 186,
      188
    envelope expansion effect,
      198
    gravitational potential,
      185-86
    Kepler's law, 185, 189
    Lagrangian points, 186,
      188
    light curve expected, 195-
      96
    luminosity of primary
      after mass exchange,
      194, 196
    mass exchange
      see Mars exchange

CUMULATIVE INDEXES

VOLUMES 5 - 9

INDEX OF CONTRIBUTING AUTHORS

INDEX OF CHAPTER TITLES

VOLUMES 5-9